Business Economics and Finance with MATLAB®, GIS, and Simulation Models

Business Economics and Finance with MATLAB®, GIS, and Simulation Models

Patrick L. Anderson
Principal, Anderson Economic Group

CHAPMAN & HALL/CRC
A CRC Press Company
Boca Raton London New York Washington, D.C.

Library of Congress Cataloging-in-Publication Data

Anderson, Patrick L.
Business economics and nance with MA TLAB, GIS and simulation models / Patrick L. Anderson.
p. cm.
Includes index.
ISBN 1-58488-348-0 (alk. paper)
1. Managerial economics—Mathematical models. I. Title.

HD30.22.A53 2004
338.5'01'13—dc22

2003069765

Direct all inquiries to CRC Press LLC, 2000 N.W. Corporate Blvd., Boca Raton, Florida 33431.

Visit the CRC Press Web site at www.crcpress.com

International Standard Book Number 1-58488-348-0
Library of Congress Card Number 2003069765
Printed in the United States of America 4 5 6 7 8 9 0
Printed on acid-free paper

Dedication

This book is dedicated to the memory of Lt. Robert Regan and firefighters Joseph Agnello and Peter Anthony Vega, who fell in the line of duty on September 11, 2001, rescuing people in the Marriott World Trade Center — including the author, who is forever indebted.

Acknowledgements

This book would not have been completed without the assistance of the following people:

David Littman, who mentored me for three critical years early in my career, and whom I consider the dean of American business economists.

Ian Clemens, who first introduced me to MATLAB and who helped develop some of the routines for gravity models, sprawl, and diversificiation presented here.

Ilhan Geckil and Chris Cotton, who worked with me on a number of projects excerpted here, including fiscal and economic model analyses of retail sales.

Christine LeNet, who both reviewed the content in the business, tax policy, and other chapters, and formatted figures and document files for publication.

Arianna Pieper and Megan Boone, who formatted, corrected, adjusted the figures, and checked the manuscript—and provided an occasional polite recommendation that I rephrase a few things.

Vlad Hlasny, Dan Li, Kevin Dick, Michael Robbins, Raymond Neveu, Jay Abrams, and Ralph Sandler, who provide invaluable review comments on earlier manuscripts.

Bob Stern of CRC Press, who was supportive, encouraging, and politely insistent on getting this project done.

Finally, my wife, Madhu, and our children Neal, Sean, and Mohra endured many evenings, late nights, and long days without me while I labored on the text, especially the past year. It has been a long journey, but not a lonely one for an author much blessed with a wonderful family.

Table of Contents

List of Tables

List of Figures

List of Code Fragments

1

How to Use This Book

Organization of the Book

This book is organized into chapters, each covering a topic. Sections in each of the chapters describe in more detail one aspect of the topic. Most readers will want to read a few chapters immediately and then return to cover specific topics as their interests warrant.

The chapters proceed as follows:

1. The first chapter describes the organization and purpose of the book as well as how to acquire the business economics toolbox prepared by the author for readers of the book. It also provides certain maxims of business economics. Almost all readers will want to review this chapter first.
2. The second chapter contrasts business economics with academic economics and discusses why simulation models in general, and MATLAB and Simulink in particular, offer advantages to the practicing economist.[1] Most readers will benefit from this chapter.
3. The third chapter describes the MATLAB and Simulink environment and provides guidelines for creating and using models in this environment. Style guidelines are presented that help make such models easier to understand and debug.
4. Chapter 4 focuses on data. We present methods to get your data into the MATLAB environment and to report it out. The chapter also contains a specific introduction to the use of Microsoft Excel as an environment for easily collecting, importing, and reporting data. Three appendices to this chapter discuss XML and structured data, creating and using custom data structures, and also importing files of various formats. Most readers will again benefit from these chapters, although those with some familiarity with the MATLAB environment will skim portions of them.

[1] MATLAB® and Simulink® are trademarks of The Mathworks, Inc., 3 Apple Hill Drive, Natick, MA 01760.

5. The fifth chapter discusses certain building blocks of economic analysis used in economics and then provides custom-programmed functions and models which implement these tools. These tools form the basis of the economics toolbox created for readers of this book. The chapter also discusses creating and using a library of such functions. This chapter introduces material not available in any other form and will be essential reading for those who wish to implement the models described later in the book.
6. The chapters that follow describe specific fields of application:
   a. Economic and fiscal impact models as well as tax policy and forecasting tax revenue are covered in Chapter 6, Chapter 7, and Chapter 8. These chapters provide practitioners in these fields with advanced tools that enable more accurate, reliable analysis. The first sections of the chapter on economic impact models identify the severe weaknesses in many common economic impact reports, adding in some examples of particularly absurd gross exaggeration in published reports. These sections close with a plea for ethics in that subfield. As much of our work has been in public policy, we use these tools on a regular basis and hope that the analytical tools will help improve the level of analysis and that our admonitions about ethics in using them will not fall on deaf ears.
   b. Regional economics is discussed in Chapter 9, particularly the proper analysis of economic diversification for states or regions and a rigorous discussion of urban sprawl. In the first sections, we present techniques that have been recently developed and provide a much better concept of risk and return in regional economies than naive diversification analyses that are not grounded in a rigorous theory of economic growth and variation among industries. In the second section, a numerical sprawl index is presented for the first time in print.
   c. Chapter 10, Chapter 11, and Chapter 12 cover the economics of a firm. Chapter 10 describes modeling a firm, focusing on the crucial question of how a firm makes money. An iterative method of matching market value and expenses based on value is introduced. Chapter 11 discusses business valuation, noting the various methods used to estimate the value of a business and the various definitions of value. The chapter differs from most other references in this field in two ways: first, we concentrate on what we consider the most important consideration in the market value of a company, namely its prospects for future earnings; and, second, we utilize the greater analytical

power of quantitative models that the MATLAB environment provides.

d. Retail sales are discussed in Chapter 13. Much of the material presented here, including the derivation of the distance-sales relationship from microeconomic foundations, is unique to this book. Other portions provide craft knowledge of this interesting field that is rarely described in print.
e. Chapter 14 briefly covers statistical process control techniques in manufacturing — a topic that is less well known than it should be among business economists and managers, given its importance.
f. Fuzzy logic models are introduced in Chapter 15. Fuzzy logic is actually quite rigorous, though the name it has been given may be inhibiting its adoption as a powerful tool for incorporating various types of information, including vaguely defined or subjective information, into a rigorous inference system.
g. Chapter 16 deals with using MATLAB and Simulink with the Internet and includes suggestions on analytical applications on the Web, sharing and displaying information, and the use of the MATLAB Web server. This is optional reading for many readers, but it contains unique material for those who wish to advance to the state of the art for analytics and data transfer on the Web.
h. The last chapter contains extensive information on the proper use of graphics to communicate data and specific MATLAB code to generate and customize a number of charts. The chapter starts with a review of the classic rules for graphical excellence and descriptions of the all-too-common graphical errors such as "chartjunk" and distortion due to improper use of 3-D graphics. There are very few texts in the social sciences which offer guidance in the proper use of graphics, and we intend this to be one.

7. Each of these chapters describes briefly the theory behind the models and then outlines an approach. These chapters can be read, skimmed, or skipped, depending on the interest at hand. A reader dealing with a finance problem, or one interested in fuzzy logic applications, could jump right to this chapter (after, of course, reviewing the initial chapters) without necessarily reading up on economic or fiscal impact models. Numerous cross-references indicate where similar or extended material appears elsewhere in the book.
8. The appendices contain troubleshooting suggestions that could save you a lot of time and trouble.

Intended Audience

This text is primarily designed for two audiences:

1. Economists, finance and valuation professionals, market researchers, public policy analysts, and other practitioners whose occupation demands the ability to model market behavior under a variety of real-world conditions
2. Teachers or students in graduate-level classes in applied economics, particularly those who cover topics such as fiscal and economic impact analysis, fuzzy logic applications, retail sales analysis, and the integration of geographic information systems, as this is one of the very few books to rigorously cover these subjects

The book may also be used by those interested in specialized topics rarely covered in economics texts, such as those listed above, and by those who have extensive background in mathematical or simulation models (such as MATLAB and Simulink) and wish to extend their knowledge.

Prerequisite Knowledge

We presume the reader has the following knowledge:

1. A good grounding in the laws of microeconomics and a familiarity with their application. Most practitioners and graduate students in economics, finance, public policy, and related disciplines should either be able to acquire this or should have already done so. Most undergraduates will need additional training.
2. A working understanding of mathematics, including (for some chapters) an understanding of the calculus used in comparative statistics. The completion of a graduate course in math for economists would be helpful, but is not mandatory.
3. For many of the chapters, a familiarity with (but not necessarily expertise in) the MATLAB software environment. Most of the applications presented here could be accomplished in other software environments, but we provide examples of applications in MATLAB. See "References on MATLAB and Applied Economics" on page 9.

4. For some chapters, an awareness (but not necessarily a working knowledge) of geographical information systems.

We suggest some resources on these topics under "References on MATLAB and Applied Economics" on page 9.

Suggested Reading Plans

The organization of this book allows you to read those portions most relevant to the work you are doing at present, in particular:

- Regardless of whether you are new to business economics or a veteran, read the brief "Maxims of Business Economics" on page 15.
- If you are new to the use of simulation models, it is best to start at the beginning. Go through the chapters that explain the MATLAB and Simulink environment, how to get your data, and creating and using library functions. Then jump to the most interesting application chapters.
- If you are focused on one type of application and have a working familiarity with MATLAB or Simulink, finish this chapter and then jump straight to the application chapters. Be prepared to review information in the earlier chapters or in the appendices.
- If you are currently using MATLAB or Simulink and are looking for suggested approaches or troubleshooting information, scan the chapters on troubleshooting in the appendices, any relevant application chapters, and the early chapters on design guidelines, getting your data, and library functions.
- If you want to learn about using MATLAB and Simulink on the Internet, jump to that chapter but be prepared to refer to other areas.

Typographic Conventions

We have adopted certain typographic conventions that will assist the reader in understanding the text, such as:

- The main text of the book is written in a normal font, like this.
- Text that is part of a computer program or is a command you type in a software environment is in a keystroke font.

- Program commands for software that uses menus are sometimes summarized by listing the menu selections that execute them with the piping separator ("|") between the menu selections. For example, to use the MATLAB menu command to open a new Simulink model, you would execute the following menu items:
  1. First, hit the File menu with your mouse or with a keyboard command (such as Alt-F).
  2. Second, hit the New submenu command on the File menu.
  3. Third, select Model from the File submenu.

We summarize this set of commands as: File | New | Model.

Changing Software Commands

Keep in mind that software companies change the menus in their products frequently, and you may need to use a different set of commands if you are using a different version of the software or on a different operating system. Also, there may be a shortcut method using a combination of keystrokes or mouse actions that can accomplish the same operation in a different way.

MATLAB Programming and Code

We have provided the actual code for a number of MATLAB programs we have prepared to illustrate the techniques described in this book. Longer sections are listed as the code fragment "A Simple Startup File" on page 7.[2] We call these code fragments because they are written in a specific software language or code, and because many of the longer code sections are shortened, the code included in this text may be only a portion or fragment of the entire file.

Main Purpose of Code

Experienced MATLAB users will undoubtedly notice ways in which the execution of these programs could be speeded up or different ways of accomplishing the same thing. The primary purpose of including the programs is to demonstrate how to perform certain tasks and not to execute those tasks with maximum efficiency. Therefore, you will note that many of the programs include extensive comment lines (which begin with a percent-sign [%] mark) that are ignored by the computer. If you are reviewing these code fragments in order to perform a similar task, pay attention to these comment lines.

---

[2] A startup file is a set of instructions run by MATLAB when the program is first run. We recommend that users customize their startup files; this is discussed in Chapter 3, this book, under "Setting Up the MATLAB Workspace" on page 30.

Those who can recommend a more efficient — or more effective — code to accomplish the same result are encouraged to send a note to the author. See the contact information at "Contacting the Author" on page 9.

Version Notes

The versions of MATLAB used during the development of this book included R11, R12, R13, and a prereleased version of R14. We believe that all of the routines will run, in some cases with minor modifications, on R12 and R13. Most will run on R11 as well. Based on our testing of the comparison business economics toolbox with the prereleased version of R14, we anticipate that all of the routines included here will also work under R14, perhaps with minor modifications. Users should expect that different versions will have some interface differences as well.

Code Fragment 1-1. A Simple Startup File

```
% Startup file for Patrick Anderson
% Place in MATLABroot\work directory
%
% ---Change directory to folder with Matlab models
CD P:\models;
%
% ---Messages---
message = ['Good afternoon, Mr. Anderson. God bless
you and your work'];
disp(message);
%
% ---Show version---
ver;
```

Accessing Files and Online Resources

Many of the MATLAB program files described in this book, including script and function m-files, and Simulink models, are available online from the author or publisher. Other information about the content of the book, including corrections of any errors, may also be posted.

We hope this book will be used for years. Note that Web sites tend to have short life spans, at least at the same Internet address (known as a universal resource locator or URL). Therefore, we are providing you with a set of Internet addresses to examine in the hope that several years from the publication date you can access the information prepared for readers of this book.

Accessing the Economics Toolbox

The author has prepared a business economics toolbox containing many of the MATLAB functions, scripts, and Simulink model files described in this book. It is the author's intention to offer this toolbox to *bona fide* purchasers of this book as an additional, licensed software product. The license for the product allows purchasers of this book to download and use the toolbox free of charge subject to the conditions stated below:

Disclaimers and Limitations

It is unfortunate, but our litigious society requires even those that provide products free to outline certain limitations and disclaimers, which we do below:

- While it is the author's intention to provide the business economics toolbox, we do not guarantee that the files will be available indefinitely, that they will work on specific computer platforms, or that a fee will not be charged for a license at some future date.
- The author may require the endorsement or acceptance of a license as a condition for providing the toolbox and to make further restrictions, including restrictions on reproduction or resale.
- Acquiring a license for the toolbox does not entitle a reader to a license for MATLAB, and vice versa.
- Neither purchase of the book nor downloading the toolbox qualifies the reader for technical assistance or help of any kind.
- Any downloaded files are provided with no warranty, including no warranty of fitness or merchantability. Users assume all risks of use.
- The author may provide the toolbox in compressed format, require verification of book purchase, charge an additional fee to unlock the code, or rely on an intermediary (such as the publisher or The MathWorks) to provide the toolbox.

Contacting the Author and Publisher

Readers should periodically check the following Web sites for related resources and the companion economics toolbox.

Contacting the Author

Patrick L. Anderson is a principal in the consulting firm of Anderson Economic Group headquartered in East Lansing, MI. The firm maintains a number of Web sites, including http://www.andersoneconomicgroup.com.

A set of pages on this site is devoted to this book, and it is our intention to post any revision notes, related works, resources, errata, and other information on those pages. If you have a suggestion about a topic to be covered in a future edition, an observation on a better approach to some problem, or you have discovered a typographical or conceptual error, please contact us via e-mail or regular mail. Check the Web site for current mailing and e-mail addresses.

Contacting the Publisher

CRC Press maintains an active Web site, which includes downloadable files and other information for many of its titles. The current address is http://www.crcpress.com.

Contacting The MathWorks

The MathWorks maintains an active Web site which includes a file exchange area, product information, technical notes, and other support information. The current address is http://www.MathWorks.com.

References on MATLAB and Applied Economics

This book is about applying economic principles to practical problems in business, finance, public policy, and other fields. The principles themselves are timeless; the applications described in this book can be done in a number of software environments. Indeed, the first principle we recommend — important enough that we call it a maxim — is to think through the problem before trying to restate it in mathematical terms.[3]

---

[3] "Think first — calculate later." See "Maxims of Business Economics" on page 15.

However, once you have thought through the problem, you will normally have to describe it in mathematical terms and do so in a software environment that allows you to complete the calculations, simulations, reporting, and graphical illustrations that are necessary. We describe in a later section the advantages and disadvantages of different categories of software that can accomplish some or all of these tasks.[4]

In this book, we provide detailed descriptions of analysis in three of these environments: mathematical and simulation models (MATLAB and Simulink), Geographic Information Systems (GIS), and spreadsheets (particularly Microsoft Excel). This book, however, does not provide a tutorial for these software programs. Therefore, we provide below suggestions for those readers who would benefit from additional information on the software products described in this book.

Suggested Resources for MATLAB Users in Business, Economics, and Finance

The following resources are recommended:

1. New users of MATLAB will probably benefit from one or more of the general reference books or booklets listed below in "General Books on Using MATLAB." These will not provide application advice though, and most examples in them are from other fields.
2. Within this book we introduce a number of basic and many advanced MATLAB techniques that are quite helpful in business, economics, and finance. A chapter-by-chapter summary of these is in "MATLAB Explanations in this Book" on page 11.
3. Those with a rudimentary knowledge of MATLAB can simply read the book in one of the ways suggested in "Suggested Reading Plans" on page 5 and determine whether any of the references listed in the chapters are necessary for their work.

General Books on Using MATLAB

- *A Primer on MATLAB*, 6th ed., by Kermit Sigmon and Timothy Davis, CRC Press, Boca Raton, FL, 1994. This is a very thin, almost vest-pocket reference on the basic MATLAB commands, syntax, and other essentials.
- *Mastering MATLAB 6* by Duane Hanselman and Bruce Littlefield, Prentice Hall, Upper Saddle River, NJ, 2001. Subtitled accurately as "A Comprehensive Tutorial and Reference." The authors also provide, for a nominal charge, a related set of utility programs.[5]

[4] See Chapter 2, this book, under "Software Environments" on page 21 and under "Limitations of Spreadsheet Models" on page 23.

[5] We describe one in Chapter 4, this book, under "A Shortcut Utility" on page 78.

- *Graphics and GUIs with MATLAB*, 3rd. ed., by Patrick Marchand and O. Thomas Holland, Chapman & Hall/CRC Press, Boca Raton, FL, 2003. This comprehensive book provides extensive information on creating graphics in MATLAB. Handle Graphics is one of the most powerful features in MATLAB. We provide examples of graphics routines that are designed for users of this book in Chapter 17, "Graphics and Other Topics," but more advanced users will want a reference book on graphics, and we recommend this one.
- *MATLAB User Guides* and other documentation distributed by The MathWorks with the MATLAB software. Many publishers of software programs provide little or no documentation or provide poorly written guides that must be supplemented by additional purchases.[6] Fortunately, The MathWorks provides extensive references that are, by and large, well written and authoritative. This should typically be the first place a user goes for explanation and guidance on using the program.[7] However, most of these references (with a few notable exceptions) are not intended for practitioners in economics, business, or related fields, and such individuals will often benefit from additional information in books such as this that focus on their fields of interest.

MATLAB Explanations in This Book

While this book is not an introduction to MATLAB, the following sections provide, in summary form, explanatory material on working within MATLAB:

- Chapter 3, "MATLAB and Simulink Design Guidelines," introduces Simulink models and provides guidance on creating them. The same chapter provides style guidelines and shows how to document your models so that others understand them.
- Chapter 4, "Importing and Reporting Your Data," describes various methods of getting data into your model. A handy appendix contains a summary of file types and the best way to use the data in such files in MATLAB. The chapter also discusses the various data types used in MATLAB. The benefits of a custom data structure are developed and an example given for later use in business valuation.

---

[6] Particularly frustrating are those publishers that provide little in the way of usable information with their software and then sell basic user assistance and introductory help books to the users saddled with the poorly documented software.

[7] Users can get guidance from: (1) within the command window, the Help function for all commands (e.g., typing Help Print will produce information on the Print command); (2) the help pages available from the desktop; and (3) the extensive technical notes and other information available on the MathWorks Web site which is at: http://www.MathWorks.com.

- Chapter 5, "Library Functions for Business Economics," reviews fundamental tools of economic theory and shows how to create MATLAB tools that apply those tools to practical problems. Specialized functions useful to business, economics, and finance are presented, including supply and demand calculations and special Simulink blocks for constant growth, varying growth, and elastic impact.
- Chapter 6, "Economic Impact Models," and Chapter 7, "Fiscal Impact Models," present Simulink models for these tasks.
- Chapter 8, "Tax Revenue and Tax Policy," describes functions that model the U.S. income tax, U.S. payroll taxes, and other taxes.
- Chapter 9, "Regional Economics," provides mathematical tools for analyzing regional economies, including newly developed methods for calculating the diversification (risk and reward characteristics) of regional economies and their recession severity risk.
- Chapter 10, "Applications for Business," includes a useful projection function, as well as a general business model and an iterative tax and market value simulation model.
- Chapter 11, "Business Valuation and Damages Estimation," uses custom data structures and functions for different valuation formulas to estimate values using multiple methods.
- Chapter 12, "Application for Finance," introduces the many MATLAB capabilities for financial modeling and includes a special GetStocks function that acquires data from an Internet provider, analyzes it using special operations used in finance, and then graphs it.
- Chapter 13, "Modeling Location and Retail Sales," explains a number of rarely documented methods to model retail sales, including gravity models. We present functions to perform distance-sales analyses and introduce optimization as a method within MATLAB to estimate parameters for models that are nonlinear.
- Chapter 14, "Applications for Manufacturing," provides an overview of an important topic barely understood outside engineering circles — statistical process control (SPC). Some fairly simple SPC functions are presented along with a few suggestions for understanding this important field.
- Chapter 15, "Fuzzy Logic Business Application," provides a brief introduction to this burgeoning field which is sometimes called artificial intelligence or expert systems. A fuzzy logic income tax

audit predictor is presented, which was run on the Internet (via a MATLAB Web server) for more than a year, encompassing two April 15 filing dates. A credit risk simulation model is also presented in this chapter.

- Chapter 16, "Bringing Analytic Power to the Internet," begins with a sober reflection on the relatively primitive amount of actual analytical power provided by most Web sites. It then describes how the use of a MATLAB Web server can significantly increase the ability to provide advanced analytics to a remote user.
- Chapter 17, "Graphics and Other Topics," is devoted entirely to the proper use of graphics. It provides examples showing how to generate test data, draw all the major type of graphs, customize them with different scale, annotation, and other features, and identify them for future use.
- Appendix A: "Troubleshooting," provides general MATLAB troubleshooting.
- Appendix B: "Excel Link Debugging," focuses on this useful but troublesome connection and how to make it work.

Books on Applying MATLAB for Business, Economics, and Related Fields

We are not aware of any other book that describes how to apply MATLAB to general business economics and finance problems.[8] However, there are many texts that focus on specific fields within or overlapping those discussed in this book. Those we recommend are:

- Our book starts with a perspective similar to the book *Handbook with MATLAB*[8a] with an application of social science principles to real-world problems. As such, this book describes a number of statistical techniques (such as the bootstrap) that are particularly applicable to real-world problems. The comparison of academic and computational statistics was the model for our comparison in Chapter 2, under "Business Economics vs. Academic Economics" on page 19.
- "Macro-Investment Analysis," an "electronic work-in-progress" by William F. Sharpe,[9] winner of a Nobel Prize in economics,[9a] was

[8] The author would be happy to be informed of new texts or older texts that escaped his knowledge; see "Contacting the Author" on page 9.

[8a] *Computational Statistics Handbook with MATLAB* by Wendy Martinez and Angel Martinez, CRC Press, Boca Raton, FL, 2001.

[9] It can be found at: http://www.stanford.edu/~wfsharpe/mia/mia.htm. See also the references in Chapter 12 in this book, "Applications for Finance," in the sections noted.

[9a] William Sharpe, Harry Markowitz, and Merton Miller won the Noble Prize in Economics in 1990 for their pioneering work in finance.

one of the creators of the mean-variance framework for analyzing investment portfolios. We describe applications of mean-variance analysis of investment portfolios in Chapter 12, this book, under "Investment Portfolio Analysis" on page 303.

- The books by Paolo Brandimarte and Neil Chriss on the analysis of options and other derivative securities using MATLAB are listed in Chapter 12, "Applications for Finance."
- *Applied Computational Economics and Finance* by Mario Miranda and Paul Fackler, MIT Press, Cambridge, MA, 2002. This book describes numerical methods and the mathematics behind them in depth and provides a number of application examples using MATLAB, custom-programmed optimization, and other routines.

Where other published works have been used in a significant way to prepare this text, we have cited them in the section describing the application. We also provide citations for reference texts covering the fundamentals of certain techniques.

Fundamental and Historical References

Finally, we have made a special effort to cite those who pioneered certain techniques or who originated important ideas that are now incorporated into the mainstream. These individuals and their contributions are too often neglected in contemporary works, especially those that are focused on software programs. Such neglect not only robs these individuals of the recognition they deserve, it also breeds a group of practitioners that lacks the knowledge of why, how, and under what conditions a certain approach works well.[10] Albeit briefly, we describe the foundations in most chapters.

Books for Other Fields

Those interested in applying MATLAB in other fields will often find multiple texts on specific applications. There are over 300 published books on using MATLAB in fields such as engineering, signal processing, mathematics, numerical methods, finite analysis, radar, communications, control systems, neural networks, mechanics, filters, earth sciences, electronics, statistics, and probability. This extensive library is one of the advantages of using MATLAB as a software environment. The MathWorks web site lists many of these books by category.

---

[10] "Those that cannot remember the past are condemned to repeat it," George Santayana's observation, is a broad statement about the human condition. In a narrower sense, those that cannot remember why a certain method was developed in the past are likely to misuse it in the future.

Maxims of Business Economics

Economists learn a lot through detailed academic work and then receive additional practical training. When prompted about their training, I have noted that professionals early in their career recite specific academic classes as well as opportunities to learn skills. The same question, put to successful business people later in their careers, often produces a surprisingly different set of lessons that has enabled them to succeed.

Most of the remainder of this book is devoted to skills in applying business economics, including quantitative techniques, analytical techniques, and presentation techniques, as well as methodological notes and cautions. In this section, however, we summarize a set of maxims that will enable economists to be successful in their careers regardless of how much math they know.

With that said, here are six maxims for business economics:

1. *Remember — it is a social science.*

   There is a joke that "an economist is someone who looks at reality and wonders whether it would work in theory." Like most jokes, there is a kernel of truth in that observation. Economics models human behavior. Economics applied to real-world situations, the focus of this book, models human behavior when such humans are placed in common situations. Applying economic thinking to actual human behavior forces one to confront the untidy, unexpected, and occasionally unexplainable behavior of fellow members of one's species. Thus, the first maxim is to remember that economics is a social science, by which we mean economics only works when it models human behavior.

   The corollary to this is: when human beings do not behave like the model, it is probably the model that is wrong.

2. *All economics is microeconomics.*

   All social science is an attempt to explain the behavior of individuals. Of course, the aggregate of those individual behaviors is the behavior of groups, including groups of consumers, producers, taxpayers, government officials, and other groups that are important to a business economist. However, when tempted to describe macroeconomic behavior that is at odds with the microeconomic foundation for individual behavior, remember this maxim: if it does not work at the microeconomic level, it will not work at the macroeconomic level.

   Recent pioneers in economics, including at least a few winners of the Nobel Prize, have documented failures of macroeconomics that stem directly from violation of this maxim.[11]

Robert Lucas summarized one set of such errors, in macroeconomics as follows:

> The prevailing strategy for macroeconomic modeling in the early 1960s held that the individual or sectoral models arising out of this intertemporal theorizing could then simply be combined to form a single model, the way Keynes, Tinbergen, and their successors assembled a consumption function, an investment function, and so on into a model of an entire economy. But models of individual decisions over time necessarily involve expected future prices. Some microeconomic analyses treated these prices as known; others imputed adaptive forecasting rules to maximizing firms and households. However it was done, the "church supper" models assembled from such individual components implied behavior of actual equilibrium prices and incomes that bore no relation to, and were in general grossly inconsistent with, the price expectations that the theory imputed to individual agents.[12]

Going back to the times before Alfred Nobel, pioneers of economics such as Adam Smith, David Ricardo, and others grounded their economic arguments on the behavior of individuals. Modern pioneers of business similarly achieve wealth and influence by providing products that individuals (persons and companies) want to buy and not by providing products that on average, or in aggregate, the economy as a whole will want to buy.[13] While your goal may or may not be winning

---

[11] The Nobel Prize winners that spring to mind include the following four:

Milton Friedman's pioneering work on monetary policy illustrated how failures in government policy, primarily monetary policy, induced individuals to behavior that deepened the Great Depression. His work on income and consumption in the late 1950s is a classic demolition of the use of macroeconomic data alone to model the behavior of individual consumers, and it established "permanent income" as the basis for sound thinking about consumption patterns ever since.

Robert Lucas, who pioneered modern "rational expectations" theory, demonstrated how traditional Keynesian economics was frequently at odds with individual incentives and, therefore, wrongly predicted how an economy would react to policy changes. It is interesting to note that one of the first applications of what became known as "rational expectations" theory was by John Muth in 1961, explaining how to model consumption, given Friedman's "permanent income" theory.

James Buchanan, who pioneered modern public choice theory, demonstrated that government officials and interest groups behave like normal, rational individual economic actors.

James Heckman focused his 2000 Nobel Prize Lecture on the importance of microdata and the tools within microeconometrics to deal with heterogeneity in the population.

See *New Palgrave Dictionary of Economics*, various entries; Nobel e-Museum at http://www.nobel.se; J. Muth, "Rational expectations and the theory of price movements," *Econometrica*, 29, 315–35, 1961; M. Friedman and A. Schwarz, *A Monetary History of the United States, 1867–1960* (Princeton, NJ: NBER, 1963); M. Friedman, *A Theory of the Consumption Function* (Princeton, NJ: Princeton University Press, 1957).

[12] Lucas, Nobel Prize Lecture, December 1995; found at: http://www.nobel.se.

[13] I should add, at this point, the political economy corollary of this maxim: "All politics is local." Attributed to the former Speaker of the U.S. House of Representatives Thomas O'Neill, this corollary is part of the underlying knowledge about human behavior from which springs the study of political economy.

the Nobel Prize, you can certainly succeed much faster by remembering this maxim.

3. *Think first — calculate later.*

   Many highly technical, mathematically intense, and impressive-looking economic analyses are completely worthless because the author did not think through the problem before producing his or her equations. The proliferation of software, including spreadsheets, statistical packages, and simulation software such as MATLAB, greatly expands the power of the economist to model behavior and provide detailed, multiperiod forecasts or analyses. Power, however, can be misused. Simply because one can extrapolate for another 10 years and produce a nice chart, does not mean the exercise is useful.

   A recurring theme in this book is the importance of thinking first before rushing to create a complicated model.

   I should also note the academic corollary to this maxim: think first — differentiate later.

4. *Know your data.*

   This is an area where business economics is quite different from theoretical economics. Analysts should always spend time understanding the source, reliability, meaning, and deficiencies in the data. Most data have some deficiencies. Some data are produced by parties with an interest that should be understood before pasting the data into an equation. Be sure that you always know your data.

5. *Be ethical.*

   There are many opportunities in the world of academic and business economics to cut ethical corners in an attempt to produce a more sensational analysis, masquerade someone else's work as your own, or simply toot your own horn inappropriately. No book by a human author can reform this essentially human deficiency. However, if you do not recognize a standard, you will certainly not follow it. Therefore, establish an ethical standard in your work and discipline yourself and your coworkers to follow it.

6. *Be humble.*

   The best thinkers start off by knowing what they do not know and admitting it. If you are reading this book, you are, in all likelihood, an intelligent person who either now or in the future will be considered an expert on some topics. Resist the temptation to claim expertise in an area in which you are an intellectual bystander. In your own field, be careful to cite the contributions of others and to identify the bases of your assumptions, the sources of your data, and the limitations of your knowledge and experience.

Conclusion

These maxims could be applied in many fields. I hope the remainder of this book imparts a great deal of practical knowledge about modeling economic behavior, and that both author and reader follow these simple maxims.

2

Mathematical and Simulation Models in Business Economics

Business Economics vs. Academic Economics

Economics is a social science. It seeks to describe the incentives operating on individuals as they work, save, invest, contribute, vote, and consume. Like any field of science, economics can be studied at the widest theoretical level, at a minute practical level, and at any point in between.

This book looks at the practical applications of economic reasoning, using techniques that are based in rigorous theory and proven in actual practice. We entitle the book *Business Economics and Finance* because the majority of the applications focus on individuals in business dealings. However, there are many topics that focus on government organizations and their interaction with individual workers, managers, and consumers.

It is worth pausing to review how this practical, applied economics differs from the theoretical economics that is taught in colleges and universities today. While there is plenty of cross-over among the subdisciplines, and most economists devote some of their time to both, it is worth considering the differences. Table 2-1, "Theoretical vs. Applied Economics," highlights some of the important characteristics of both.[14]

Why Practice Business Economics?

As Table 2-1 indicates, academic economics normally deals in a smoother world in which data are available or easily assumed, model structures are specified often without argument, and rigorous statistical inferences are common practice. Why, then, practice applied economics with its messy or difficult-to-obtain data, lack of specification of models, and stricter budget constraints?

The answer to that question is much like the answer to the question "Why climb the mountain?" Because it is there! There are a huge number of practical questions facing individuals, their employers, their governments, and the

[14] This comparison is inspired by a similar one prepared by W.L. Martinez and A.R. Martinez in *Handbook of Computational Statistics* (Boca Raton, FL: CRC Press, 2002), in which they compare academic statistics with computational statistics.

stores and service providers that sell to them. These questions normally cannot be solved with theoretical economics alone; they must be approached with applied economics. This book deals with applied economics and shows how advanced use of mathematical and simulation software can help answer the tough, messy, and immediate challenges of today.

Why Study Academic Economics?

The challenge of practical business economics should not dissuade one from learning in an academic setting. Many, though not all, of the techniques

TABLE 2-1

Theoretical vs. Applied Economics

Characteristic	Theoretical Economics	Applied Economics
Data	Usually assumed to exist in exactly the right periodicity	Must be acquired, often with significant effort; will contain errors, breaks, and different periodicities
Data sources	Data are assumed to exist; unobservable variables may be used in equations	Desired data often do not exist. Observed data must be used from private, government, and nonprofit organizations, and experiments
Analysis	Simple to complex algorithms; often developing theory using mathematics	Simple to complex algorithms; often testing applications of existing theory using statistics
Questions	Can be posed by researcher, based on interest; abstraction allows for intense focus on specific questions	Usually posed by events beyond the control of the researcher; often multiple questions are posed
Time and budget constraints	Time constraints are largely based on academic calendar; academic budgets include time for pure research	Time constraints are largely based on events; budget constraints established by clients
Inference practices for parameters	Rigorous statistical inference, given strong assumptions and assumed data	Practical limitations, given potential for specification error and data problems
Assumptions about data and error process	Strong, often unverified assumptions regarding data and errors; error processes often assumed to be clean	Relatively few assumptions about data and errors; error processes often messy
Knowledge of underlying structure	Often strong assumptions about underlying structure, based on theory	Structure may be unclear, or is typically inferred from actual behavior
Key focus of specification	Optimal analysis	Robust analysis
Importance of location and geography	Often ignored or assumed away	Normally quite important and may be the focus of inquiry

described in this book originated in academic settings. Even the mountain climber who climbs mountains because they are there probably went to a mountain-climbing school.

This book borrows some techniques of academic economics in laying out data and sources, giving proper credit to those that have developed or written about a problem before, and in conveying information to others interested in this field. While practical problems require practical solutions, one should still remain true to the essential philosophies of social science.

Why Use Mathematical and Simulation Software?

Given some data and a set of equations that summarize a particular market, an economist will normally use some type of computer software to collect the data, display it, and calculate equations using it. There are many types of software available today that allow for automatic calculations of a series of equations. Each has its intended purpose and is normally quite good for that purpose. Before describing how a true simulation modeling environment can be used to great advantage, let us review the categories of software that can be used in applied economics.

Software Environments

The following illustrative list of software environments includes specific products as well as their intended purposes:[15]

- *Spreadsheet software* such as Microsoft Excel, Quattro Pro, Star Office, etc.: The design of these packages renders them particularly useful for typical accounting tasks, and their feature set can be extended to include various statistical and financial functions.
- *Statistical software,* including SAS, SPSS, Minitab, Stata, Splus, etc.: These are designed primarily to estimate parameters for equations, given a structure and data.[16]

[15] Various products mentioned here are trademarked or copyrighted by their respective owners. The list is illustrative rather than exhaustive. Many readers will know of other examples, including examples available under the GNU public license and shareware or freeware licenses.

[16] These packages normally allow for "nonparametric"statistical tests. However, even nonparametric statistics typically require an assumed model structure, even if it is only assuming that one event causes another. By contrast, we consider "time series" methods (including ARIMA and VAR methods) as "nonstructural" because they do not require the specification of an underlying model structure.

- *Econometric software,* including RATS, TSP, Shazam, Eviews, etc.: They are similar to statistical software but are designed around time-series economic data.
- *Geographic information systems* (GIS) such as MapInfo, ArcView, TransCAD, Maptitude, the mapping toolbox for MATLAB and free or open-source software such as Geode and GRASS: These software environments allow analysis to include spatial variables. A feature of this book is the explicit inclusion of geographic information in the analysis.
- *Mathematical software,* including MATLAB, Octave, Maple, Gauss, and Mathematica: These contain an enormous reservoir of mathematical routines and can be programmed to produce virtually any function or graphic.
- *True simulation models:* These incorporate systems of dynamic equations that can be solved iteratively over multiple periods. These include custom-programmed models in specific fields,[17] certain applications of mathematical, statistical, and econometric software (including some listed above), and the Simulink software developed by The MathWorks, which is based on MATLAB.[18]

Simulation Models vs. Spreadsheets

Spreadsheets are probably the most common analytical program used in business economics. Spreadsheets have powerful advantages, including low costs, intuitive use, and an ability to easily print the data used in its calculations. For most accounting work and many other uses (including uses we will describe in later chapters), you cannot beat a spreadsheet package.

However, for analytical work requiring dynamic interaction among variables — meaning that one variable affects another, which affects two more, and the latter affects the results of the next-period variables — spreadsheets fall far behind. Although you can "trick" a spreadsheet into performing some advanced analysis and even limited dynamic equations, such uses are beyond their intended scope.[19]

---

17 For example, the rainfall-runoff TOPMODEL first developed in 1979 and written in Fortran (found at http://www.es.lancs.ac.uk/hfdg/topmodel.html); Agricultural Non-Point Source (AGNPS), developed by the USDA Agricultural Research Service to predict soil erosion and nutrient transport within agricultural watersheds and written in C (found at: http://www.geog.uni-hannover.de/phygeo/grass/agnps.html); and numerous others, particularly in electronic devices and software design.

18 MATLAB® and Simulink® are trademarks of The MathWorks, Inc., which can be found at http://www.MathWorks.com.

19 For example, some years ago the author created a multiperiod model of the educational market, which was used to forecast the migration of students to private schools should a school-choice program involving those schools be approved by voters in a large industrial state. The

Limitations of Spreadsheet Models

Spreadsheet software has serious limitations as an analytical tool. These limitations include:

1. A spreadsheet is inherently a two-dimensional device. Many problems can be solved in a 2-D environment, including almost all accounting problems. Indeed, *spreadsheet* is a term borrowed from accounting. However, most problems in business are not two-dimensional.
2. While spreadsheets easily display the *data* used for calculations, they do not easily display the *formulas* used to calculate that data. If all the calculations are simple additions and the proper formatting is applied, this limitation is largely overcome. However, when anything more complex than simple addition or subtraction occurs, spreadsheets often obscure the calculations.[20]
3. Most problems involving multiple variables that interact with each other over time cannot be modeled correctly using a spreadsheet.[21] Even if the analyst creates multiple worksheets showing different scenarios in an attempt to overcome the 2-D limitations of the spreadsheet itself, the number of potential scenarios overwhelms the ability of the software to calculate and then express the results.

   With two dimensions, a spreadsheet can typically handle either a small set of variables over time, or a large set of variables at one time — but not both. The sheer size of a spreadsheet model

---

model, implemented in Microsoft Excel, took nearly a dozen individual spreadsheets, each with many columns, to project the fiscal impact. The modeling task took at least a month. Explaining how it was done took many pages in the resulting report. More recently, using the tools and skills explained in this book, a similar project was completed in half the time with more accuracy. See Patrick L. Anderson et al., The Universal Tuition Tax Credit, Midland, MI: The Mackinac Center, 1998; available at http://www.mackinac.org.

[20] Even multiplication is difficult to show adequately. For example, consider how taxes are shown on accounting statements. Unless the tax is a straightforward multiple (such as a flat-rate sales tax), the tax liability is normally shown without providing the reader the ability to review the tax rate, base, or calculations.

Of course, it is *possible* to see the formulas in a spreadsheet, but the variables used are typically cell reference, and a special command or set of keystrokes is often required to reveal the formulas used to generate results in each cell of a spreadsheet.

[21] It is possible to display the results of one scenario involving multiple, interactive variables over more than one time period. This is what accounting statements for a firm provide. However, if you look at a set of income statements and balance sheets for a corporation and ask "What if sales were higher, commissions lower, and personnel costs sharply reduced," you run immediately into the limitations of a 2-D environment. Although you could display this alternate scenario, the environment itself would not keep track of the relationships among these variables or allow an efficient manner of generating a large number of alternate scenarios involving true interactivity among the variables.

necessary to model a multiperiod, multivariable problem can be overwhelming.[22]

4. Although not a defect in the software itself, the widespread use of spreadsheets induces an intellectual laziness regarding how variables interact together. It is much too easy to look at a column of numbers, notice that they add up correctly and then conclude that the number at the bottom of the column must be correct.

Comparison with Simulation Models

Consider how a true simulation modeling environment, such as MATLAB or Simulink, overcomes some of the limitations of spreadsheet models:

1. A simulation model is n-dimensional. You can include as many variables as you think are important and relate them in multiple ways. You are not restricted to thinking in rows and columns.
2. An enormous variety and number of operators and functions can be used in simulation models. Furthermore, these operators and functions are clearly visible when creating the model, rather than being buried underneath a cell in a spreadsheet. This allows — in fact, requires — you to specify exactly how variables interact.
3. Simulation models are designed to handle a large number of variables which interact and change over time. Such an environment can therefore model more effectively the actual workings of markets with many participants.
4. Using a simulation model environment forces the analyst to explicitly identify the variables, and describe their interaction over time. Indeed, part of the relative difficulty in using the environment is the requirement to explicitly describe all these factors in a manner that is absolutely consistent. While within a spreadsheet environment you can easily mix apples with oranges, it is more difficult to do so within MATLAB.[23] While you can make errors using simulation models, gross specification errors are less likely.

---

[22] As an illustration, note the size of the spreadsheets used to forecast business income statements. It is not uncommon for these to be 10 columns wide by 40 rows long, and many are much larger. Even allowing for half of the cells to be used for formatting, notes, or white space, that is 200 actual data cells. Each cell, in turn, has either data input or a formula, and one can only infer what that is from the context.

[23] For example, a common error in spreadsheet environments is mixing balance sheet items (which are stocks) with income statement variables (which are flows). It is quite easy to, for example, add expenses to assets to produce a bogus concept that, nonetheless, is added up correctly. If you attempt to do this within MATLAB or Simulink, you may receive error messages, as the matrix dimensions may not match. Remember that in a spreadsheet, if you can put numbers in a column, the program will add them up!

Why Use MATLAB and Simulink?

Once you step beyond spreadsheets and statistical software, the available choices dwindle. In our experience, mathematical modeling software provides enormous advantages to those who must model real-world markets.

Using MATLAB

We have found that MATLAB and its companion simulation software, Simulink, offer unparalleled power, reliability, transparency, and flexibility to analyze complex economic phenomena and graphically display the results, as described below:

Power: The ability to use almost every mathematical technique now in use and custom-program many more gives the MATLAB user extraordinary capabilities.

Reliability: MATLAB code is close to a high-level computer language such as C. It therefore does not have the undocumented features that are common in operating systems and popular office software, particularly the versions that are designed for home or light office use on Microsoft Windows platforms.

Transparency: All MATLAB functions, formulas, and data can be accessed by the experienced user. This means there should be no "black box" calculations. Even intermediate and temporary data in the midst of being calculated in a function file can be viewed through the debugger.

Flexibility: MATLAB was not originally designed for economics — but then, it was not originally designed for the many fields in which it is now used. However, the building-block nature of the commands allows a savvy user to create almost any conceivable application.

Using Simulink

In addition to the advantages of MATLAB, there are additional advantages in using Simulink for economists, business analysts, and others interested in solving problems in economics, finance, public policy, and business. They are as follows:

1. Many economic questions are better answered with a simulation model than simply guessing, attempting to construct an analytical example, or using spreadsheets to mimic part of the underlying process.

2. The entire environment allows for rigorous, exact, multiperiod calculations using a very large array of mathematical calculations.
3. The Simulink environment allows for the model to be built and displayed graphically, enabling a user to see directly how each variable affects another. While the graphical approach is equivalent to sets of equations (indeed, it is a set of equations), the model schematics are often easier to understand.
4. Extensive graphical and diagnostic tools are available for reviewing and reporting on the model and the data.
5. Given the model's graphical output, the methodology can be presented in a transparent manner, allowing reviewers to assess directly and independently how the results were obtained, what data were used, and how the calculations were performed.
6. The environment is designed as a true simulation model and includes parameters, algorithms, and a software interface designed for that purpose.

Drawbacks of MATLAB and Simulink

In the impressive list of attributes above, there are two notable absences: "easy to use" and "inexpensive to acquire." However, economics is about trade-offs. Through this book and others recommended here, we can help the interested reader quickly use MATLAB to perform advanced analyses and create impressive graphics.

It is still harder to perform simple calculations in MATLAB than it is to perform the same tasks in a spreadsheet. It is sometimes harder to perform rudimentary analysis (such as a present value, straightforward financial statement analysis, or exploratory data analysis) in MATLAB than in other software. Reporting results, especially of simple calculations, also takes more work. However, the power and other advantages have convinced us that the environment is the best available for many uses in demanding analytical tasks such as those presented in economics, finance, public policy, and business.

When To Use Mathematical or Simulation Models

Table 2-2, "When to Use Simulation Models," summarizes our advice on when to use simulation modeling, spreadsheet, and statistical or mathematical software.

When To Use GIS and Database Software

We integrate the use of GIS in business economics problems in which location is a major factor. Although neglected in the economics literature for the past

TABLE 2-2

When To Use Simulation Models

Task	Spreadsheet Software	Mathematical or Statistical Software	Simulation Modeling Software
Add up columns of numbers, such as in most accounting tasks	Recommended		
Calculate discounts, simple multiplication and division of multiple numbers	Recommended (for one- or two-period analyses)	Recommended (for multiple-period analyses)	
Test out various specifications of a model, given the data		Recommended, especially if statistical inference is required	Possible use, especially when specification is complicated
Project variables many periods into the future		Recommended (for simple models)	Recommended (for simple and complex models)
Considering multiple scenarios, given a model with interaction among variables and changes in a handful of variables			Recommended

century, location is an essential — and sometimes primary — element of many problems in business economics.

We describe explicitly the use of GIS in this book in a few instances. However, for many other applications, we incorporate spatial analysis without explicitly describing the use of GIS. We encourage the reader to consider how detailed information on location, distance, proximity, and other spatial characteristics affect the analysis. In many cases, GIS can assist the user in quantifying the effects of these variables.

Similarly, we describe the importance and acquisition of data in several sections of this book. For many users, relational databases will be of great use in storing and retrieving the data. For others, the time and expense necessary to use such software will be prohibitive, and other methods will be more than adequate.

An Admonition

Using a true mathematical modeling environment overcomes the limitations of spreadsheet models and allows much more powerful analysis. However, you can make errors using MATLAB just as you can using Excel. Therefore, we offer the following warnings to users of mathematical and simulation models:

1. *Document your models and analysis.* This helps to avoid making errors in the first place and improves your ability to catch them in the future. We suggest methods of doing so in Chapter 3, "MATLAB and Simulink Design Guidelines," and in Chapter 4, "Importing and Reporting Your Data," as well as in other chapters.
2. *Use style guidelines.* In this way, it is easier to understand the structure of models — either the ones recommended in this book or the ones you develop yourself. We devote Chapter 3, "MATLAB and Simulink Design Guidelines," to this topic.
3. *Build models so that they report intermediate results and maintain data consistency.* "Garbage in, garbage out" applies to the most sophisticated models as well as the most simple ones. Knowing your data at the start and following it all the way through will ensure that you do not fall prey to the dreaded "garbage in, garbage out" syndrome.
4. *Remember the maxims of business economics.* These maxims, included in Chapter 1 under "Maxims of Business Economics" on page 15 are more important than any software. For example, the admonition "Think first, calculate later," applies whether you have an abacus or an elaborate simulation model. Having a powerful hammer requires you to be a careful carpenter!

3

MATLAB and Simulink Design Guidelines

The Importance of Good Design

It is easy to underestimate the importance of good design and good design guidelines for economic models. In this chapter, we first define "design," and then provide guidelines for designing MATLAB and Simulink models. We also suggest simulation parameters that will work well for most models used in business economics. This chapter provides quite specific advice for those working in the MATLAB environment; for those working in other software environments, it suggests good practices that will need to be adapted to the specific environment.

What is Design?

The design of a model includes the applied analytical techniques used; the style used to organize the model; the methods of storing, naming, and reporting data; and the display of results of that analysis in tabular or graphical formats. We cover these topics as follows:

1. The fundamental architecture of models is covered in the application chapters.
2. The style guidelines, including methods of naming variables, ordering equations, laying out equations (or their graphical expressions such as in Simulink), and describing the model, are covered in this chapter. These guidelines are generally followed in the applications shown in the other chapters.
3. Methods of collecting and storing data, identifying the sources of the data (sometimes known as metadata), and reporting it are covered in this chapter and in Chapter 4, "Importing and Reporting Your Data."
4. Methods of accurately displaying the most important data to maximize understanding and minimize distortion, misleading impressions, and distractions are covered in Chapter 17, under "Rules for Effective Graphics" on page 426.

Why Use Good Design?

There are several reasons for carefully considering how to design models, graphics, and tables:

1. Good design makes it easier for the creator of the model to understand his or her work, improve it, and use it.
2. Good design makes it much easier for others to understand a model, make suggestions, or ask intelligent questions about it.
3. Good design makes errors and weaknesses in logic more noticeable and therefore more likely to be corrected.
4. Good design makes the results more credible.
5. Good design makes the results more intelligible.

Setting Up the MATLAB Workspace

The first place to start with good design is the design of the workspace. Just as it is difficult to find and complete work in a disorganized office, it is easy to work in a well-organized software environment.

We suggest the following ways of organizing the MATLAB software environment.

1. Create your own startup file, which should accomplish the following :
   a. Display any greetings or messages unique to the user. An example of this is in Chapter 1, under "A Simple Startup File" on page 7.
   b. Change the default directory to the one in which the model files are normally stored.
   c. Read in any parameters that are used repeatedly, such as simulation start and end dates for Simulink models. See "Simulation Parameters and the SIM command" on page 49.
   d. Input any optional settings.

   A suggested startup file is included in the Business Economics Toolbox that is a companion to this book.
2. Store your model files and any related data files in an entirely different directory, drive, or location other than the MATLAB program files. You will probably install, delete, and reinstall various

versions of the software many times, and keeping program files separate from data files is always good practice.

3. Back up your model and data files on a regular basis.
4. Within the location you have selected for your model and data files, make separate directories or other subdivisions for special toolboxes, projects, or data. For example, purchasers of this book that have downloaded the Business Economics Toolbox will want to place it in a separate directory within this location.
5. Document all the data files you create, even if it involves simply labeling the files properly so they can be distinguished by project or type.
6. Document all your custom-programmed files, including Simulink models and MATLAB script and function files (m-files). See "Documenting Your Models" on page 53.
7. Use consistent style guidelines in your models and m-files, even if they are not those recommended here!

Note that different operating systems and different versions will require somewhat different syntax, commands, and installation procedures. However, following these simple suggestions will make your work more efficient.

Organizing MATLAB Models

We will call a model in MATLAB a collection of commands that accomplish an analysis. Such a model can be quite simple and need not be dynamic, multiperiod, or multivariable. One of the powerful features of the MATLAB environment is the ability to open, review, edit, and save existing command files, as well as to create new ones. In this section, we briefly overview these files and suggest guidelines for creating, using, and describing them.

MATLAB Scripts and Functions

Most of the work done in the MATLAB environment is performed by command files that end in the suffix .m, such as print.m. These files may have been included as part of the MATLAB software, created by the user, or created by a third party. Because of the suffix, these files are often collectively called m-files.

There are two types of m-files — scripts and functions. Scripts are commands that take no argument — they perform exactly the same tasks each time. Scripts operate on the data already in the workspace, which can be directly accessed. An analogy in the physical world are commands such as "go to sleep" or "blink your eyes." These commands mean essentially the same thing to a person, no matter what the context.

An example in the physical world is the command "run." It only makes sense if you provide an additional parameter, namely a direction in which to run. Functions, on the other hand, are designed to operate on the arguments provided to them and can return specific output. In addition, functions operate within a separate workspace. Only the data you provide them (through the arguments native to the function) can be used in function calculations. Similarly, only the output specifically generated by the function will be available in the workspace afterwards.

Functions can be very simple or very complex. For example, the command `abs(7)` calls the `abs` function to calculate the absolute value of the argument to the function 7. There is one argument and one output. Other commands take multiple arguments and return multiple outputs.

Callbacks

A command file that runs a set of commands when prompted by another command is sometimes called a "callback." Callbacks that are customized to a specific model can be very useful. We discuss their use for Simulink models in "A Master Simulation Callback" on page 52. They are also used to execute tasks specified by a user's interaction with a graphical user interface (GUI), although we do not cover this topic in this book.

Design Guidelines for m-files

This is not a book on programming. However, there are some style guidelines that we follow in creating MATLAB commands, including both scripts and functions:

1. Describe your m-file. MATLAB reserves a specific part of an m-file for this purpose. The H1 or Help line which follows the line that begins the file will be displayed in certain listings of the files in a directory. Make sure you include a description here when you create an m-file.

   Describe the various sections of your m-file using comment lines.
2. Make commands and models as simple as possible. Do not overengineer your model or make it more complex than it already is.
3. Make commands robust. A robust method is one that works well under different circumstances. Try to make your applications

robust enough to survive small human or computer errors — and to stop when major errors are committed.

4. Use hierarchy to organize. A well-designed model is often modularized — it is put together in different building blocks. Using such an approach makes it much easier to troubleshoot and to run consistently well. It is also a tremendous cost savings as further revisions are likely to be requested when just one or two sections of the command need to be updated.

   MATLAB has several attributes that make it easier to use a hierarchical, modular approach. In Simulink these include the ability to organize portions of a model into a subsystem, and in MATLAB the ability to use functions and subfunctions which operate in a separate workspace.[23a]
5. Whenever possible, separate data, parameters, and calculations. This makes your programs more flexible, powerful, and transparent.
6. Make sure that the input data, important intermediate results, and output data are provided to the user. This again makes the program more transparent and easier to use over time. As an aid to this, consider the recommendations for the use of data structures in Chapter 4 under "Using Structures to Organize Data" on page 74.
7. In models that perform multiple tasks, consider including commands that cause information on the progress of the tasks to be displayed in the workspace. "Master Simulation Callback" on page 58 is an example of an m-file in which different sections are described by comment lines and which causes information to be provided to the user at different stages.
8. Carefully consider how to name your m-files, so that they do not confuse the user, duplicate another command filename, or become either too specific or too general.

Initializing a Model

We recommend that a model be *initialized* by a separate program. Such a program will often call up the data, and request any additional parameters from the user. It may also run the required functions (after having supplied the data), and then save and report the results.

Initializing a MATLAB Model

Typically, models created within MATLAB are initialized manually when first developed. Over time, a well-developed model can be initialized in a script

[23a] A "workspace" is a section of computer memory.

m-file. We sometimes call this a "Master [project name]" m-file since it will do all the necessary steps to open the model, input the data, run the simulation, and save and display the output.

Initializing a Simulink Model

We recommend that a Simulink model be initialized by a special callback program for this specific use. See "A Master Simulation Callback" on page 52. We also provide such a callback in the Business Economics Toolbox that is a companion to this book.

Introduction to Simulink Models

One of the major advantages of working within the MATLAB software environment, especially with Simulink, is the ability to comprehensively model and then simulate complex, dynamic economic phenomena. By dynamic we mean changing over time. Most problems in business economics and finance involve multiple variables that change over time. Thus, simulation models are a natural and powerful tool for these problems.

Simulink is not the only environment that allows simulation modeling, and indeed we present other applications of dynamic models in this book. However, Simulink has at least two significant advantages over most other environments:

1. Economic models are typically described as a system of equations involving many variables and some fixed parameters which change over time. These same equations can be modeled in Simulink but in a graphically intuitive way, like using building blocks. This is often easier than describing the same model using equations.
2. Simulink models can be constructed in a hierarchical fashion. A number of equations — all constructed and displayed graphically — can be organized into a separate *subsystem* within the overall model. These subsystems, in turn, can be organized to provide an understandable top-level view of the entire model. Constructing and visualizing a model in this manner makes it much easier to understand both the "big picture" and the specifics.

In addition to these distinctions, Simulink models can use the vast library of MATLAB functions, as well as specialized libraries for specific purposes.

The Simulink Model Schematic

After a brief review, it is fairly easy to read a Simulink model. We say "fairly easy" because it is much easier to understand than the same systems described as a set of equations, but harder to describe than, say, calculating the sales tax on a hamburger.[24] Because a Simulink model is illustrated in a way that shows how one variable affects another, we sometimes call the illustrations a *schematic* of a model.

It is easiest to think of a Simulink model as an active flowchart with variables flowing from one end of the page to another.[25] The variables used in the model are introduced on the left side of the page in boxes that contain their definitions. Other boxes perform all types of functions and operations. Lines connect these boxes indicating that the results of their calculations are sent as signals down "wires" or "pipes" to the next block. By following the wires, you can see directly how one variable affects another and, ultimately, how all the inputs result in the output.

Time and States in a Simulink Model

Simulink models are actually a set of mathematical difference equations displayed as graphical entities in a set of systems and subsystems. Time is an essential element in a dynamic system. For each time period, variables are fed into the model, used in calculations, and then the results are reported. The results from one period are available for use in the next period. Because a change in a variable in one time period can cause further changes in the next, running the model involves simulating the effects of all the variables on each other over multiple time periods.

While it is straightforward to think of Simulink models as a set of self-executing blocks in a flowchart, this is not technically correct. The equations calculate states of the model for each tick of the clock. Each block then provides information to the model at the next tick of the clock. Some blocks retain their states after each tick of the clock and can use that information at the next tick. Others, such as a simple multiplication block, do not.[26] Many blocks use parameters that affect the operations they perform.

---

[24] Describing the income tax levied on the earnings needed to buy that hamburger, however, is probably a different story.

[25] We rely on some of our style guidelines here. A user could design a Simulink model so that variables affect each other haphazardly with blocks all over the page. However, that would defeat at least one of the purposes of using this approach — an easier-to-understand presentation.

[26] See *Using Simulink, Version 5 Guide* (published by The MathWorks), Chapter 2.

Example Simulink Models

We start with a few examples which will illustrate how a Simulink model can perform complex, multistage calculations.

Example Economics Model

In Chapter 5, "Library Functions for Business Economics," we illustrate a simple model that projects sales given a change in price. The model incorporates a number of specialized business economics blocks that are described in that chapter. The house net benefit model described next is more complex but does not use any specialized blocks, so we introduce it first. Some readers may want to look ahead to Chapter 5 to see the other model which is described in "Example Simulink Model" on page 114.

House Net Benefits Model

Figure 3-1, "Simulink model example: house net benefits," illustrates a Simulink model that calculates the costs and benefits of owning a home. The model incorporates both the costs of ownership (including interest, insurance, and maintenance) as well as the benefits (the rental equivalent of living space, the tax deduction for mortgage interest payments, and any capital gain or loss on sale).

This is the type of analysis that is frequently done — people commonly talk about "making money on a house" — but rarely done properly. To properly account for all these variables over a multiple-year time frame is a serious undertaking.

Top View

In the top view of the model, the flow information starts in the upper left of the schematic in Figure 3-1 in the box labeled "home value subsystem." In this subsystem, we calculate two variables: the initial value of the home and the current value. Both these variables are needed to calculate the total costs and benefits.

We also see that, once calculated, the values for initial value and current value are used in other subsystems. Visually, it appears that the values of these variables go through the wires or pipes to the next subsystem. The visual appearance is consistent with reality because the values that emerge

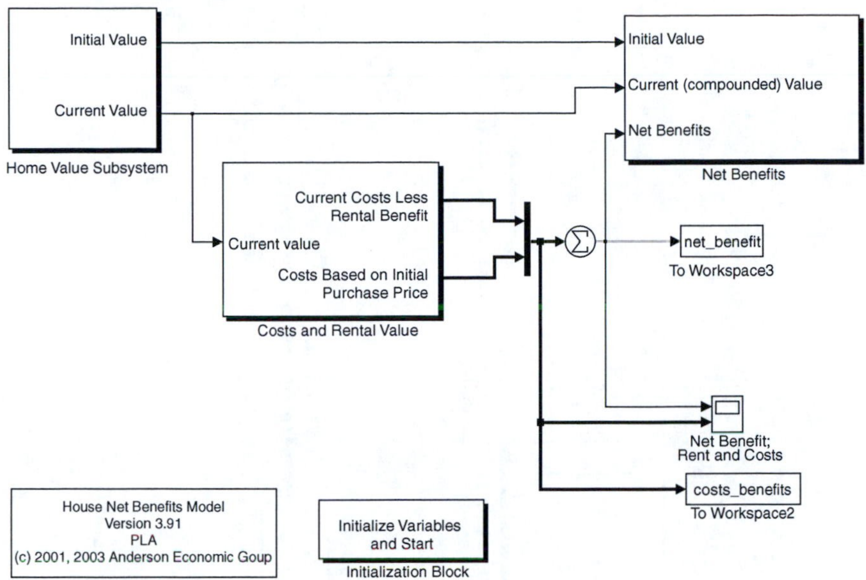

FIGURE 3-1
Simulink model example: house net benefits.

from the set of equations represented in one subsystem flow into equations represented by another subsystem.

The results from the home value subsystem are used in two other subsystems: (1) Net Benefits and (2) Costs and Rental Value. The results of the equations in these systems are also shown being summed (in the block showing the Greek letter *sigma*), being collected into variables that will be saved in the workspace for later review, and being represented graphically in a "scope."

Costs and Rental Value Subsystem

The equations within the Costs and Rental Value subsystem are illustrated in Figure 3-2, "Costs and rental value subsystem." You will note that the pipes running into a subsystem are illustrated within that subsystem as inport (input) blocks, which should be labeled with the proper variable name. Those running out are called output or outport blocks.

Other Costs Subsystems

Subsystems can nest inside one another so that the user can build multiple levels of a model and work or illustrate one level at a time.

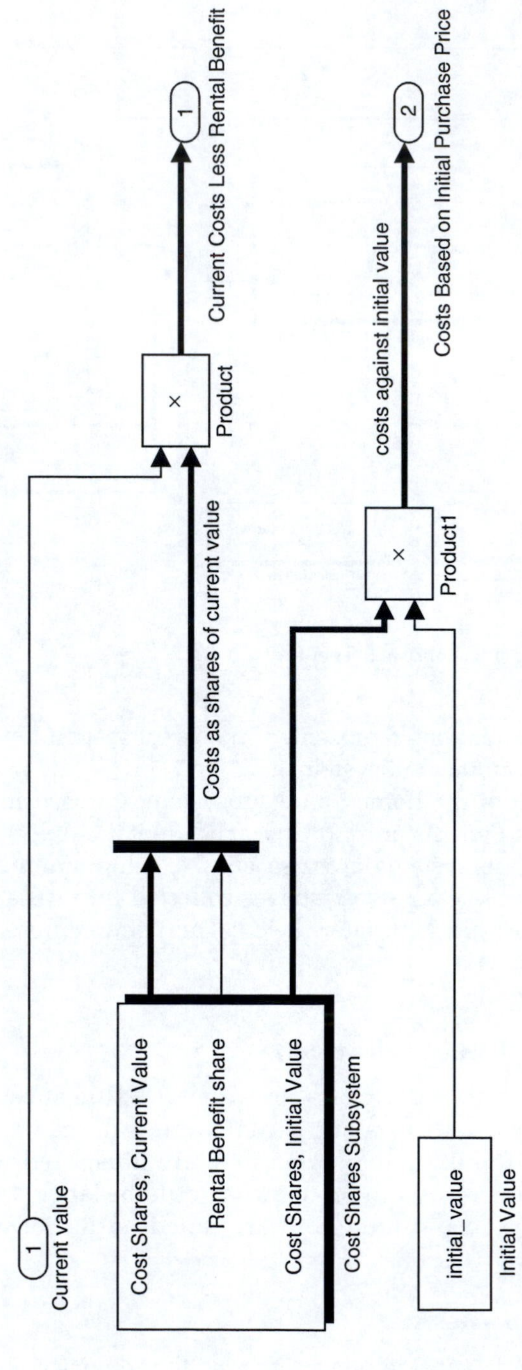

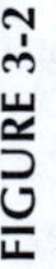

FIGURE 3-2
Costs and rental value subsystem.

The Cost Shares subsystem (which appears within the Cost and Rental Value subsystem) is illustrated in Figure 3-3, "Simulink model example: cost shares subsystem." Within that subsystem is still another layer which is the subsystem illustrated in Figure 3-4, "Simulink model example: cost subsystem. "

Output Blocks

Returning to the top-level view in Figure 3-1, "Top-level view of Simulink model," we see the results of multiple calculations are the variables net benefit and costs_benefits. These are deposited into the to_workspace variable blocks that list the variable names. Using these blocks you can inspect, analyze, and make further calculations with the resulting values.

Multiplexed or Vector Output

You will note that certain lines are wider than others. The wide lines indicate that more than one variable is being carried in the same "pipe" or "wire." This is known as multiplexing. A tremendous advantage of Simulink models is the ability to carry and preserve the values of multiple intermediate variables in this manner and easily sum these up into a single variable. The addition block, with the Greek letter *sigma* illustrated, sums up these intermediate variables into a single variable, which has a single value each time period.

Simulink Model Design Considerations

Style Guidelines for Models

To ensure that others can understand your Simulink model, and can follow the logic and mechanics of the model, you should adopt a set of style guidelines for your Simulink models. Such guidelines have the following purposes:

1. Ensuring that inputs, outputs, models, subsystems, and other building blocks are each distinctively labeled and that each class of building block is displayed differently
2. Minimizing the number of distracting, confusing, or spurious bits of information displayed

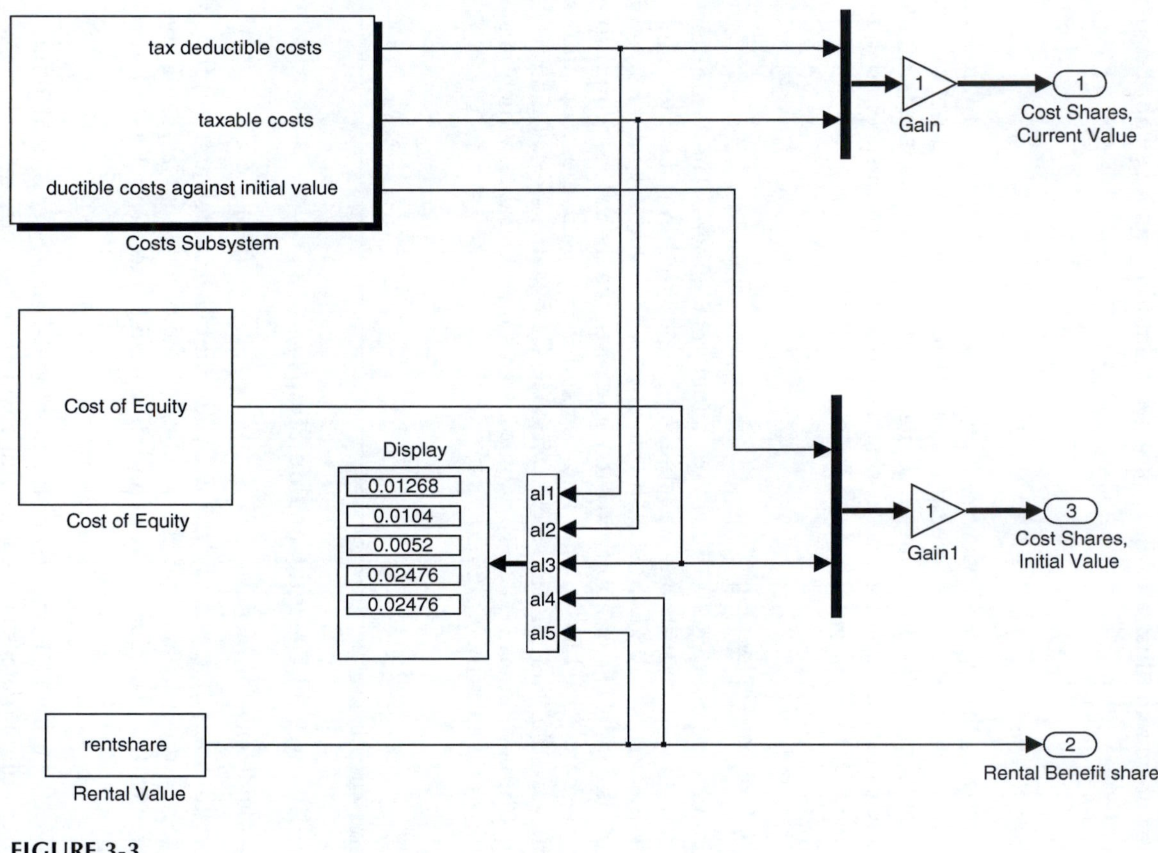

FIGURE 3-3
Simulink model example: cost shares subsystem.

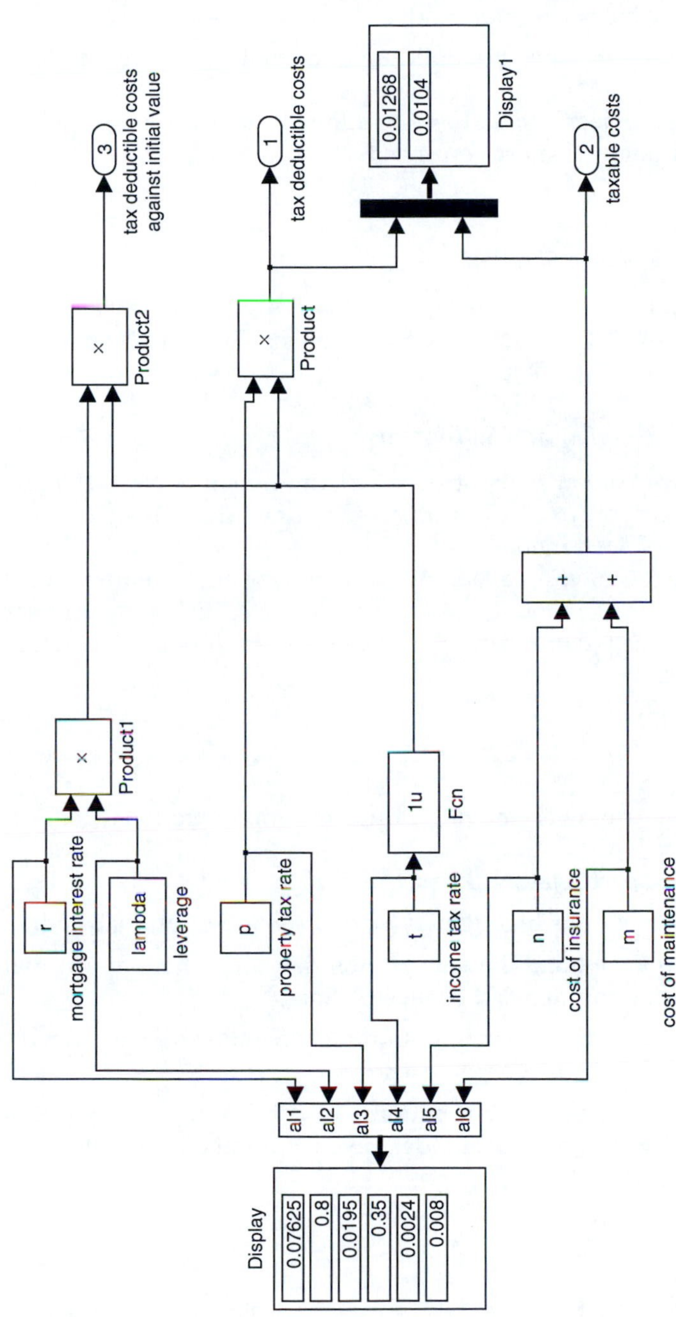

FIGURE 3-4
Simulink model example: cost subsystem.

3. Allowing the user to logically group the systems, subsystems, and building blocks of your model
4. Improving the speed of building and troubleshooting models

Based on these objectives, we have developed a set of style guidelines that you can use or adapt to meet your needs.

Major Style Guidelines: Simulink

The major graphical styles we recommend and use for Simulink models are described in Table 3-1, "Major Style Elements for Simulink Models."

Style of Model Overview and Subsystems

We discuss how to organize the structure of Simulink models in "Organizing Simulink Models" on page 45. In this section, we discuss style guidelines for the display of such models.

Simulink models have an overview pane that covers the entire model and subsystems within the model that can be illustrated in separate panes. We suggest slightly different design guidelines for the top-level view and for the subsystems.

Top-Level View

The top-level view of every model should include the following:

- A title properly stating the purpose of the model.
- Major subsystems highlighted by drop-shadows and labeled.
- Connections among the subsystems that are, where appropriate, made using multiplexed connector lines.[27]
- If used, a clearly labeled initiation block would show up in red on the computer screen of the user.
- A description box at the bottom describing the purpose of the model. The author of the model, revision number, and date should be included.

[27] In the example shown in Fig 3-1, we could have used multiplexed lines to connect the major subsystems. However, in order to show the different streams of data flowing from the first subsystem, We chose to reveal the number and titles of each by de-multiplexing the signals before bringing them into the subsystem on the right. The Simulink blocks that perform the combining of multiple lines and then separating them are called Mux and Demux.

TABLE 3-1

Major Style Elements for Simulink Models

Element	Style	Comments
Major subsystem	Large box, drop shadow	This style highlights the most important portions of the model and makes use of one of the most important advantages of Simulink: the ability to easily grasp how data in a model flows from one group of calculations to another
Description box	Small box in bottom of top-level model view	Should describe the title of the model, the author(s) and sponsoring organization, and the revision number or date. The Model Info Simulink block can be used for this purpose
Initialization block	A small block labeled "initiation block," which runs an initiation script when selected and clicked; this is typically outlined in red	Runs the initiation script that sets default values, reads in data, and provides parameters used in the simulation
Titles	Bold letters at the top of the frame that state the purpose of the model or subsystem	Use a consistent font and type size, and place in the same position in each view

Subsystems

- Subsystems need not have initiation blocks or description boxes.
- Subsystems can have more text when needed to explain the model.
- Subsystems can include scopes, graphs, and other modules that are useful in debugging or checking results but which would clutter up the top-level view.

Example: Top-Level View

Figure 3-5 shows the top-level view of an economic impact model that is notable for its simplicity.

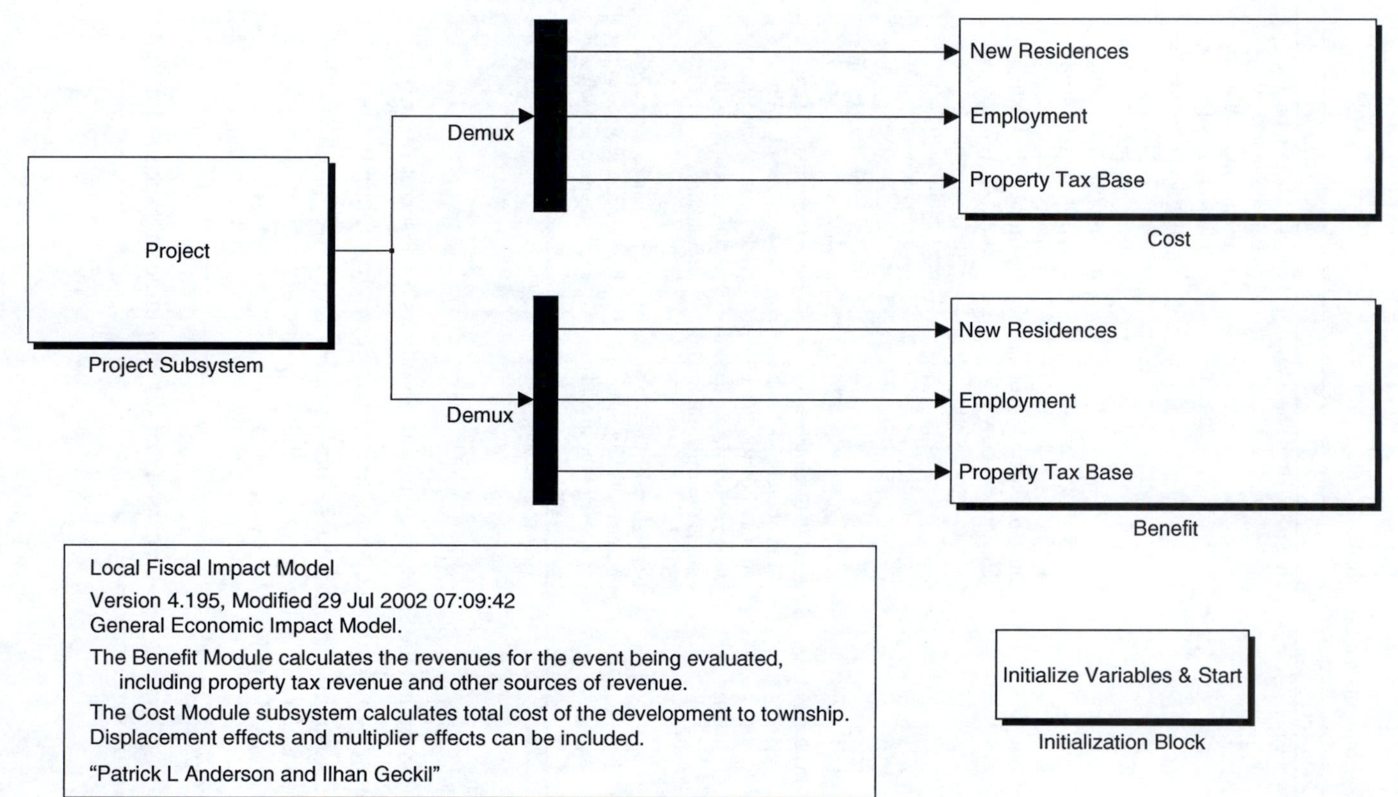

FIGURE 3-5
Top-level view of Simulink model.

Organizing Simulink Models

Models should be structurally well organized, for a number of reasons, including:

- A well-organized model reflects well-organized thinking.
- As no system is perfect, good organization helps you catch errors and debug quicker.
- Good organization helps convince your model's users (or consumers of your model's output) of the integrity of your approach.

Given this motivation, we recommend the following guidelines for organizing your model:

1. Identify the major factors that drive the other variables and combine them into one or more sets of inputs.
2. Collect similar inputs into major subsystems that can be summarized by a single concept. Mask the subsystems so that the top level of the model shows a properly labeled subsystem containing related variables, rather than a tangle of signals and individual variables.
3. Collect similar output variables together in masked subsystems with inputs created for input and intermediate variables.
4. Place the input variables on the left side of the top-level model view and output variables on the right. This makes the model read the same way as sentences in English and other languages.[28]
5. Connect the subsystems together, to the extent possible, using signals that are multiplexed into big pipes. Where possible, avoid overlapping the signal pipes or crossing them.

The top-level view in Figure 3-5 illustrates these guidelines well.

[28] These are style guidelines and therefore suggest conventions that fit well within the culture in which the primary audience of the book will likely live and work. For those in different cultures and subcultures (many organizations have strong subcultures of their own), these should be modified.

The Art of Model Organization

While these written guidelines provide rules to follow, the practice of organizing models is as much art as science. We often find that models are organized, reorganized, and then re-reorganized for presentation. Following a few simple rules at the outset helps minimize the trauma of these changes.

Style Guidelines for Specific Blocks

In addition to the above guidelines for the organization and display of your model, we also recommend a consistent style for specific blocks. Our recommendations are summarized in Table 3-2, "Styles for Specific Blocks."

Initializing a Model

In well-designed models, there are often a number of variables, parameters, and callbacks that need to be run to initialize the model or make it ready

TABLE 3-2

Styles for Specific Blocks

Element	Color of Block	Comments
Inputs and outputs (inports and outports)	Blue	Inputs and outputs should be identified clearly. Some users may prefer to have their outputs (outports) a different color; if so, we suggest orange
Constants	Green	The blocks are always growing the same data
To workspace	Magenta	Highlights data that will be sent to the workspace. You may wish to set parameters associated with this block and then copy it several times
Scopes and displays	Light Blue	These do not affect the working of the model, and making them a light color allows the eye to avoid concentrating on them
Mathematical operators (sum, multiplication, powers, etc.)	Black	These are the basics of the model and should be arranged for clarity, not highlighted

for operation. As an analogy, think of driving a car. Before you can safely and effectively drive a car, you should have gas and oil in the engine, air in the tires, a driver's license, a set of directions, and knowledge of traffic laws. If we could initialize each car and driver with all these before they started the car (especially the safety instructions), we would all be better off.

When a simulation model is used to analyze complex economic phenomena, a careful initialization brings the right data to the model at the right time.

Initializing a Simulink Model

Initializing a Simulink model establishes the variable and parameter values and runs any preparatory callbacks. An initialization block makes this easy by allowing the model user to simply click on the block to initialize the model. Figure 3-6 illustrates such a block.

Creating an Initialization Block

The following instructions are based on the most current version of MATLAB and Simulink at the time this text was written.[28a] Later versions or versions used on different operating systems can be expected to have minor changes in the interface.

Use the Simulink Library Browser, select the subsystem block, and copy it into the top-level view of the model. You can go inside the subsystem to remove any sample variables or connections. Then (if you wish to follow our style guidelines) add a drop shadow and color the foreground red using the right-click functionality of your mouse.

Display: To make the block display the desired text, select the subsystem and use the Edit | Mask Subsystem command. Once inside the mask editor, you will find tabs for separate portions of the editor: Initialization, Icon, Parameters, and Documentation.[28b]

Type the following in the icon portion of the window:

Code Fragment 3-1. Display Command in Masked Subsystem

```
disp('Initialize Variables & Start' );
```

Help Information: To add help information, you may wish to add in the "Mask Help" portion of the Documentation window the following:

[28a] This procedure was confirmed with release 13.1.

[28b] You can also remove any inports or outports to give a clearer appearance.

Code Fragment 3-2. Help Information in Masked Subsystem

```
Double click to start simulation.
```

Initialization: For the Initialization portion of the block you should add the following:

1. A description of the block in the field allotted for Mask Type.
2. Any commands needed for variables that must be available only within the space allotted for calculations made within the masked block. For initialization commands this is usually nothing.

The MATLAB help system provides excellent guidance on this topic.

Setting Parameters for Initialization Blocks: Highlight the block, then use `gcb` to find the name. Then use the following code line to set the name of the block:

Code Fragment 3-3. Setting Parameters for Blocks

```
set_param(gcb, 'name', 'initialization:')
```

It will then show up properly on the model browser.[29] See Figure 3-6, "Initialization block in a Simulink model."

Running Initialization Scripts on Command

To make the block call an initialization script you have already written, take the following steps:

1. Right-click on the block and issue the Block Properties command.
2. In the panel marked "callbacks," select Open Function and insert the name of the initialization script you wish to call. This will cause the script to run whenever you double click on the block.
3. While in this property editor for the block, add descriptive text in the "general" panel and select the attribute you want duplicated.
4. Block Annotation in the panel:

[29] This can also be done using the "Block Properties" menu for the subsystem, as described below.

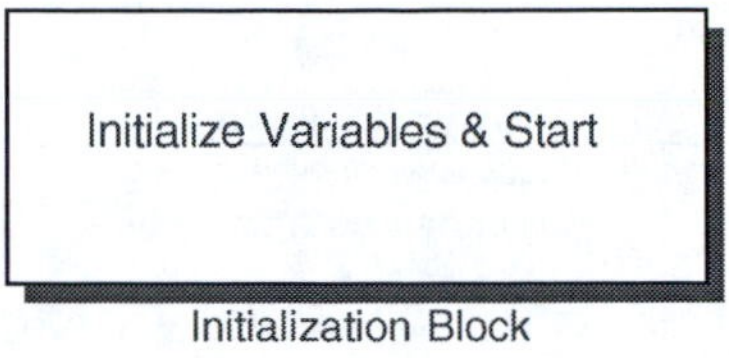

FIGURE 3-6
Initialization block in a Simulink model.

Code Fragment 3-4. Self-Describing Variable for Blocks

```
%<description>
```

This will cause the phrase written as the description of the block to show up in the model window. As an alternative, you can also specify a pre-load call back, using the model properties dialog box.

Simulation Parameters and the SIM Command

The simulation model may not run if any one of a number of parameters is not set correctly. The following section describes the parameters to set for each simulation model, and how they are set.

Time and Solver Parameters in Simulink

One of the more confusing aspects of Simulink can be its time and solver parameters. For many challenges in business economics, these can be set quite simply as discrete units of time that are consistent with the conventional description of the problem at hand. Variables that are measured only at specific intervals are known as *discrete* variables within the mathematically precise MATLAB environment. Most business variables such as sales, revenue, profits, employees, etc., are typically measured in discrete intervals.

Simulink has the power to deal with problems in mathematics that require continuous variables and solvers that find solutions for more difficult (in a matrix algebra sense) problems. This power is often not required for business economics problems. However, a few applications demonstrated later in this book — such as iterative models in Chapter 10 and dynamicroprogramming in Chapter 11 — will make use of it.

TABLE 3-3

Key Simulation Parameters

Parameter	Suggested Settings	Comments
Start time	Current year or period (e.g., quarter or month); may use Tstart variable	Business problems often have time as the key index
Stop time	End of forecast period; may use Tstop variable	Use same units as for start time
Solver	Fixed-step, discrete	Most business problems take place in a time period that can be indexed using specific time intervals
Fixed-step size	Unit of time, such as year, month, or quarter; may use Tstep variable	Must match the start and stop times

Setting Parameters through a Menu

Simulink allows key parameters to be set through a menu, and this is the easiest way to get a single model started. The key parameters are described in Table 3-3, "Key Simulation Parameters."

Simulink has a menu that allows the manual setting of these parameters. The menu can be reached by the Simulation | Parameters keystroke sequence or by manually selecting the menu item.

This *ad hoc* method may be sufficient for many purposes, but there is a quicker method for those that routinely model business economics phenomena. We describe a method of automating the process of setting these parameters using the SIM command with the selection of specific options.

The SIM Command

The SIM command causes MATLAB to simulate a model using certain parameters defined by the command user or set by default. To review the options and general command syntax, you can review the Help information for the command.[30]

We recommend that users explicitly set the options described below and provide code fragments showing how to set them.

Workspace and timespan: Workspace and timespan are important arguments. The workspace is the (virtual) subset of the universe in which variables exist. If the variables you create are in the base workspace, then you need to work in that workspace. Normally, that is what happens. However, computer operating systems usually set up multiple workspaces so that variables in one do not leak into another. For example, MATLAB sets up separate workspaces for functions.

[30] This can be obtained by typing help sim at the command prompt or by looking through the Simulink user guide.

Workspace for standard models: The source workspace is the workspace in which the model seeks variables. The default for this is the base workspace. This will normally work well. However, sometimes leaving it in this mode will result in errors. In such cases, try setting this at current.

Workspace for Web server models: The source workspace is often different for workstation models and those run on a Web server. For Web server models, use the current workspace.

Saving Parameters in a Structure Variable

The following line will set just this one parameter in the structure called sset:

Code Fragment 3-5. Setting the Source Workspace

```
sset = simset('SrcWorkspace','current');
```

Timespan: The timespan is the set of time periods beginning with the starting time period and incrementing each period until the end. This is a critical setting for simulation models since the entire model is recalculated each increment. We recommend that you establish the starting, stopping, and step times in your simulation m-file. A code fragment that accomplishes this for a discrete system follows:

Code Fragment 3-6. Setting Time Parameters

```
Tstart = 2003;

Tstop = 2013;

Tstep = 1;

Years = [Tstart:Tstep:Tstop];
```

This fragment also creates the variable Years that can be used to summarize the output for analysis or publication. In this case, we added an apostrophe after the vector definition to convert it to a column vector.

Obviously, if you are using a continuous system or running data of a different periodicity, you would change the Tstep variable and probably rename the Years variable as well.

Model: The model is the most obvious parameter. Make sure you have it correctly named and you have it enclosed in single quotes. By convention, you do not include the file extension.

Setting the Parameters

In general, we recommend you formally set the parameters for a simulation model in a special structure variable. You can do this by including the following code fragment in your main m-file:

Code Fragment 3-7. Setting Options and Simulating Model

```
...
sset = simset('SrcWorkspace','current');
sim('model', [Tstart Tstop], sset, [])
...
```

The first line sets the options parameters for the simulation, particularly the source workspace, and saves it in a structure called sset.

The second line simulates the model using the starting and stopping time and the parameters set previously in the option structure sset.

Note that if you use the left and right brackets ("[]") instead of a date range, the simulation will default to the standard value.[31] The brackets shown in the code fragment (at the end of the second line) are for a different argument indicating external inputs. Using open and close brackets for this argument indicates no external inputs.

A Master Simulation Callback

We incorporate a number of the recommendations described above in a master callback for Simulink simulation models. This script could be called by an initialization block like that shown in Figure 3-6, "Initialization block in a Simulink model" on page 49 or directly run.

The master callback has sections that perform the following tasks:

1. Initialize the model including setting key parameters such as the timespan and model name.
2. Get data for the variables which may be done from an outside file.
3. Perform any initial calculations if necessary.
4. Simulate the model.

[31] You can review these using the simulation parameters menu command. They may be set by a startup file or a preferences file.
Of course, check the help documents of your specific version to see if minor syntax changes are needed from the examples shown in the text.

5. Graph the results.
6. Optionally, call a second callback that runs base and alternate case simulations.

The script file "Master Simulation Callback" is in the appendix to this chapter.

Using the Master Simulation Callback

The callback will run the example economics model illustrated in Chapter 5, "Library Functions for Business Economics," starting with "Example Simulink Model" on page 114.

Documenting Your Models

Documenting your model is an important and often left-to-the-end step. The following steps help document your model. They can be accomplished while building or adjusting it.

Masked Message Boxes

Message boxes help you know what is going on in your model and communicate that to other users. An easy way to create a message box in Simulink is to create a masked subsystem.

This is straightforward:

1. Create a subsystem in your model using the Simulink menu.
2. Edit the subsystem.
3. In the dialogue box created by the Edit|Mask command, select the Icon tab.
4. Type in the drawing box the disp or display command with the variable you want displayed as the argument.

This method has limitations such as the inability of disp to take more than one argument. The following examples show how you can use the simple method when needed and find creative ways around the disp limitation.

Trigger Boxes

Many times you want to model behavior that starts when some other variable reaches a certain level or on a certain date. You can visually remind users of your model with trigger boxes.

For example, if you have a trigger date in your model, you could define a variable called trigger that is equal to the trigger date. Then, using a masked subsystem, specify the icon for the subsystem.

In the icon dialogue box for a masked subsystem type:

```
disp(trigger)
```

The trigger date then appears, magically, right in the box when you look at the model.

Callback Messages

We suggest that complex simulation models be initialized by a callback program which loads the variables and sets the simulation parameters. The same program can simulate the model, perhaps using two sets of variables.

This approach will be more advantageous to the user and easier to debug if you include messages to the user in the callback program. This is quite easy as shown in the following example:

Code Fragment 3-8. Messages in a Callback

```
% load variables
load myvariables
disp('variables loaded...')
whos;
```

In the example, the callback loads the variables and after doing so displays a short message in the command window indicating this step was successfully accomplished. Then follows a whos command which will cause all the variables in the workspace to be individually listed as well.

Complex String Arrays

Sometimes a simple one-line character string is not sufficient for your message. In such cases, you create a character array that includes more than one string. This allows you to display multiple lines.

You can also change character arrays to cell arrays, which allows for more precise addressing of the elements.[32] The following code fragment illustrates this.

Code Fragment 3-9. Character Arrays

```
%       create character array
s =     char('Hello', 'Neal!');
%       create similar cell array
scell = {'Hello', 'Neal'}
%       convert character array to cell array
schar = cellstr(s)
%       get information on string functions
help strfun
%       display arrays
disp(s)
disp(scell)
disp(schar)
```

The code fragment above will create three different variables containing the same information. The following fragment shows the result of the display commands for the three variables.

Code Fragment 3-10. Character and Cell Array Displays

```
Hello
Neal!
'Hello'    'Neal!'
    'Hello'
    'Neal!'
```

[32] We will discuss character arrays (strings) and related functions shortly.

Using Strings in Message Boxes: You can use these different methods of storing and displaying strings in Simulink message boxes. First, create a character array (for example, "s") with the required text information. Then place the following command in an icon dialogue box or later in a callback or similar script file:

```
disp(s);
```

MATLAB will then create a two-line display such as that in the code fragment above.

String Commands

The use of strings to document models can be tricky because the character array data type is different from the data type of most numbers. Therefore, mixing strings with numbers requires careful handling including the use of special commands.

While we will not provide a tutorial on the different data types in this book, we suggest a few quick references for practitioners.

1. Use the MATLAB help resources for string functions as shown in "Character Arrays" on page 55, as well as for strings in general. This can be accessed by simply typing `help strings` in the command window.
2. Use the `num2str` or `int2str` commands to convert numbers or integers to strings, especially when you want to include a number in a sentence.
3. Use cell arrays to create multiple-line texts. We use this extensively in graphics for titles and text messages. See Chapter 17, especially, "Methods of Specific Plots" on page 427.

A Custom Business Economics Simulink Library

A powerful feature of the MATLAB environment is the ability to create and use custom function, script, and model files. In this section, we describe how to create a library of special business economics model files which can then be used repeatedly in Simulink models.

This book's companion Business Economics Toolbox contains a library of such models.

To Set Up a Library

Use the File | New command after having already started a new model.

To Put Blocks in the Library

Copy blocks from other models into the new library. Be sure to unlock the library before editing or adding anything and then save it.

If you are adding blocks from other libraries, you must Edit | Break Library Link in order to have a separate block saved in the library you are editing. Many frustrating difficulties are caused by not saving the blocks in this manner, as blocks can appear in the libraries but still create unresolved library link errors when you attempt to simulate a model.

To Make the Library Appear in the Simulink Browser

Find the m-file slblocks.m (use the which slblocks command), and put it in a folder which is in your path. We suggest creating your own Library or Tools folder to keep it separate from folders installed by MATLAB automatically.

Edit the slblocks.m file to make reference to the file name and screen name you select. For example:

Code Fragment 3-11. Library Blocks

```
% Define the library list for the Simulink Library
% Browser.

% Return the name of the library model and the name
% for it

% These are MATLAB structures, so use the format
% "Browser(n).Library

% and then "Browser(n).Name" to identify the file
% name (Library) and the screen name (Name). The "n"
% is a number identifying the order in

% the structure.

Browser(1).Library = 'Economics';

Browser(1).Name    = 'AEG Economics Models';

Browser(2).Library = 'Test2Lib';

Browser(2).Name     = 'Test Libraries';
```

This fragment will cause two libraries to appear on the Simulink Library Browser, AEG Economics Models and Test Libraries, each associated with a specific file name.

Troubleshooting

If you have unresolved library links, carefully review, step by step, the source of each block and the referencing of it by MATLAB.[33]

Appendix: Master Simulation Callback

The following code fragment can be adapted to initialize, run, and display the output from Simulink models.

Code Fragment 3-12. Master Simulation Callback

```
%    Base Simulation Callback m-file

%    Initializes and Runs Simulink Models

%  In this example, the model is:
%  economics_example.mdl;

%  May be adapted to initialize and run other
%  simulink models

%

% PLA June 5, 2003

% (c) 2003 Anderson Economic Group,

% part of Business Economics Toolbox

% Use by license only.

echo off;

%-----------------------------------------------------
%   I. Initialize Simulation Model
%-----------------------------------------------------
```

[33] See also MathWorks solution 9009 for more details. These and other technical notes can be accessed through The MathWorks Web site.

```
%-----------------------------------------------------
%   Time Parameters (may be in .mat file);

Tstart = 2003;

Tstop = 2012;

Tstep = 1;

Years = [Tstart:Tstep:Tstop]';  %make column vector;
% use in graphing

% ---------------------------------------------------

% Set model name

modelname = 'economics_example';

%-----------------------------------------------------
%   II. Get Data

%-----------------------------------------------------
%   Get data

% Option 1: from GUI (workstation) or from HTML (Web
% server),

% Option 2: get data from .mat file

%   Load workspace (.mat) file, if necessary

%   (make sure file is located in reachable
%   directory.)

%  load sample.mat;

% Option 3: Enter base data in callback:

%  Input Variables and Parameters, and echo in
%  workspace

display('Echoing Key Variables and Parameters.');

echo on;

% Input1 = 100;

% Input2 = 100;

% Input3 = 100;

% ----Inputs to Initialize economics_example

Elasticity = -1.0;

InitialIncome = 1000000;
```

```
growth = .01 + rand(size(Years))*.03;
% 1% secular growth trend, plus random information
AverageGrowth = mean(growth);
% calculated mean growth
% ----end inputs for economics_example
display('Data read in.')
echo off;
%-----------------------------------------------------
%   III. Initial Variable Calculations (if needed)
%-----------------------------------------------------


%-----------------------------------------------------
%   IV. Simulation
%-----------------------------------------------------
%   Simulate Model
%   Sim command takes model, timespan, options, and
%   external input arguments.
%   Options structure and Timespan should be set
%   above.


%   Simulation Parameters


%   These should be entered in the 'simulation
%   parameters' part of the Simulink model or through
%   the SIMSET command.
%   Source workspace should be set to 'current' for
%   Web server applications;
%   'base' is default and works for workstation
%   applications.
%
%   The following structure (named here "sset") can
%   be used to capture multiple parameters,including
%   model, timespan, options (including workspace),
```

```
%   and external inputs.
sset = simset('SrcWorkspace','current');
sim(modelname, [Tstart:Tstep:Tstop], sset, []);
display('Model Simulated.')
%   Show variables in workspace, if desired
%   Delete for Web server applications
whos
%-----------------------------------------------
%   V. Graph Results
%-----------------------------------------------
display('Graphing Results.');
%   Define variables used for graphing
millions = 1000000;
billions = 1000 * millions;
% Assemble variables for graphing purposes
% (using example variables; substitute for different
% model)
Y1 = [Income./millions];
Y2 = [Sales./millions];
%   Call specific graphing commands, or call graphing
%   callback here.
figure(21), plot(Years, Y2),
legend('Constant Growth', 'Varying Growth');
title('Income With Varying and Constant Growth
Rates', ...
    'fontweight', 'bold');
Xlabel('Years'); Ylabel('$ Millions');
figure(22), bar(Y2),
legend('Sales');
title('Economics Example: Sales Before and After
Policy Change', ...
```

```
    'fontweight', 'bold');
Xlabel('Years'); Ylabel('$ Millions');
% label years
j = num2str(Years);
jk = cellstr(j);
set(gca, 'Xticklabel', jk);
%--------------------------------------------------
%  VI. Run Base and Alternate Simulations (if
%  needed)
%--------------------------------------------------
% base_alt_callback
%----------------------------end--------------------
```

4

Importing and Reporting Your Data

The Importance of Data

Most analysts fall victim to the "falling in love with your model" syndrome at one time or another. As the author of a whole book on models, we confess at least a passing infatuation with a model or two. However, it is time to have a sober discussion of the fundamental importance of *data* and the necessity for good methods of getting the data in and then reporting it out.

We are all familiar with the adage "garbage in, garbage out." This timeless critique warns us against using even a good model with bad data. In this chapter we describe methods of getting the data into your model, documenting the data, and reporting it in such a way that it informs your readers. But first we warn against the perils of using bad data — perils that have increased with the widespread availability of inaccurate, poorly documented, and sometimes just plain bad data on the Internet.

This chapter includes guidance on:

1. The critical importance of knowing your data, including warnings about data that are available on the Internet.
2. Getting data from a variety of sources.
3. The limitations of Web sites and the HTML format in which most Web-based information is presented, along with the burgeoning use of the inherently superior XML format.
4. Inputting data from the commonly used spreadsheet program Microsoft Excel and using Excel to report the results of your analysis.
5. Organizing data into *structures* for efficient storage and retrieval.
6. Documenting your analysis, using built-in MATLAB tools like the diary function, the report generator, and other means.

There are also three appendices to this chapter:

- "Appendix I: XML and Structured Data" on page 83
- "Appendix II: Custom Data Structure Creation" on page 87
- "Appendix III: File Importing with MATLAB" on page 93

Know Your Data

The Critical Need

This book is primarily designed to explain advanced techniques for analysis. While the author and the reader of this text share an interest in analytical methods, all analysis depends on information. Unless you have good information, the most sophisticated analytical tools are expensive toys.

"Garbage in, garbage out" is a reminder that is far too easily forgotten. The Internet appears to provide a cornucopia of information, and newspapers are filled with scientific-sounding assertions and impressively precise numbers. Before we start listing sources of data, we must consider the following axiom: any analysis is only as good as the data that goes into it.

Below, we offer two sets of caution: the first set on data in general and the second a specific set of warnings about data appearing on the Internet.

General Cautions on Use of Data

1. *Know the data before you use it.* Knowing your data means knowing the source, evaluating its reliability, and understanding the concepts, the units, the time intervals, geographic scope, and other factors that make data useful.[34]
2. *Describe properly the original source of your data.* The original source of the data, even if you obtained the data from an intermediary, should be cited. In particular, the original source of your data is probably not a Web site (unless the Web server generates its own data, such as usage data). Information that is displayed on a Web site comes from some source. Always cite the source, as well as the Web site from which it was downloaded.

[34] See Chapter 1, "How to Use This Book," especially the section "Maxims of Business Economics" on page 15.

For example, information cited simply as www.mygoofynews.com or "Big City News" provides little confidence.[35] On the other hand, "Data on cartoon characters from the Dowagiac Cartoon Club," found at: http://www.mygoofynews.com tells the reader who actually provided the data and how they could confirm it.

3. *Be careful with information from advocacy groups and self-interested parties.* Such information often contains buried judgments that severely limit its usefulness. In particular, much survey data — though not all — are based on questions that push the responses one way or another. Some advocacy groups do produce consistently reliable data. Others simply do not. Before you use it, check.

   In addition, be aware that government agencies are often interested parties. Whether in government or private industry, the employees of an organization paid to oversee a program will neither be inclined to carefully document its failings nor advertise those failings to their customers. Sometimes the most interesting government-provided information does not appear on the Web site, is released late, has confusing definitions, or is buried under a mountain of disclaimers and press releases.

4. *Remember that information in a news story may not be accurate.* The use of information in a news story, even from a reputable news organization, means the reporter and editor thought the information was interesting and that it met their reliability standards. Often, neither the reporter nor the editor have the background, training, or time to adequately review the information. Furthermore, news organizations are profit-maximizing firms. Their revenue is based on drawing in readers and viewers.

   Be sure to check the original source of the data, or properly indicate in your analysis that you used data reported by a source but not verified by you.

A Warning on Internet Data

It has become commonplace to assert that nearly everything is available on the Internet.[36] In fact, much of the most useful information is not on the Internet. By "useful," we mean information that is reliable, well-formed, and necessary to the task at hand. The vast majority of information on the Internet fails one or more of these tests.

---

[35] As an example of how little confidence this inspires, note that political campaign ads attacking other candidates often "cite" the "source" of their information using this style. The viewer then has no ability to judge the actual source and little ability to even confirm there is a source.

[36] For example, in the cult car-tuner film *The Fast and the Furious*, the leader of a gang of racers tells the upstart that he knows he had been in prison because almost everything is available on the Internet, including the upstart's criminal record. It turns out that the upstart is actually an undercover cop. The moral of the story — at least one of them — is that one should not trust information that almost anyone can post.

While the Internet seems like a gold mine of information, it should really be described as a gold-plated mine shaft, leading to both gold and sludge. With this in mind, consider the following warnings about Internet data:

1. Information that appears on Internet sites is often mislabeled, improperly described, incomplete, or just plain wrong.
2. Unless there is a revision date on the Web page, do not assume that the information is current.
3. If data is republished on one Web site but not completely described there, always check to see the original publisher or creator of the data. Reputable data providers usually provide links to those sites.
4. Almost anybody can establish a Web site and post information on it.
5. There should not be two standards for reliability. Ask the same questions about data gathered on the Web as data gathered from any other medium.

Getting Data into a Model

In this section, we will discuss some tools for getting data into your model, including:

- A standard Web browser, along with MATLAB, Excel, or other analysis programs
- Excel Link (which requires Microsoft Excel, a spreadsheet program)

Getting Data from the Internet

The growth of the Internet has been a boon to users of data. However, there are still many limitations in the availability, format, and reliability of data posted on the Web. To help you get your data in usable format, we offer a few useful tips and tools in the following pages, all based on a standard Web browser and Internet connection.

Limitations of HTML

We describe in Appendix I to this chapter the interesting history of the language that made the World Wide Web possible.[37] This language, Hypertext Markup Language (HTML) allowed for innumerable hypertext links to establish a web of connections among pages of information. However, this great strength in linking was not matched in other characteristics, notably the maintenance of structure among data.

The limitations of HTML described above have become more evident as Web usage has grown. This has motivated the development of the Extensible Markup Language (XML). XML is intended to maintain the structure of data. Information in XML is "marked up" to allow the recipient to understand the types of information that is being transmitted. For example, an XML document may contain *tags* that distinguish the title, author, text, tabular data, date, and any other type of information for which an XML description has been created. By contrast, HTML will merely display some of this information in different fonts, colors, and spacing.

For recommendations on creating applications for the future that incorporate the advantages offered by XML see Appendix I to this chapter. However, for the other portions of this chapter we will assume you are working with HTML pages and want to get the data from the Web pages into your intelligent hands.

Getting Data From HTML Pages

For this discussion, we assume that either you or more likely your source of data has not been provided data in XML or other structured format. We assume instead that you are staring at a Web page that contains the data you want and noting that it is trapped in an HTML table.[38] How do you get it into MATLAB or another analysis program? Here are some tactics:

1. *Seek other formats.* See if the data source also provides the tables in another format. Many U.S. government sources provide data in multiple formats, including text files and files formatted to be read directly by spreadsheet programs such as Excel, Quattro, or Lotus 1-2-3.
2. *Use a spreadsheet Web query.* If the Web page has a stable address (i.e., an absolute URL such as: http://www.andersoneconomicgroup.com/index.html) and the HTML contains data in a table format, you can use a Web query from

[37] See "Appendix I: XML and Structured Data" on page 83.

[38] If you have yet to find the Web page at which to stare, see "Using Excel Link" on page 70.

spreadsheet programs such as Microsoft Excel.[39] A Web query finds the required Web page and copies all or part of the data into the spreadsheet program.

Web queries can be tricky as they depend on another server, another program, and instructions that may be incomplete. Always check your data against the original source when using a Web query! Despite these cautions, Web queries frequently save a lot of time and reduce the number of transcription errors that arise from retyping the data.

3. *Try a multistage cut-and-paste.* We have found you can sometimes trick a set of software programs into accepting your data, by using a multistage cut-and-paste procedure. This works as follows:

   a. Copy the HTML table into a spreadsheet program. It may just sit there, all in one cell, or it may nicely separate into columns and rows.

   b. Copy the cell from the spreadsheet into a word processing program. Some of these programs, such as Adobe FrameMaker, contain options for how to paste the object into their format. These options may include text, RTF text, unicode text, HTML, and others. Try the most-formatted option first and then work down to plain text.

   c. After experimenting with these in different order, you will often eventually get your data into a neat, properly formatted table that can be imported directly into MATLAB.

4. *Use MATLAB's Data Import Wizard.* MATLAB itself has a fairly competent data import wizard which can open many spreadsheet files. This is an option for those who do not have Excel Link or find it difficult to use.

Example — Election Data: After the 2000 presidential election produced a contested result in Florida, we analyzed the election data obtained largely through online sources. The voluminous election data was first located on different Web sites, including Web sites maintained by news organizations.[40] As these pages tend to change quickly, we saved the URL of the data page, used a web query within Microsoft Excel to download the data into Excel, and edited and formatted the data within that environment. At that point,

---

[39] A Web page generated from a database program (which may have a very long, indecipherable string in the address pane of your Web browser) may not have an absolute URL and may not work with a Web query. Even if it does produce data in your spreadsheet, it may be the wrong data! (See the car depreciation data example on page 67.)

[40] As suggested under "General Cautions on Use of Data" on page 64, we viewed more than one source of data and carefully noted the source as well as the revision date.

we could import the data into MATLAB using Excel Link for analysis. Summaries of our analysis are available online.[41]

Example — Automobile Depreciation Data: We found interesting data on the historic depreciation values of automobiles at the Website of Edmunds.[42] However, the data was in HTML tables. Copying the data directly into Excel produced one big cell — almost useless on its own. Using a Web query, given that the table was produced by a database program on the server, produced incorrect results.[43]

To address this problem, we went to the Web page generated by Edmunds from our original search and copied the table into Excel. We then recopied the Excel cell into Adobe FrameMaker. Voila! The data we wanted, in a nice table! Of course, the "nice table" was in a document preparation software program and it also had neat punctuation marks such as hyphens, commas, and dollar signs. Instead of trying to clear that up in MATLAB, we edited those out using the Search-and-Replace command. It could then paste easily into Matlab.

Direct Internet Data Acquisition

MATLAB in recent versions also has limited ability to directly access a Web page (including an FTP site) and access the information there. Coupled with the ability to search within the information, this provides a limited data-mining technique.

The example in "URLREAD for Direct Internet Data Access" below illustrates a compact script for getting information from a known site on the Internet and searching it for specific words or phrases.

There are obvious limitations in this approach; in particular, tabular data will be mixed up in the same long string as all other information on the site. You will typically have to use other techniques to transfer tabular data in complete form once you have located it.

Code Fragment 4-1. URLREAD for Direct Internet Data Access

```
% example of "urlread" command; checks web site,
% downloads content,

% searches for instances of specific words.
```

[41] See Patrick L. Anderson, "Statistical Analysis of the Florida Presidential Election Results" and related Web pages on Michigan Results; available at: http://www.andersoneconomicgroup.com/Projects/policy/elections/stat_anal_fl.htm; "Michigan Undervote Could Outweigh Florida's," *The Detroit News*, December 13, 2000.

[42] The Edmunds site is at: http://www.edmunds.com. Also, see Chapter 12, "Applications of Finance," especially "Depreciation" on page 301.

[43] As noted above, Web queries are not reliable when used with Web pages generated by database programs. This is not the fault of the sponsor of the site, who is often attempting to provide data to a large number of users, most of whom are (unfortunately) not economists.

```
% Note: must have active internet connection.
s = urlread('http://www.andersoneconomicgroup.com')
s1 = findstr(s, 'Michigan Unemployment')
s2 = findstr(s, 'Unemployment')
% --end--
```

Using Excel Link

One of the most useful features of MATLAB is its interface with Microsoft Excel, using the Excel Link Toolbox.[44]

Excel is a powerful, inexpensive, and nearly ubiquitous software for assembling and reporting data. Certain government data sources are provided in Excel format, and many businesses use it. Furthermore, Excel itself contains powerful file transformation routines that can be used to import data from other formats.[45] Thus, some familiarity with the Excel Link Toolbox can make the real-life chore of getting the right data easier.

The following discussion describes how this feature can be used in regular workstation sessions. Throughout we assume that the user is using a Microsoft Windows environment and an Excel 97 or later version.[46] Even for applications that may later be ported to a Web server, this is a useful strategy for development and testing. Please refer first to the MATLAB product instructions for Excel Link if you are unfamiliar with it.

Importing Your Data: A General Strategy

We often employ the following general strategy for importing data:

1. Collect the initial numeric data in an Excel worksheet. The worksheet should list the title of the variable, note its source, note any

[44] Excel is a trademarked product of Microsoft Corporation.

[45] Numerous observers have criticized Microsoft for embedding features in its software that make it more difficult to use with non-Microsoft software; I have also weighed in on this controversy. See, for example, Patrick L. Anderson, "Microsoft is Indeed Cyber-Bully," *The Detroit News*, November 11, 1999. Indeed, the U.S. government's antitrust lawsuits against Microsoft resulted in a judicial ruling that the Windows operating system had been designed to make it difficult to use non-Microsoft Web browsers.

Computer users today have multiple options, including running their workstations and servers on competing operating systems such as Linux or Unix. My inclusion of a discussion on Excel Link in this book does not constitute an endorsement of Microsoft's operating systems. It does constitute an endorsement of using Microsoft Excel as a useful front end for obtaining many data sources, inputting it easily into MATLAB, and reporting or communicating it to others, including those who (by choice or by c ircumstances) use Microsoft products.

[46] We have successfully used this feature with Excel 2000 and 2003. Please see the discussion on running Excel Link under different versions of MATLAB in "Appendix B: Excel Link Debugging," on page 455.

limitations or cautions on its use, and include the number for the datum. This can be done in a simple table format.

2. Embed in the table the necessary Excel Link commands to put the variables into the MATLAB workspace. The first active column contains the name of the variable and the next columns contain the values in the data vector. The third column contains a definition statement in the form of variable_name = [amount1 amount2]. This can be entered manually or by using the concatenate command in Excel.[47]
3. Manually execute these cells to ensure that they properly send the correct data.

   See "Sample Excel Sheet for Inputting Data" below. Note that the MLEvalString command in the last column must refer to the cell that contains the definition statement.
4. Once the worksheet is completed and all the data entered, save and reopen the worksheet. This automatically transfers all the data into the MATLAB workspace.
5. Save the data in the MATLAB workspace to a .mat file.
6. Use the .mat file for MATLAB sessions.
7. When data needs to be updated or revised, one normally needs only to replace the numeric values, save and reopen the worksheet, and then resave the variables in the MATLAB workspace.

Table 4-1, "Sample Excel Sheet for Inputting Data," illustrates this strategy.

Reporting Your Data

A complementary use of Excel is to report data produced in your MATLAB sessions. This normally is done by using the getmatrix buttons in the Excel

TABLE 4-1

Sample Excel Sheet for Inputting Data

Variable Name	Amount1	Amount2	Definition Statement	Evaluation Statement
Revenue	100	200	Revenue = [100 200]	MLEvalString(D2)

[47] For example, the following statement will grab the variable title and the values and then form the proper definition statement:
"=CONCATENATE(A2,"=[",B2," ",C2,"];")." After one such statement is created properly, you can use the Edit | Fill command to create others in the same column, creating variable definitions for more rows of data.

toolbar. However, for repeated use of similar data, this process can be automated in the same fashion as importing data.

The method for reporting your data is the same as for inputting the data with the following changes:

1. Define the range of cells into which you want to place the data calculated in MATLAB. Put the ap propriate headings and related formatting items in place as well. The data cells will be blank.
2. Put a getmatrix command in a nearby cell of the Excel spreadsheet which includes the name of the variable in MATLAB and the range of cells allotted for the variables.
3. Execute the command when the data are available.
4. Check to ensure that the correct data are in the correct cells. It is very easy to misspecify a cell range and end up with data in the wrong place.

Debugging Excel Link

See "Appendix B: Excel Link Debugging," on page 455, which lists a number of tactics for debugging this useful but tricky utility.

Reporting Your Results

We discussed above the methods for getting data into your model. In this section, we discuss a few methods for getting data out of your model, including:

- The diary command in MATLAB
- The MATLAB Report Generator
- Careful reporting in a function or script

The Diary Command

An easy method of recording your results is with the diary command. It can be invoked from the command line or from where it is embedded in an m-file. It automatically generates a record.

MATLAB Report Engine

Another way to report data and document a model is by using the MATLAB or Simulink report engine. These are separate toolboxes that can be purchased and provide powerful capabilities to produce reports in various formats.

The report engines can be difficult to use at first. The reader should first consult the MATLAB help information and try a few sample reports. After gaining some familiarity with the tool, the following suggestions will be helpful:

1. Make liberal use of text headings, such as chapters and section titles, to separate out portions of the report.
2. Make liberal use of the paragraph feature which inserts into the report text explaining the analysis being performed. Remember that your reader will generally not know as much about your model and data as you, so explain both.
3. Think carefully about the order and syntax of commands that are executed by the report generator. If you wish to show, for example, the results of an analysis in both table and figure form, you must generally read in the data, perform the analysis, and only then generate the tables and figures.
4. When attempting to capture figures in a report, use the "figure loop" feature, remembering to include the "figure snapshot" feature within the loop. A related tip is to make sure that each figure you wish to capture is open in a separate figure window.

Careful Programming

You will note that in many of the m-files reprinted in this book there are well-placed commands that require MATLAB to report on the actions in a model, or to summarize data in a variable and echo it in the workspace, or to save the data into a structure. Such a discipline is very helpful when trying to debug a model, explain it to another person, or simply use it effectively.

Recommended Practices

We recommend the following practices when creating or editing m-files:

Put comments in the m-files that identify what the next section of code will do.

Use commands like `disp` `display` and `sprintf` to cause messages to be sent to the command window. These messages can simply report on the program

("... now simulating model") or inform the user of the values of certain variables.

Create output variables for your functions that collect the needed information. The use of structures allows you to collect disparate types of information, including text, into one variable in an efficient manner. See "Data Arrays and Data Structures" below.

Using Structures to Organize Data

One of the more powerful features of MATLAB is its ability to handle multiple classes of data.

Data Arrays and Data Structures

Most of the time we deal directly with scalars, vectors, and matrices of data. These are *arrays* of data. They are often supplemented with *strings* of text, which are also known as *character arrays*.

The following simple commands create a string array and then create scalar, vector, and matrix arrays of numeric data:

Code Fragment 4-2. Entering String and Numeric Data Arrays

```
message = ['The following two-dimensional arrays
contain numeric values:'];

a = [1];

b = [1 2 3 4];

c = [1 2 3 4; 5 6 7 8];
```

Uses and Limitations of Arrays

For straightforward calculations, even extraordinarily complex ones, the use of vectors and matrices will often be sufficient.

Users of spreadsheets will find these data types easy to understand, display, and import into MATLAB. However, if any of the following situations exist, you may need more powerful data types:

- You have many variables.
- You have the same variables for each of several entities.

- You want to collect descriptive information which may include text and numbers into one variable.
- You want to pass groups of data back and forth to functions, or wish to organize or store information for several entities.
- You find it easier to manage large amounts of data by first conceptually organizing information and then storing it according to group.

Most business economists and others dealing with the issues discussed in this book will find themselves in these situations.

Cell Arrays

Cell arrays are variables in which objects of many types are put into "cells." Cell arrays are like a warehouse with many rooms, each of which could be located by listing the room and the bin number. In MATLAB, cell arrays are distinguished by, among other things, their use of curly brackets. For example, in the following code fragment we create a cell array of variables and then extract them.

Code Fragment 4-3. Cell Arrays

```
% example of cell array usage

% create cell array (empty)
info = cell(2,2)

% put in strings (note curly brackets)
info{1} = 'first bit of information'

% put in vector
info{2} = [1 2 3 4 5]

% put in matrix
info{3} = [20 30; 40 50]
```

```
info{4} = 'last bit of information'

% now, get information out; address with curly
% brackets
info{4}

% now, get information with plain addressing (note
% format)
info(4)
```

Structures

Structure data variables are another format in which you structure a variable so that it holds, with a rigorous logical structure, a number of "fields" that can each have a variable of a different type.

Structures are similar to databases in that the entries are denoted as fields within an overall structure name. Fields can be added, deleted, or addressed using standard MATLAB commands.

Structure Syntax

Structures, like databases, are addressed using the database name and then a field name. The syntax, used by MATLAB delimiter so that any known variable within a structure can be addressed in MATLAB using the name of the structure and the name of the field. A similar convention is followed in other environments such as XML.[48]

In the following code fragment, the structure variable book is created and fields for the title, subtitle, author, and publisher (all strings) are inserted into the structure, followed by the numeric array (in this case a scalar) of the number of chapters.

[48] We describe later in this chapter rules for structured documents called the Document Object Model or the DOM. These rules are the basis for the XML data standard. In anticipation of that material, note how the variable.field syntax allows for different elements of a whole topic, which are related by the topic but are not of the same data type, to be grouped together in a logical fashion. MATLAB structures do not strictly follow DOM rules, but they do follow a larger convention that becomes quite helpful when the number of variables becomes large.

Code Fragment 4-4. Structure Syntax

```
book.title = 'Business Economics and Finance'

book.subtitle = 'Using Matlab, GIS, and Simulation
Models'

book.author = 'Patrick L. Anderson'

book.publisher = 'CRC Press'

book.number_of_chapters = [17]


book


book =


book.title = 'Business Economics and Finance'

book.subtitle = 'Using Matlab, GIS, and Simulation
Models'

book.author = 'Patrick L. Anderson'

book.publisher = 'CRC Press'

book.number_of_chapters = 17

publisher: 'CRC Press'
```

Methods of Packing and Unpacking Structures

MATLAB includes commands that create structures, "pack" information into them, and then "unpack" the information. Packing variables into a structure means creating the field for the variable and then putting the variable into that field. Think of packing as analogous to carefully packing a suitcase so that the shoes, socks, and all other items go into the right compartment.

These commands include:

- You can directly create a structure or, add a field to an existing structure by naming it and assigning information to it.
- struct is a command that builds a data structure from variables in a workspace and names the fields in the structure.
- cell is a command that builds cell arrays.

- cell2struct and struct2cell are commands that take the elements of a structure or cell array and create a new variable in a different format.
- rmfield, setfield, and getfield are commands that remove, set, and get a certain field within a structure.

For example, you can create a structure and pack it with variables as follows:

Code Fragment 4-5. Creating, Packing, and Unpacking a Structure

```
% put in strings (note curly brackets)
infostruct.intro = 'first bit of information'
% put in vector
infostruct.data1 = [1 2 3 4 5]
% put in matrix
infostruct.data2 = [20 30; 40 50]
infostruct.extro = 'last bit of information'
% now, get information out; address directly
infostruct.data2
% now, get information on entire structure
infostruct
%--------------------------------------------------
% Packing all at once
%--------------------------------------------------
infostruct2 = struct('first_info', info{1}, 'data1',
infostruct.data1);
infostruct
```

A Shortcut Utility

You can always create structures and perform related tasks using the built-in MATLAB commands. We also know of a shortcut method to packing and unpacking structures that assumes you name the fields in a structure the same as the related variable names. This utility function, mmv2struct (vector to structure), was created as part of the Mastering MATLAB Toolbox by the authors of *Mastering MATLAB* 6.[49] You will see it used occasionally in the m-files described in this book.

Example: Data on Ten Businesses

For example, if you want to have 10 different variables containing information on 10 different businesses, you could create 100 different variables, each named differently. Or, you could create 10 structures, one for each of the businesses, and have 10 consistently named fields within these structures.

A Method for Organizing Data

The following method has proven effective in assembling large amounts of information into useful structures which can be saved, reused, and stored.

1. Get the information into the MATLAB workspace. This can be done in any of the normal methods, using Excel Link, importing from other sources, direct input, or loading data stored in .mat files. If you are organizing data from multiple entities or concepts, it may be useful to have the data from only one concept at a time in the workspace.
2. Pack the variables into structures, using one of the following techniques:
   a. Directly create them, using MATLAB struct or other commands.
   b. Use the mmv2struct function described above.
   c. Create a script m-file that checks to see if all the needed data are in the workspace, collects the data, and packs it into a structure with all the necessary fields. An example is shown in Appendix II: Custom Data Structure Creation on page 87.
3. After a structure is created, save it into a .mat file, using the save command with the -append option.
4. Clear your workspace.
5. Perform these steps again, as many times as needed.

Example: Valuing Multiple Businesses

The following example shows how this method of organizing data can be used for a project in which multiple businesses are evaluated for the purposes of estimating the market value. We describe the methods of valuing

[49] D. Hanselman and B. Littlefield, *Mastering MATLAB 6* (Upper Saddle River, NJ: Prentice Hall, 2001), Chapter 38. This and other references to MATLAB are discussed in Chapter 1, this book, under "General Books on Using MATLAB" on page 10.

Also see Chapter 1, "General Books on using MATLAB," for information on obtaining toolboxes prepared by the authors of these texts.

businesses in Chapter 11, "Business Valuation and Damages Estimation," where we make use of the data structures created here.

We approach the project according to the steps described in the preceding section.

1. We assemble the information on the subject businesses, including sales history, profit margins, cost of capital, projections of future performance, and other relevant information in a worksheet. We then get that information into the MATLAB workspace by using Excel Link.
2. We pack the variables for each business into a data structure. This can be done using standard commands such as those described in the previous section. However, in this example we make use of a custom script file called data_bd, which we describe in Appendix II: Custom Data Structure Creation on page 87.
3. We save the resulting structures to a .mat file, repeating these steps for each subject business and appending the structure for each to the file.

Code Fragment 4-6. Saving Structures to a .mat File

```
whos;                  %Check to see what is in
                       %workspace.
data_bd;               %Script to check data and
                       %create structure; script will
                       %ask user for name of structure;
                       %in this example, input "dutch".
save distributors.mat bdistA -append;
                       %Saves to "distributors. mat" the
                       %structure variable
                       %"bdistA," appending it to existing
                       %contents.
clear;                 %Clears all variables in workspace.
```

Data Organizing Tools

Organizing your data becomes increasingly important as the complexity of the project increases and as the number of projects multiply. The ability to organize data varies tremendously across different software environments.

There are certain attributes we desire in a software environment that organizes data, such as:

- A hierarchical structure which natively organizes details about a concept into a larger whole
- The ability to directly address a specific field or entry in a table without wading through all entries
- The incorporation and maintenance of *metadata*, or data about data, which allows for the documenting of data sources, inclusion of notes, and other essential information

Discussion: Spreadsheets and Databases

Spreadsheets are quite deficient in these attributes. They are inherently flat tables with no structure and with no direct connection between metadata and the data. While you can trick a spreadsheet into acting like a database, it is inherently a flat table.

Relational databases are a large improvement over spreadsheets. They have structure, and a skilled user can document and enforce data integrity among related elements. However, the skill required is fairly high for such use; there is a reason why DBA (database administrator) is a common acronym while SA (spreadsheet administrator) is not.[50]

MATLAB has specific commands that allow it to directly query many relational databases. A Database Toolbox is available that allows MATLAB to query databases, import data, export data, and perform other tasks with relational databases. We do not describe this toolbox in this textbook, but you may wish to consider this toolbox if you regularly access large databases.

Even if you do not wish to directly interact with a relational database, you can normally use data stored within it by exporting the data in one or more forms. Most databases can export information in flat files that can be read by MATLAB, and most can now export it in the XML format.[51]

---

[50] Just to check, the author asked a search engine to find Web pages containing "DBA" and "database"; over 236,000 were found, many of which specifically referenced a database administrator. By contrast, 51,000 instances of "spreadsheet" were found also containing "SA." Most of these turned out to be foreign words, other acronyms, or the surprisingly common use of novel contractions like that in the phrase "what's a spreadsheet?"

[51] For a discussion of XML, see "History of HTML, Other Formats, and XML" on page 84.

A Note on GIS Databases

In addition, the Mapping Toolbox includes commands for accessing data from Geographic Information Systems (GIS) databases, including the TIGER files maintained by the U.S. Census Bureau. We do not describe this toolbox in this textbook, but the use of GIS in analyzing retail sales and other elements of the economics of location and geography has been described in Chapter 13, "Modeling Location and Retail Sales," on page 327, and Chapter 9, "Regional Economics," on page 189.

Creating Custom Structures

Within MATLAB, we often wish to use the benefits of a relational database, specifically the ability to structure and organize data. These benefits can often be obtained by designing an application from the beginning to use data stored in MATLAB structures.

In the preceding section, we described the general tools used in creating, packing, and unpacking structures.

Additional Methods

The methods described above do not exhaust the possibilities for exchanging data and reporting on the results of an analysis. Below we briefly mention three sets of methods; the first set will be available to most users with minimal or no additional cost. The second set will require substantial customization and additional programming. The third set is described in Chapter 16, "Bringing Analytic Power to the Internet."

Command History, Diary, and Notebook

1. For all users, the `diary` command allows for the commands to be collected in a file along with most of the command-line output. This allows for the results of a session to be replicated later on.
2. For all users, the command history window collects the commands you have typed in a session from which you can quickly create a script file. (The command history window appears at the user's option in MATLAB 6 and later versions; type `desktop` and then use the View menu to open it.)
3. For those using Microsoft Word, the `notebook` command will collect input and output in a Word file. Even if you do not use Word, the file format may be accessible by your word processing software.

Custom Interfaces

You may wish to consider the following custom interfaces which will require additional software, are platform-specific, and may also require the assistance of a specially trained programmer:

1. Using MATLAB as an engine to perform calculations demanded by a C or Fortran program, which is available on multiple platforms.
2. Creating MEX files from MATLAB m-files, which then can be run on other machines.
3. For those using Microsoft Windows computers, a custom Active-X interface can be created based on a client/server model. MATLAB has built-in functions that enable this approach.
4. For those using Microsoft Windows, a custom DDE (dynamic data exchange) interface can also be created which implements a peer-to-peer communication.

Good descriptions of these possibilities are provided in other books on MATLAB.[52] See also the MATLAB user guides.

MATLAB Web Server

In Chapter 16, "Bringing Analytic Power to the Internet," we describe running applications on Web servers which can extend the potential reach of MATLAB analytics to anyone with a Web browser and an Internet connection.

Appendix I: XML and Structured Data

Limitations of HTML

One of the problems of the World Wide Web is the limitation built into the basic HTML language used to display text, numbers, and graphics. HTML means Hypertext Markup Language, which means that the main purposes of HTML documents are to provide hypertext links to other sources of information (the ubiquitous links feature of Web pages) and markup capabilities.

Adding links was a huge innovation, and it made the Internet more than an interesting device for dedicated academics and technophiles.

However, the markup side of HTML was never designed to do much more than simply display information. Therefore, HTML pages that contain tables

[52] See, in particular, D. Hanselman and B. Littlefield, *Mastering* MATLAB 6 (Upper Saddle River, NJ: Prentice Hall, 2001), Chapters 34 and 36.

of data can display the data for a reader but normally cannot provide the hierarchy, source information, or easy communication with other programs.

History of HTML, Other Formats, and XML

A brief history of HTML and the emergence of newer standards will motivate the discussions that follow. HTML emerged from an older and much more sophisticated standard — Standardized General Markup Language or SGML. SGML was created in the 1960s, and the SGML standard was codified by ANSI in 1986.[53] It is a truly international standard, and the ISO organization also maintains an SGML specification.[54]

SGML is quite sophisticated and is used by organizations with complex document production needs, including the U.S. Department of Defense and the U.S. Internal Revenue Service. Indeed, this book was created in a software format that is based on SGML.[55] In 1990 an application of SGML was created for the purpose of allowing scientists to communicate with each other easily.[56] This HTML language now dominates the World Wide Web that exploded across the globe in the following decade. Thus, the majority of data we encounter on the Internet is still in flat files or in HTML tables.

Use of XML Today: Its Use in MATLAB

XML is already widely used and most Web browsers already accept XML pages. There are a huge number of references on XML available, and at least a few books that describe the implications of this new technology for managers.[57]

Indeed, portions of the MATLAB Help documentation are maintained in XML format. Thus, most MATLAB users are already using XML perhaps without being aware of it.

Example: The info.xml File for the Economics Toolbox

We provide an example file in XML format below, which is part of the Business Economics Toolbox that is a companion to the textbook. In Code

[53] ANSI is the American National Standards Institute whose Web site is at: http://www.ansi.org.

[54] ISO is the common name for the International Organization for Standardization and publishes numerous standards. Probably the best known are ISO 9000, dealing with quality assurance and management, and ISO 14000, dealing with environmental quality in production. Its Web site at http://www.iso.org offers an SGML specification for a fee.

[55] This book was created in Adobe FrameMaker which maintains the SGML structure of a document. Adobe's Web site is at: http://www.adobe.com.

[56] HTML was created by Tim Berners-Lee at CERN in Geneva. The chronology here is based on the excellent reference to the history of SGML, HTML, and XML, *XML for the Absolute Beginner,* provided by the nonprofit Organization for the Advancement of Structured Information Systems (OASIS), whose Web site is: http://www.xml.org.

[57] There are numerous references on XML programming. However, to understand how XML can be used in business, we recommend the book by Kevin Dick, *XML: A Manager's Guide*, 2nd. ed. (Boston, MA: Addison-Wesley, 2002).

Fragment 4-7, "The info.xml File for the Economics Toolbox," on page 85, we have stored information on aspects of the toolbox which can be accessed by MATLAB users through the launchpad.[58]

Notes on Reading XML: We will not describe XML syntax here. However, for ease in reading the info.xml file, we introduce the following rules:

- To be well formed, each element should begin with a tag that describes the type of element (for example, <listitem>) and then end with a tag that similarly terminates the element (for example, </listitem>).
- Comments are begun with a bracket and exclamation point. (For example, <!-- See Matlab Help for file ...)

XML and the Document Object Model

XML follows certain rules called the Document Object Model or the DOM. This topic is described in Chapter 16, under "Structured Data and the Document Object Model" on page 415.

Recommendation: Plan for Future Data Exchange

It is likely that the reliance on HTML will change in the future, although the change will inevitably involve some clash of standards. We recommend that those developing applications that they expect to use in the future consider the XML standard when planning for future data exchanges via the Internet.

Code Fragment 4-7. The info.xml File for the Economics Toolbox

```
<productinfo>
<MATLABrelease>13+</MATLABrelease>
<name>Economics Toolbox</name>
<type>Toolbox</type>
```

[58] To use the launchpad, open the MATLAB desktop (if it is not already open) and specify that you want to see the launchpad. (Use the View menu.) Right-click on any of the toolboxes and click Refresh. MATLAB lists any toolbox directory in the path with an info.xml file in the resulting graphical hierarchy of available features.

The launchpad should include, under the Toolboxes hierarchy, a listing of certain commands, help functions, and other attributes of the Economics Toolbox.

```
<icon>$toolbox/MATLAB/icons/MATLABicon.gif</icon>
<list>
<listitem>
<label>Testplots</label>
<callback>testplot</callback>
<icon>D://models/economics/economics_toolbox.jpg</
icon>
</listitem>
<listitem>
<label>Help--not currently implemented</label>
<callback>doc econ_toolbox_help/</callback>
<icon>$toolbox/MATLAB/icons/book_mat.gif</icon>
</listitem>
<listitem>
<label>Book Publisher (CRC Press) Web Site</label>
<callback>web http://www.crcpress.com -browser;</
callback>
<icon>$toolbox/MATLAB/icons/webicon.gif</icon>
</listitem>
<listitem>
<label>Book Author (Anderson Economic Group) Web
Site</label>
<callback>web http://www.andersoneconomicgroup.com -
browser;</callback>
<icon>$toolbox/MATLAB/icons/webicon.gif</icon>
</listitem>
</list>
</productinfo>
<!--See MATLAB Help for file input-output, regarding
XML files, for discussion of info.xml file. -->
```

```
<!--Prepared by Anderson Economic Group LLC for
purchasers of Business Economics Using MATLAB,
published by CRC Press. Use by license only. -- >

<!--Note that MATLAB appears sensitive to indents and
spacing in these xml files. Use "smart indent". -->
```

Appendix II: Custom Data Structure Creation

We provide below a custom script file for use in business valuations in which the same data will be collected for multiple businesses and then used for calculations.

The data structure below is highly customized for this purpose; indeed, it is customized for valuation of a specific type of business, namely beer distribution. We discuss in Chapter 11, "Business Valuation and Damages Estimation," the importance of using methods appropriate to both the industry and the business. Therefore, we provide no "one size fits all" routines.

While this routine will not collect the data necessary to estimate the value of, say, a retailer or manufacturing firm, the structure of the program can be used for most projects in which a set of disparate data are collected for each of several subjects. You will note the following sections in the program:

1. Naming the data structure
2. Checking key variables for both existence and reasonableness
3. Checking the overall number of inputs
4. Packing the variables into a data structure, which includes the metadata necessary for a user to later identify the source and intended application of the data

Variables in This Custom Structure

The purpose of describing this method in this chapter is to explain the method of creating a data structure, not to discuss the data. However, for those readers unfamiliar with the variables (or those who are waiting to read the chapters on business and valuation), we describe briefly the purposes of the variables in the script.

- The name of the case is requested for each business for obvious reasons.

- Data on sales history, costs, and profits are requested for each business.
- Cost-of-capital data are also requested; the wacc variables are various estimates of the weighted average cost of capital for a firm.

The data and its uses are discussed in Chapter 11, under "Example: Beverage Distributors" on page 272.

Code Fragment 4-8. Creating a Valuation Data Structure

```
%DATA_BD
% Creates valuation data structure from workspace
% variables.
% Resulting data structure can be used in VAL_BD
% (for beer distributors)
% or can be modified for use in other valuation
% tasks.
% Note that data must already be in workspace for
% command to work.
%
% uses MM toolbox (mmv2struct); can use STRUCT
% instead.
% See also VAL_BD
%
% PLA June 9, 2003
% (c) 2003 Anderson Economic Group,
% part of Business Economics Toolbox
% Use by license only.
%------------------------------------------------
% I. Get intended filename and any notes
%------------------------------------------------
filenamestring = input('Enter name of file: ', 's');
%------------------------------------------------
```

```
% II. Check individual data variables for
%     existence and reasonableness
%---------------------------------------------------

disp('Checking for necessary data in workspace
...');

% Insert placeholders, or warn user, for certain
% variables

if exist('name_of_case')

% use entered data

else

   warning('Missing name_of_case variable.');

end

% price to retailer, if necessary to calculate

if exist('ptr')

   % use entered data

   messtemp = ['Using entered PTR ', num2str(ptr),
               '.'];

   disp(messtemp);

   clear messtemp;

else

   ptr=sales_rev_base/sales_case_base;

   disp('Using calculated PTR.');

end

% check amount

if ptr > 30

   warning('Calculated price to retailer exceeds
$30. Recheck input data.');

elseif ptr < 10

   warning('Calculated price to retailer less than
$5. Recheck input data.');

end

% salable assets-------------------------------------
```

```
if exist('SalableAssets')
   %use entered data
else
   SalableAssets = 25000;
   warning('No Salable Assets Entered. Using
placeholder value of $25,000.');
end
% discount rate on liquidation
if exist('discount_rate_liq_sales')
%use entered data
else
   discount_rate_liq_sales = .25;
   warning('No discount rate on liquidation of
assets entered. Using placeholder value of 25%.');
end
% growth rates and wacc-----------------------------
if growth_mid >1
   warning('Growth rates should be entered as
decimals, such as ".05" rather than "5%".');
elseif (growth_mid) < 0
   disp('Growth rate entered as negative; check
value.');
end
if wacc_low >1
   warning('Cost-of-capital and growth rates should
be entered as decimals.');
elseif (wacc_low*discount_rate_close) < 0
   warning('Cost-of-capital and discount rates
should be positive values.');
end
% substitution parameters---------------------------
if exist('eps_sub_ratio')
```

```
   %use entered data
else
   eps_sub_ratio = 0.65; %AEG estimate from past
work.
   warning('No ratio of net earnings on substituted
brands entered. Using placeholder value of 67%.');
end
if exist('sub_rate')
   %use entered data
else
   sub_rate = 0.9;
   warning('No substituion rate entered. Using
placeholder value of 90%.');
end
% Forecast horizon for FCF methods
if exist('numperiods')
   %use entered data
else
   numperiods = 10;
   warning('Number of periods for forecasting cash
flow not entered. Using placeholder value of 10
years.');
end
%---------------------------------------------------
% III. Check number of inputs
%---------------------------------------------------


keyinputs = {sales_rev_base, sales_case_base, ptr,
sales_trend, ...
   SalableAssets, discount_rate_liq_sales,
sales_decay_rate,
   discount_rate_close, ...
   earnings_pretax_share, earnings_pretax, ...
```

```
   growth_low, growth_high, growth_mid, numperiods,
...
   case_mult_low, case_mult_high, ...
   income_tax_rate, eps_sub_ratio, sub_rate,
ptr_sub, ...
   wacc, wacc_low, wacc_high, lost_profit_alt};
if size(keyinputs, 2) < 24
   warning('May be missing field of input
structure.');
end
clear keyinputs;
%---------------------------------------------------
% IV. Create structure
%---------------------------------------------------
% note string variables are added here, while
% checkdata section
% checks to see number of numeric array variables.
filenametemp = mmv2struct(name_of_case,
sales_rev_base, sales_case_base, ptr, sales_trend,
...
   SalableAssets, discount_rate_liq_sales,
sales_decay_rate,
   discount_rate_close, ...
   earnings_pretax_share, earnings_pretax, ...
   growth_low, growth_high, growth_mid, numperiods, ...
   case_mult_low, case_mult_high, ...
   income_tax_rate, eps_sub_ratio, sub_rate, ptr_sub, ...
   wacc, wacc_low, wacc_high, lost_profit_alt);
namingcommand = [filenamestring, ' = filenametemp'];
eval(namingcommand);
clear filenametemp;
message = ['Structure ', filenamestring, ' in
workspace.'];
```

```
disp(message);

%-----------------------------------------------------

% V. Append to .mat file (optional)

%-----------------------------------------------------

% Appending to .mat file

% After creating a number of structures, you may
% want to add them to a .mat file. If so, the

% following command may be a guide:

%

% save valinformation.mat samplebd1    %first time;
% creates .mat file

% save valinformation.mat samplebd2 -append
% subsequent times; appends data to existing file

% where "valinformation.mat" is a filename,
% "samplebd" is the variable, and -append

% indicates the variable should be appended to the
% existing contents of the .mat file.

%---End-----------------------------------------------
```

Appendix III: File Importing with MATLAB

In this appendix, we summarize information on a variety of data types.[59]

Delimited Files

Definition and Sources: Most spreadsheet, word processing, database, and other software will write a comma-delimited file, often with a .txt extension.

Helpful MATLAB *Functions*: If you have a delimited file, look at DLMREAD and CSVREAD.

Example Data Sources: ftp://ftp.ny.frb.org/prime/Prime.txt; http://www.bea.gov/bea/dn/dpga.csv; ftp://ftp.bls.gov/pub/news.release/cpi.txt; http://www.nber.org/nberces/bbg96_87.txt

[59] I am indebted to Michael Robbins for suggesting this compendium, for ideas on handling formatted data, and for providing example sources on the Internet.

Common Problems: These include files that do not have the same number of delimiters on each line, headers, skipped lines, etc.

Common Solutions: If you cannot make these files consistent, you must treat them as formatted files (see "Formatted Files" on page 95). Sometimes using an intermediate software environment (such as a text editor, word processor, or spreadsheet) will shorten the process. Many of these have powerful capabilities to make repeated corrections and can strip out extraneous formatting marks quickly.

Web-Based Data: HTML

Helpful MATLAB Functions: If you have an identified page with a static URL, try URLREAD. It will copy the data on the page into a string.

From within certain spreadsheet programs, try a Web query as described in "Getting Data From HTML Pages" on page 67.

Example Data Sources: http://www.newyorkfed.org/autorates/selectedrates/display/ selrates.cfm

Common Problems: Once the data from the URL are copied into a string, it must be treated as sloppy data such as HTML data. This will often require additional editing like that suggested for formatted or delimited files.

Common Solutions: Look at STRFUN for a list of functions that may help you parse your data from the resulting string. See also the advice below for formatted files and above for delimited files. These data are often quite messy.

Web-Based Data: XML

Helpful MATLAB Functions: Try XMLREAD. As the use of this type of data is growing rapidly, check for future enhancements and third-party products. See "Appendix I: XML and Structured Data" and "History of HTML, Other Formats, and XML" on page 83.

You might also try a Web query, as described in "Getting Data From HTML Pages" on page 67.

Compressed Files

Helpful MATLAB Functions: You may need a utility function such as PKunzip if you download a compressed file.[60]

Example Data Sources: http://www.newyorkfed.org/rmaghome/economist/hobijn/identsrc.zip; http://research.stlouisfed.org/ fred2/ downloaddata/FRED2_xls.zip

[60] Some compression utilities are widely available or bundled with other software or operating systems. The program PKunzip is available from PKware at: http://www.pkware.com.

Common Problems: Once the file is uncompressed, you must then use one of the other methods on this page to extract the data.

Formatted Files

Helpful MATLAB Functions: If your file is delimited in a manner that can be understood by CSVREAD or DLMREAD, use these functions. If not, try the more basic commands FGETL, FGETS, FSCANF, and TEXTREAD.

Example Data Sources: Almost all data provided by a machine is formatted in some manner. Humans provide unformatted data, which is why a handwritten note is more interesting than a billing statement.

Common Problems: Oftentimes data that are formatted are not consistently formatted.

Common Solutions: You may also have to use SPRINTF, SSCANF, STRREAD, FREAD, and REGEXP. Try also the tips under "Delimited Files." Try getting the software to produce data in a different format.

Databases and Data Feeds

Helpful MATLAB Functions: If you regularly connect to a database or see the benefits in doing so, consider the MATLAB Database Toolbox or Datafeed Toolbox. There are also many powerful methods that will connect MATLAB to certain databases, including ActiveX, JAVA, ODBC, and COM.

Example Data Sources: There are many municipal, academic, and commercial databases available. Many data feeds are available as well.

Common Problems: Database access difficulties include speed issues, type conversion, time series issues, and query design.

Common Solutions: The Mathworks has a number of technical notes online for users with specific problems. Search for one that applies to your issue.

Excel Link

Helpful MATLAB Functions: For excel files look at XLSREAD, the Excellink Toolbox, and the free ActiveX connectivity examples. In this book, see in this chapter "Using Excel Link" on page 70 and "Appendix B: Excel Link and Debugging," on page 455.

Example Data Sources: http://www.bea.doc.gov/bea/dn/cwcurr.zip; http://research.stlouisfed.org/fred2/downloaddata/FRED2_xls.zip

5

Library Functions for Business Economics

Common Analytical Tools in Economics

Economists and all business analysts that use economic reasoning rely on a mental library of analytical tools. These tools are not well implemented in any single software environment. Therefore, economists typically use a range of tools, such as spreadsheets, statistical software, graphics utilities, and different modeling languages to complete their tasks.

MATLAB does not provide a complete environment for all these tasks. However, it provides much more power to perform analysis than spreadsheets, statistical software, and modeling languages alone, and can often accomplish tasks in one software environment that might otherwise take two or three.

Library of Tools in MATLAB and Simulink

In this chapter, we recommend a number of tools for use in the MATLAB environment, some of which we have programmed specifically for the field. We call these a *library* of tools partially because the metaphor of shelves of books is somewhat appropriate (a toolbox metaphor is also fitting) and partially because MATLAB provides for the creation of special libraries of m-files and Simulink blocks.

Microeconomic Theory and Practical Applications

In this chapter we first discuss the economic theory that underpins the techniques, and then present the techniques themselves, along with notes for practical application.

Comparative Statics

Many of these analytical tools are derived from comparative statics, the study of how changes in one variable affect others in a static model. Such tools have been developed by some of the great minds of the 20th century, including Alfred Marshall, John Hicks, Jon Von Neumann, and Paul Samuelson.[61]

We present below some of the building blocks of comparative statics in economics, first discussing theory and then presenting implementations using the MATLAB environment.

Estimating Demand Schedules

Consider the ubiquitous demand curve. It is based on axioms and implications of theory that establish the following rule: assuming that all other variables stay the same, an increase in the price of a product or service will result in lower demand for that product.[62]

In business economics, we will not derive the theories of microeconomics. Instead, we will apply them to real-world challenges. Numerous such challenges in business economics and related fields require a knowledge of consumer demand. We recommend the following steps for acquiring data and for specifying, estimating, and implementing a demand function.

Step One: Acquire Data

The first step is very important. The required data must not only demonstrate equilibrium prices (prices at which supply and demand intersect), but also

---

[61] Alfred Marshall's (1842–1924) Principles of Economics (London: Macmillan, 1890, 1920) introduced innovations such as the ubiquitous supply–demand charts, with price and quantity as the axes. He also defined the concept of elasticity of demand in the 1885 article "On the Graphic Method of Statistics" for the *Journal of the [London] Statistics Society,* although other economists before him had described similar concepts.

Sir John Hicks helped develop the notion of demand, substitution, and expenditures, and represented the "Slutzky" decomposition of demand changes into income and substitution effects. The theory was originally stated by Russian economist Eugene Slutzky. John Von Neumann helped develop the utility theory and economic behavior under "game" conditions.

Paul Samuelson's *Foundations of Economic Analysis* (Cambridge, MA: Harvard University Press, 1947) is the seminal work for establishing mathematical analysis of microeconomic questions within modern economics. He was awarded the 1970 Nobel Prize in economics for developing the "Static and Dynamic Economic Theory."

See Hal Varian, *Microeconomics,* 1st ed. (New York: Norton, 1978), especially the notes to chapters on the theory of the firm, theory of the market, and theory of the consumer; Economics New School Web site, History of Economic Thought, at http://cepa.newschool.edu/het/home.htm, various entries; *The New Palgrave Dictionary of Economics* (New York and London: Stockton and Macmillan, 1987); the biographical entry for Paul Samuelson at the Nobel Web site at: http://www.nobel.se/economics; the biographical entry for John Von Neumann at the University of St. Andrews (Scotland) site at: http://www.history.mcs.st-andrews.ac.uk/history/References/Von_Neumann.html.

[62] Meticulous microeconomists will note the theoretical existence of *Giffen goods,* in which a reduction in price of sufficient quantity would result in a *decrease* in demand. Potatoes are often offered as an example. An autographed copy of this book awaits the first researcher that identifies, acquires data for, and estimates the demand curve for a Giffen good.

changes in demand during times when prices change. Unless the data provide instances where demand is higher and lower, there can be no demand schedule or curve derived from the data. Similarly, there must be data that distinguishes changes in demand for a specific commodity from such changes for other commodities. This is the essence of the identification problem in econometrics.[63]

Assuming that we acquire such data, we will go to the next step.

Step Two: Estimate Demand Equation

Once the data have been acquired, we can proceed to specify a demand equation and estimate its parameters. MATLAB provides extensive statistical functions for the estimation task, including the Curve-Fitting Toolbox. Examples of demand equation specifications with good theoretical foundations include the following:

Distance in a transformed demand equation: One portion of the true price of a product or service is the cost of commuting to the store and back. This commute could be as short as a quick exit off the highway and back on (for example, when stopping for gas on an automobile drive) or several round-trip visits (for example, when selecting a college or choosing an automobile dealership). Thus, a distance–sales relationship is fundamentally a demand equation, in which a portion of the price is transformed into a different variable, namely distance. We can evaluate this more rigorously by decomposing the total price of a good into two components: the transaction price and the distance price:

$$p = p^t + p^d \tag{1}$$

where p is the true price of the good and the other terms are the transaction price (the price paid in the purchase transaction for goods and services unrelated to distance) and the price related to the distance between the consumer's location and the retail location. We insert this price into the general demand function:

$$x = f(y, p) \tag{2}$$

where y is income and p is the relative price of the good to other goods.[64] If we differentiate the demand function with regard to price, we get:

---

[63] See J. Kmenta, *Elements of Econometrics* (New York: Macmillan, 1971), Chapters 11 and 13; Kmenta also cites F. M. Fischer, *The Identification Problem in Econometrics* (New York: McGraw-Hill, 1966).

[64] Note that under the assumption of the same transaction prices at retailers in the area, the distance component becomes the key varying portion of the relative price.

$$dx = \frac{\partial x}{\partial y} dy + \frac{\partial x}{\partial p} dp \tag{3}$$

At this point we make some assumptions that simplify this relation. Assume first that the transaction price is the same at retailers located throughout the area. This will be true, or nearly true, in many cases.[65] Second, assume that income does not change during the time period of the data, an assumption that will again be consistent with the facts in many cases.[66] The assumptions and the resulting demand equation can be summarized as:

$$dy = 0$$

$$dp = dp^t + dp^d$$

$$dp^t = 0 \tag{4}$$

$$dx = \frac{\partial x}{\partial p} dp^d \tag{5}$$

Thus, under certain assumptions that will often be fulfilled, the demand equation can be transformed into a distance–sales equation.

Example Distance–Sales Analyses: We provide examples of demand equations in which the price variable is *distance* in Chapter 8, on page 173.

Indirect Addilog Demand Model: This model is based on an indirect utility function of consumers, in which their desire for goods is based on a power function of the relative price of goods and their income.[67] A linear form of the model, in which x_i/x_j is the relative demand for goods i and j and p_i/p_j is the relative price, is:

$$ln(x_i / x_j) = a + b\, ln(y) + c\, ln(p_i / p_j) \tag{6}$$

Note that the parameters a, b, and c in this equation represent combinations of parameters in the original consumer utility function. The decomposition of those parameters may or may not be important in

---

[65] In some cases, this assumption is violated by the behavior of retailers who charge premiums or offer discounts based on their locations to compensate for the distance costs of the consumers.

[66] This is generally the case with cross-sectional data. However, for data that cover multiple periods, changes in income and prices must be included in the analysis.

[67] See H. Varian, *Microeconomics* (New York: Norton, 1978), Chapter 4.

applied work.[68] However, it is important that business economists evaluating demand at least consider the foundations for specifying the equation in a certain fashion.

Linear Demand Model: This is the simplest model and the one most likely represented in textbooks. It also has the weakest theoretical basis. However, under certain limited conditions it is the most practical.

The most basic linear model is of a familiar form:

$$x_i = a + bp_i \tag{7}$$

Most analysts should include relative prices and income, using this form:

$$x_i = a + by + c(p_i / p_j) \tag{8}$$

Semi-Log Demand Model: A variation of the linear-in-the-logs model described above is:

$$ln(x_i) = a + by + c(p_i / p_j) \tag{9}$$

Other Demand Models: There are a variety of other specifications for demand functions using logs, semi-logs, powers, rational, and other combinations. The Curve-Fitting Toolbox provides a straightforward manner of specifying and estimating multiple versions. The conditions mentioned under "Practical Basis for Linear Demand Models" on page 101 are likely to hold true as useful approximations of reality.

Directly Specified Demand Models: In many cases, you will have a prior knowledge, often from previous research, that suggests a specification of a demand model. Consider carefully whether the previous analyses were based on data that are actually comparable to the market participants you are modeling and, if not, make allowance for behavior to be different.

Practical Basis for Linear Demand Models

The practical and theoretical basis for a linear demand equation is based on the following:

1. The observed demand is usually not of individual consumers but of a large number of *aggregate* consumers. Therefore,

[68] For example, it may be important to compare consumer demand for certain goods over time or across different data sets, specifications, and estimation techniques. A well-thought-out analysis will allow such comparisons and provide the analyst with the ability to compare his or her results with those of others.

restrictive utility functions used in microeconomic theory for individual consumers are not the direct underpinnings of models of aggregate consumer demand. This allows the demand functions for aggregate consumers to be modeled as if there were a large group of consumers for whom there is a continuous demand function.[69] This is especially appropriate when studying demand for a narrow product category targeting a narrow class of consumers.

2. Given the large number of different factors that affect numerous consumers, an appeal may be made to the central limit theorem to justify an assumption of normal errors when estimating such an equation.[70] Note that such an appeal *requires* a large sample size or other information on the distribution of small-sample variables.[71]
3. Although the underlying demand function of aggregate consumers is probably not linear, the *observed range* of data may be approximated by a linear function which may include variables that are products, logs, or other combinations of other variables.[72]
4. In practical applications, the relative price can be approximated by deflating the nominal price by a general price index,[73] and the

---

[69] An extensive discussion, including mathematical derivations, is in A. Mas-Collell, M.D. Whinston, and J.R. Green, *Microeconomic Theory* (New York: Oxford, 1995), Chapter 4. See also H. Varian, *Microeconomic Theory,* 3rd ed. (New York: Norton, 1992), Chapter 9.4, "Aggregating across Consumers."

[70] The central limit theorem is one of the most important in statistics and econometrics. It may be summarized as: the distribution of the mean of a very large number of independent random variables converges to a normal distribution. See, e.g., G.W. Snedecor and W.G. Cochrane, *Statistical Methods*, 8th ed. (Iowa University Press, various editions 1937–1989), Section 7.8.

A more careful mathematical derivation builds from the sample of any distribution with a finite mean and variance; almost all economic and business variables of interest satisfy this quality. The theorem holds that, if X_i are the members of a random sample from a distribution with a mean μ and positive variance σ^2, then the random variable $Y = \frac{\sqrt{n}\cdot(X-\mu)}{\sigma}$, where n is the sample size, and $\bar{X}$ is the sample mean of the X_i, has a limiting distribution that is normal, with mean zero and variance one. See R.V. Hogg and A.T. Craig, *Introduction to Mathematical Statistics*, 4th ed. (New York: Macmillian, 1978), Section 5.4.

[71] As the name implies, the central *limit* theorem describes the *limiting* distribution of a sample of variables. It does not mean that a small sample of variables about which we know very little is distributed normally.

[72] The desirable properties of standard ordinary least squares estimators for the parameter vector β in the standard model $Y=X\beta+\varepsilon$ will still hold when the expectation of the dependent variable Y, E(Y), is linear in the unknown parameter β. This allows for cases where the explanatory variables X can be the natural logs of other variables or even products or powers of other variables. For a discussion of these and other practical applications of the linear model, see H. Theil, *Principles of Econometrics* (New York: John Wiley & Sons, 1971), Section 3.9, "Limitations of the Standard Linear Model."

[73] Be careful not to indiscriminately use the U.S. CPI, which has a well-known tendency to exaggerate underlying price inflation due to a persistent substitution bias and an overweighting for many consumers in energy prices and housing costs.

budget constraint can be relaxed when the price of the commodity is very small relative to overall income.[74]

General Cautions on Estimating Equations

Always consider the following cautions when specifying and estimating equations in an economic model such as a demand equation:

1. Microeconomic theory is based on well-behaved utility functions. These utility functions and other axioms typically prohibit purchases of negative quantities or negative prices, ensure consumers live within a budget constraint, prefer more to less, and like to pay less rather than more. A model should not violate those assumptions unless there is a good reason to believe that ordinary axioms do not apply.
2. Note that many models fail to follow a number of the assumptions listed above. In particular, linear models will allow negative prices and quantities; models that do not include income ignore the budget constraint; and use of nominal prices (especially with time-series data) violates the assumption that relative prices are the basis for consumer choices. You may be able to confidently use a model that, across all real numbers, violates one or more assumptions. One may do so only if the data and theory strongly support the model, the data are well outside the implausible range, and the range of the model is restricted.
3. The fact that data can be loaded into a regression model and the model is estimated with high "goodness-of-fit" statistics (such as R^2) means very little. It is much better to use a model for which there is strong theoretical and practical support than to simply run regressions until good results are obtained. It is occasionally useful to remind oneself that "correlation is not causality".[75]
4. If demand is based on multiple variables and a model is made using only one, then the model is *misspecified*. There may be serious consequences to this, including the estimation of parameters

---

[74] This is especially true when using cross-sectional data. For time-series analysis, changes in income are quite important, as is deflating for price changes and taking into account other significant changes (such as taxes) that affect disposable income after purchase of essential commodities.

[75] The prevalence of easy-to-use statistical software is, in this sense, a bane to solid thinking. I recall the pioneering University of Michigan econometrician Saul Hymans showing me how a single linear regression equation could be estimated, laboriously, using a desk calculator. He noted that, in times when such labor was required, economists thought very carefully about their model.

that are *biased*.[76] Furthermore, omitting variables from an equation when such variables are observed by the market participants and affect their behavior will typically result in biased estimates.[77] There are econometric techniques that sometimes minimize the effect of such biases. However, it is much better to properly specify the model in the first place. Again, excellent goodness-of-fit statistics provide no benefit when the "fit" is biased.

5. Consider the measurement accuracy of your variables and whether any of them include a certain amount of random error. Any error in the dependent or independent variable results in the estimated parameter having higher variance. In general, standard econometric techniques continue to work well when the explanatory variables in a regression equation have a certain amount of randomness, as long as the random behavior is not correlated with that of the dependent variable. However, when that randomness is correlated with the behavior of the dependent variable, the resulting estimates are most likely biased.[78]

Step Three: Model Demand Equation

Using the data and estimation results from the previous two steps, we can model consumer demand for a specific commodity. MATLAB provides a number of methods, the most general being the creation of a function.

Demand function: We can program a MATLAB function to calculate demand based on one or more inputs.

For example, the following function produces a demand schedule for a commodity, using a single (price, quantity) pair and one additional parameter (slope of the demand curve). The linear demand curve used by this function and illustrated in the graph it produces is very simple. Most

---

[76] Bias in parameter estimates means that the method systematically produces an estimate that differs from the actual value. An unbiased estimate will still usually differ from the underlying value — it is, after all, an *estimate* — but has the desirable property of having an expected value equal to the actual value. A related concept is *consistency*, which applies to estimators that, as the sample size increases, are expected to converge to the actual underlying value. Estimators that are both biased and inconsistent, therefore, must be expected to provide the wrong answer no matter how much data is available.

See the following notes for further explanation.

[77] For example, Varian shows how estimating production using one variable — when another, omitted variable is also important and observed by producers — can result in overestimating the effect of the included variable. This effect can occur in estimating returns to education (because education is chosen by the same person who is earning the returns to it) and other phenomenon. See H. Varian, *Microeconomic Theory*, 3rd ed. (New York: Norton, 1992), Section 12.7.

[78] In the standard linear model $Y=\alpha+\beta X+\varepsilon$, the assumption that X is nonstochastic can be relaxed and still retain the desirable properties of the ordinary least squares estimators if X is independent of the error term ε. However, if the error term and X are correlated — as often happens in misspecified equations and correctly specified systems of simultaneous equations — the least squares estimators will be biased and inconsistent. See, e.g., J. Kmenta, *Elements of Econometrics* (New York: Macmillan, 1971), Section 8-3.

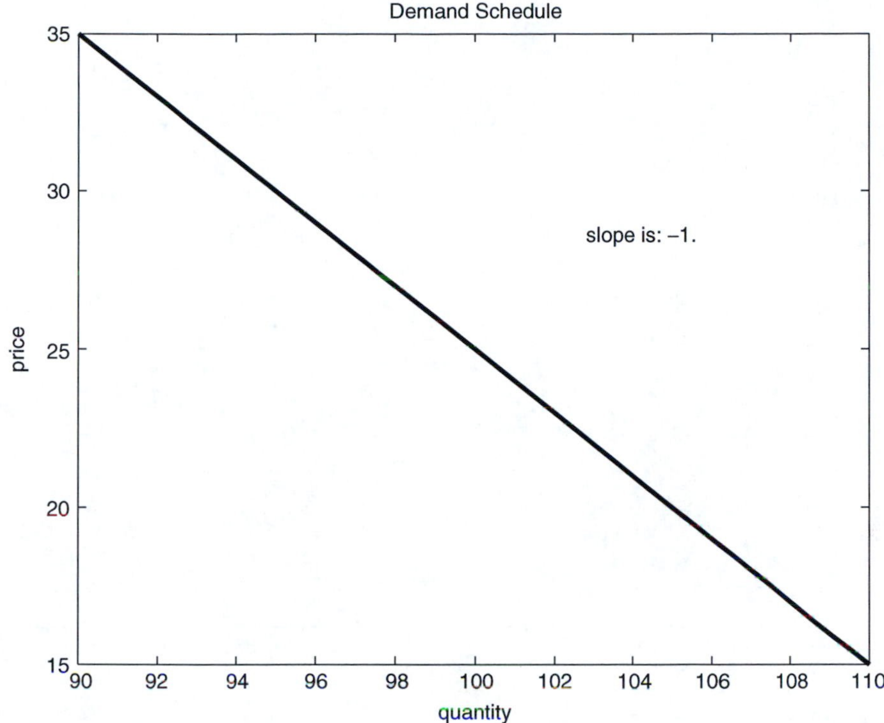

FIGURE 5-1
Linear demand function.

applications will require a more sophisticated demand function. However, it shows how demand can be modeled with a function within MATLAB and then illustrated with a graph.

See Figure 5-1, "Linear demand function," Code Fragment 5-1, "Simple Linear Demand Function," on page 120.

Supply Function

We can also model a supply schedule. A MATLAB function for this is printed in Code Fragment 5-2, "Supply Function" on page 122. The function takes as arguments a price for a supplied quantity, an elasticity of supply, and a minimum price at which producers will supply goods or services.

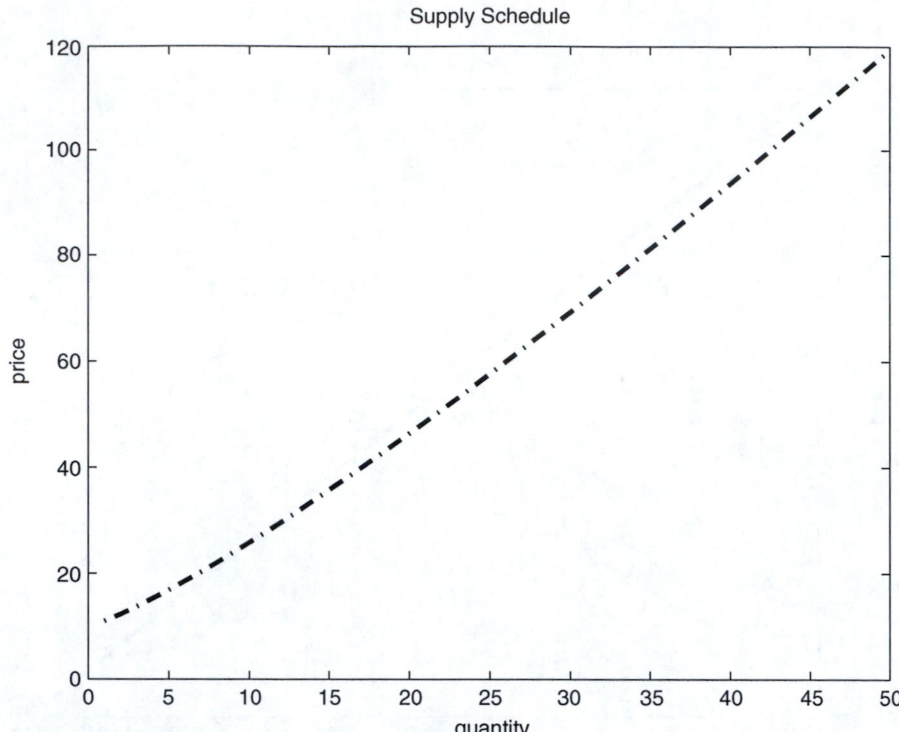

FIGURE 5-2
Supply schedule.

The function produces the quantity supplied for the price (which may be zero if it is too low) and a structure that contains the input parameters, plus the complete supply schedule.

The function also graphs the schedule, as illustrated in Figure 5-2, "Supply schedule." Finally, it includes a graphical comparison of the schedule, the elasticity parameter, and the actual arc elasticities at each interval. See Figure 5-3, "Elasticity of supply schedule."

Desirable Qualities of the Supply Function

This particular supply function is interesting and can be applied directly to many real-world problems. In particular:

- It has a "nearly constant" price elasticity of supply, as described in the next section, which is defined by a single parameter.
- It explicitly incorporates a minimum price (below which producers will not make new product) and makes the supply very sensitive to this initial price.

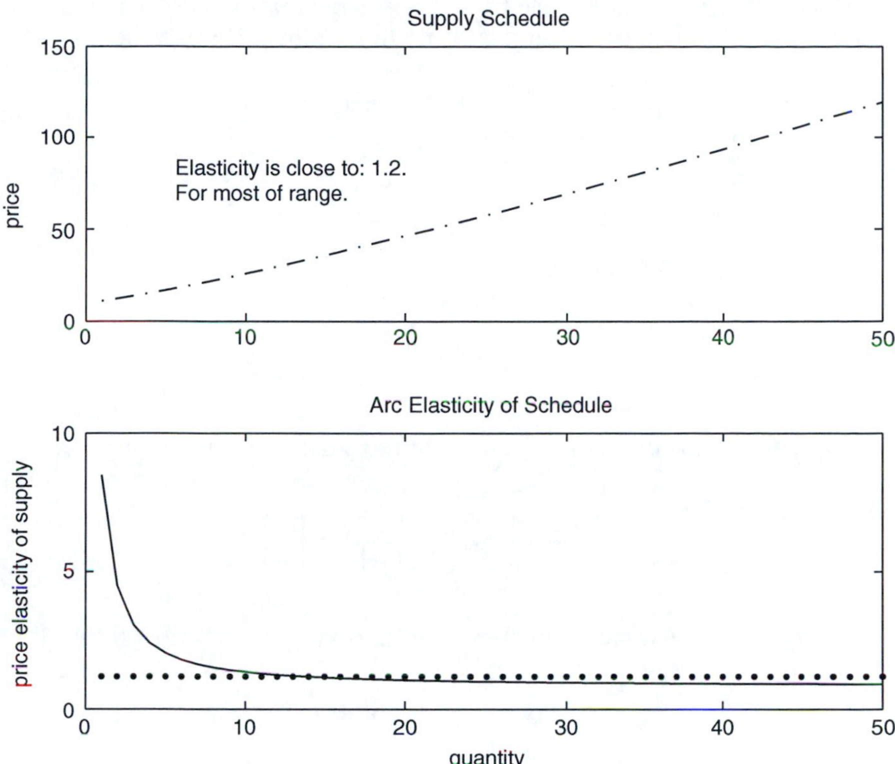

FIGURE 5-3
Elasticity of supply schedule.

- It allows the user to vary the price elasticity of supply.
- The function itself is written in a modular fashion. The actual program itself is very short. The program relies on subfunctions that handle the work and graphics.
- The function returns a data structure which records useful information about assumptions and data produced from the function. This makes it easy to record information from a session for reviewing later.

Elasticity of Supply

We derive the price elasticity of supply for this function below. A similar derivation can be used for demand and supply functions with similar forms.

The basic supply function expresses the quantity demanded as a function of price. We include an explicit minimum-price argument to the function and transform it into an inverse function in which price is determined from quantity, an elasticity parameter, and the minimum price parameter:

$$p = q^k + p^{min} \text{ from} \tag{10}$$

$$q^k = p - p^{min} \tag{11}$$

Differentiating with respect to quantity:

$$\frac{dp}{dq} = kq^{k-1} \tag{12}$$

Multiplying by the q/p ratio to calculate the price elasticity of supply:

$$\frac{dp}{dq} \cdot \frac{q}{p} = k\left(q^{k-1} \cdot \frac{q}{p}\right) = k \cdot \left(\frac{q^k}{p}\right) \tag{13}$$

Recalling from the equation above the definition of q^K and substituting it in:

$$\frac{dp}{dq} \cdot \frac{q}{p} = k \cdot \left(\frac{p - p^{min}}{p}\right) \tag{14}$$

Thus, the elasticity simplifies to the parameter k multiplied by a factor that, as prices grow, converges to one. For a good part of the usable range of the function the factor is slightly less than 1, so the elasticity of the schedule is slightly less than the elasticity parameter used in the function.[79] This is illustrated in the lower panel of Figure 5-3, "Elasticity of supply schedule."

Putting It Together

Once you can model supply and demand, you can attack many of the problems in typical economic thinking in a rigorous manner. For example, you can put them together as in Figure 5-4, "Demand and supply graph."

[79] This is sensitive to the range of plotting and the minimum price. If the minimum price is large relative to the range used for plotting, the elasticities in that range may be significantly lower than the elasticity parameter.

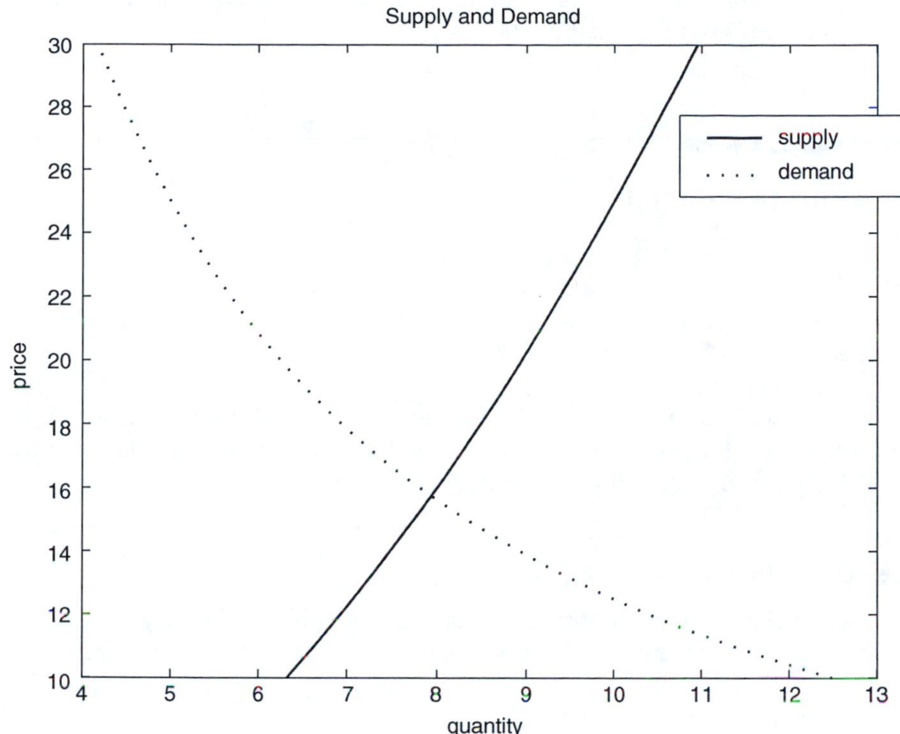

FIGURE 5-4
Demand and supply graph.

Graphics Techniques

The vast graphics capabilities of the MATLAB environment make it especially attractive for the business economist. MATLAB's *Handle Graphics* allows the experienced user to control almost every aspect of a graphic from the figure window to the axes, lines, annotations, and the tick marks.

However, that power has its price. The price in the case of MATLAB graphics is complexity. While most basic graphs can be created quickly, graphics ready for final production or publication will normally require specific instructions.

We provide numerous examples of graphics in this book, often with the m-file code to generate the graphic. A systematic discussion of graphics is in Chapter 17, "Graphics and Other Topics," on page 421. That chapter includes a discussion of the principles of graphical excellence, in "Tufte Principles" on page 421; a script m-file that generates a set of sample graphics in "Methods of Specific Plots" on page 427. This testplot.m file is included in the companion Business Economics Toolbox.

Libraries in Simulink

In addition to creating a library of MATLAB functions and scripts, you can also create libraries of blocks for use in Simulink. As business economists are likely to repeatedly use similar tools, the creation of specialized blocks can greatly speed up building and testing models.

We have created a small library of Simulink blocks which we use in the models described in this book. The following is a step-by-step procedure for establishing a business economics library into which you can place blocks developed by us, as well as your own blocks.[80]

Set Up a Library

From the Simulink model browser, use the File | New | Library command. A new library window will open. Save the library file, preferably in a "library" folder where you keep your other MATLAB files. For example, you might save a *BusEconUser* library.

To Put Blocks in the Library

Copy blocks from other models into the new library. If you are permanently adding blocks from other libraries, you must break the link to the existing model or block library in order to have a separate block saved in the library you are editing.[81] Many frustrating difficulties are caused by not saving the blocks in this manner, including "unresolved library link" errors.

Making the Library Appear in the Simulink Browser

Find the m-file slblocks.m and copy it to a folder in your path.[82] I suggest creating your own Library or Tools folder to keep it separate from folders installed by MATLAB automatically.

[80] As always, the instructions here are based on the software version available at the time of writing; check the user guide or help for later version.

[81] Try right-clicking on the block and looking under the "link options" submenu, or use an Edit | Break Library Link command. Refer to the Simulink user guide or help documents for additional information.

[82] There may be multiple files of this name, which can be found with the ('slblocks.m', 'all') command.

You can edit the slblocks.m file to make reference to the file name and screen name you select, or you can use Code Fragment 5-3,"Declaring An Economics Block Library," on page 127 as a template.

Troubleshooting Libraries

If you have unresolved library links or other difficulties, try the following steps:

1. Check to ensure that the slblocks.m file exists and correctly identifies your library files.
2. Carefully review the source of each block and the referencing of it by MATLAB.

See the Simulink user guide, the help files, or MathWorks solution 9009 for additional information.

Modifying an Existing Library

You can add blocks, edit the ones in the library, and otherwise modify a library once you have created it. To edit existing blocks, you must unlock the library before editing and then save it afterwards. Be careful when editing a library block, as the changes you make will then apply to other blocks that are linked to it and may appear in other models. In such cases, you may wish to preserve a legacy block and create a new one with a different name.

Special Library Blocks

Simulink comes with a wide array of functions, operators, signal generators, and other building blocks of a simulation model. Some of these are commonly used in business economics work. These are described below.

Blocks for Growth Models

Typically, one models a market by first identifying the current state of affairs and then allowing key variables to change. We discuss in this section techniques for describing the initial state, as well as for causing changes in that state.

Economic Base Series

We often start an analysis by assuming that consumers are spending a certain amount a year on a specific commodity, or that income is growing at an underlying rate, or that interest rates are trending to the sum of the inflation rate, plus a "real" component.

This is straightforward in Simulink, using a combination of blocks that are incorporated in the program and with blocks that are customized to perform specific calculation. The key blocks included with the program are described below. Custom blocks are described in the following section, "Custom Business Economics Blocks."

Constants

A constant block is included in Simulink. You can reference a number (e.g., 9) or a variable for which you have defined a number (e.g., sales2000).

Ramps

Simulink has a ramp block that quickly creates a series that starts at a certain number and grows linearly after that.

Periodic Signals

Simulink has an enormous capability for creating complex periodic signals. Blocks that create such signals include sine wave, signal generator, repeating sequence, and pulse generator.

Clocks

There are two clock blocks that generate time signals, which can be customized to produce time in almost every increment, starting at any time period.

Random Numbers

Simulink has two random number generators which can be used to simulate price or supply fluctuations.

Other Signals

Data from the workspace or from a specific data file can be used as inputs to a Simulink model. In addition, Simulink includes step functions and chirp signal generators. These can simulate a disturbance, or be used as switches.

Custom Business Economics Blocks

We have created a small set of blocks that are especially useful in business economics and are used in the models described in this book. They are described below.

Compound Growth

Simulink does not have a native block for generating a common basis for economic phenomena, namely a series that starts at a base number and grows at a certain rate per year.

To address this need, we created a compound growth block in the Business Economics Toolbox that you can plug into a Simulink model. The block will take a base amount and grow it a certain amount each period. Over a number of periods, this will produce the compound growth that is common in the analysis of business, economics, and other phenomenon.

Varying Growth

A related question arises when growth is not consistently occurring at the same rate. For this more complex set of calculations, we created a varying growth subsystem.

This block is implemented differently, in that you are asked to identify the initial amount, the time series vector that corresponds to the periods during which this initial amount will change, and then a factor of growth rates with one growth vector for each time period

Elasticity

A more complicated analytical tool used by economists is the concept of *elasticity* to describe how a change in one variable affects a change in another. A common consideration in business economics is how changes in prices affect changes in consumer demand. The price elasticity of demand is a convenient way of describing the percentage change in consumer demand caused by percentage change in price.

We created a block in the Business Economics Toolbox that changes the values of one variable on the basis of changes in another variable, using an elasticity factor. Using this block, you can project the path of sales, assuming changes in price and assumed price elasticity of demand. This block might be more precisely defined as "elastic impact" rather than elasticity, as it applies changes to a relationship governed by elasticity rather than calculating it.

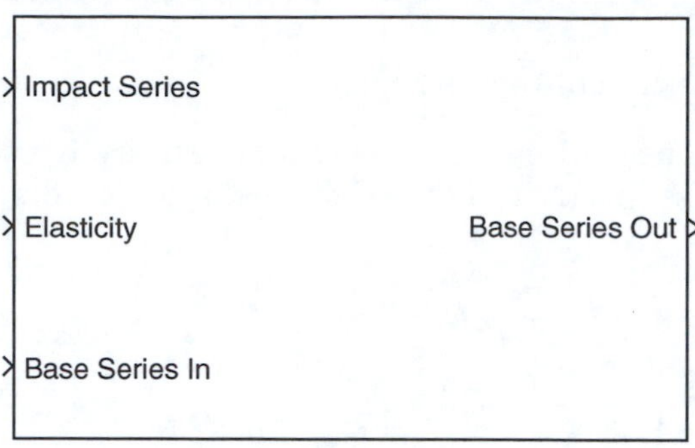

FIGURE 5-5
Elastic impact block.

The block has three inputs: the base series (the series that you will measure both before a change in another variable and after), the impact series (the variable that is changing), and the elasticity factor. Keep in mind that price elasticities of demand are normally negative. See Figure 5-5, "Elastic impact block."

Using the block is straightforward: connect the variable that changes (such as a price, tax rate, or other policy variable) to the "impact" port. Connect the elasticity (which could be a constant or a changing variable) to the "elasticity" port. Finally, connect the base series variable to the "in" and "out" ports.

In such a system, for example, the base series of "sales" will increase by 5% when the impact series of "price" drops by 10%, if the elasticity is set at –0.050.

Example Simulink Model

We have incorporated the blocks in the Economics Toolbox described above into an example model, which will illustrate a Simulink economics model and also how these various blocks work.

Purpose of the Model

The purpose of this model is to simulate sales in a market before and after a policy change that affects the price of the goods or service in question. This problem is a fundamental one in both theoretical microeconomics and applied economics, and is faced by policymakers in government as well as business leaders every day.

For this model we assume that sales are determined by income and price.[83] We will simulate the effect on sales as income changes over time and as tax policy changes affect the price of the goods or services.

Top-Level View of Model

The overview of the model shows the constants on the left: initial income and two growth variables. We would normally use only one growth variable, but for this example we include two to illustrate the use of both constant and varying growth blocks. See Figure 5-6, "Example business economics model."

The large subsystem to the right of the constants is the base case income subsystem, which takes the income and growth variables and calculates income for each period. The policy change and impact subsystem takes the income variable, and within the subsystem takes the policy variable (a tax rate) and an elasticity parameter, and calculates sales before and after the policy change. The last subsystem takes the variables in and saves them properly.

These individual subsystems are illustrated in Figure 5-7, "Base income subsystem," Figure 5-8, "Policy change subsystem," and Figure 5-9, "Income and sales subsystem."

Running the Example Model

Master Simulation Callback

We run the model using the master simulation callback that was described in Chapter 3, "MATLAB and Simulink Design Guidelines" at "A Master Simulation Callback" on page 52.

That callback script initializes the model, loads the variables, runs the simulation, and graphs the output. The key output is the projection of sales before and after the policy change. A graphical illustration of this is produced by the callback and illustrated in Figure 5-10, "Sales projection with policy change."

---

[83] See "Estimating Demand Schedules" on page 98 for a discussion of demand schedules. For this simulation model we assume that we have already properly modeled the demand for the particular good or service and obtained a price elasticity of demand as well as a relationship between income and demand. We are making further simplifying assumptions that the relative price in a demand function is captured by the after-tax price in this model and that the fundamental price and income relationships do not change during the simulation time period.

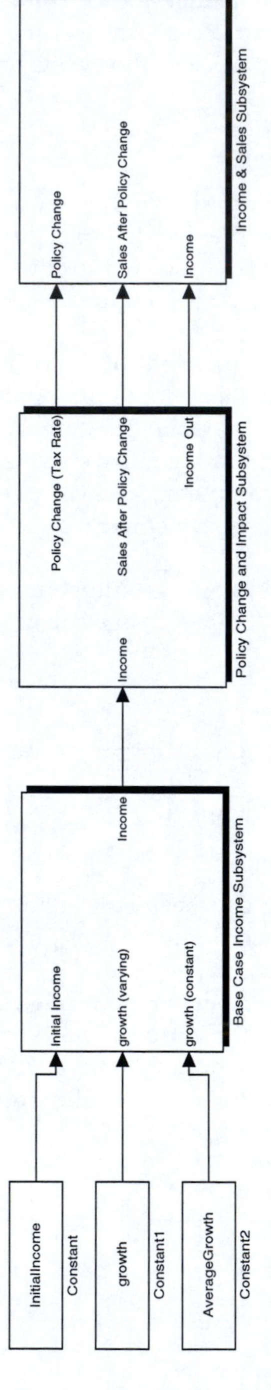

FIGURE 5-6
Example business economics model.

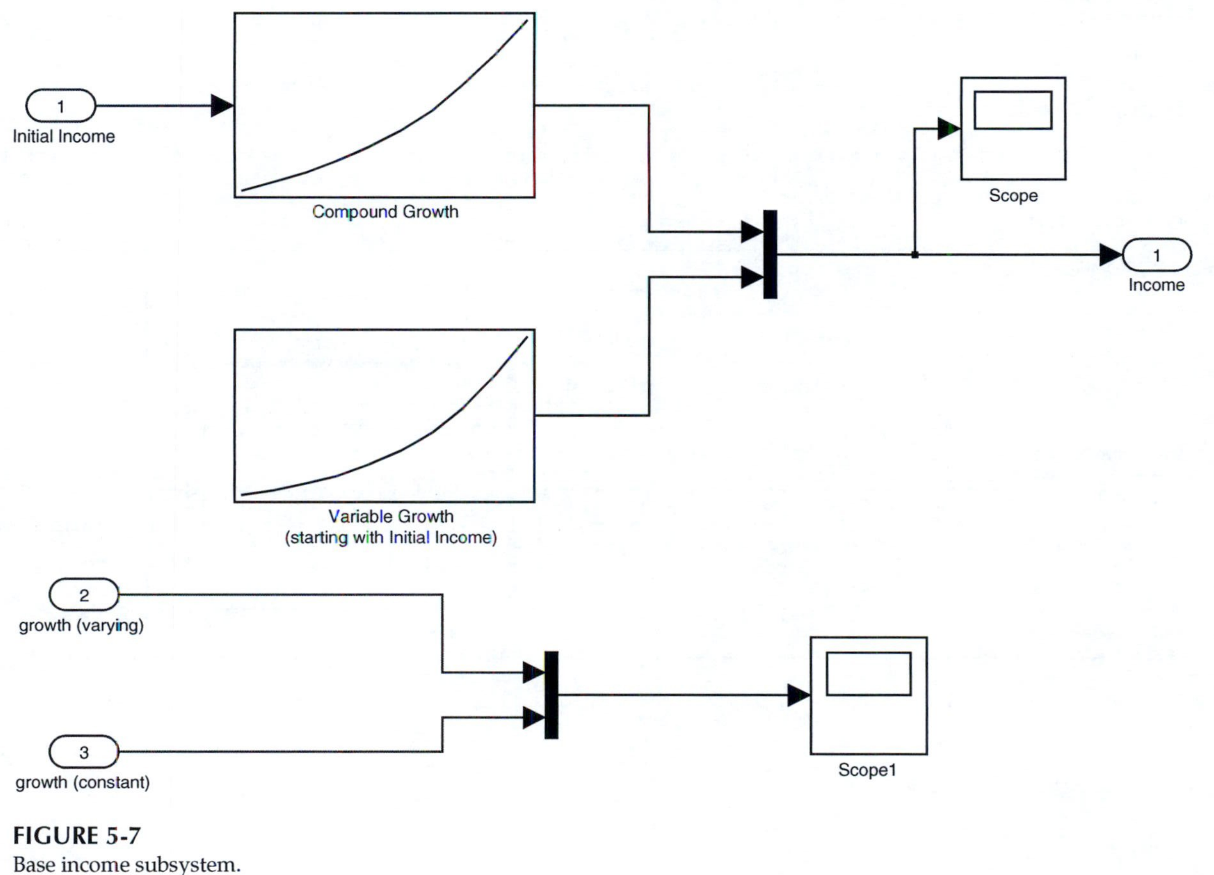

FIGURE 5-7
Base income subsystem.

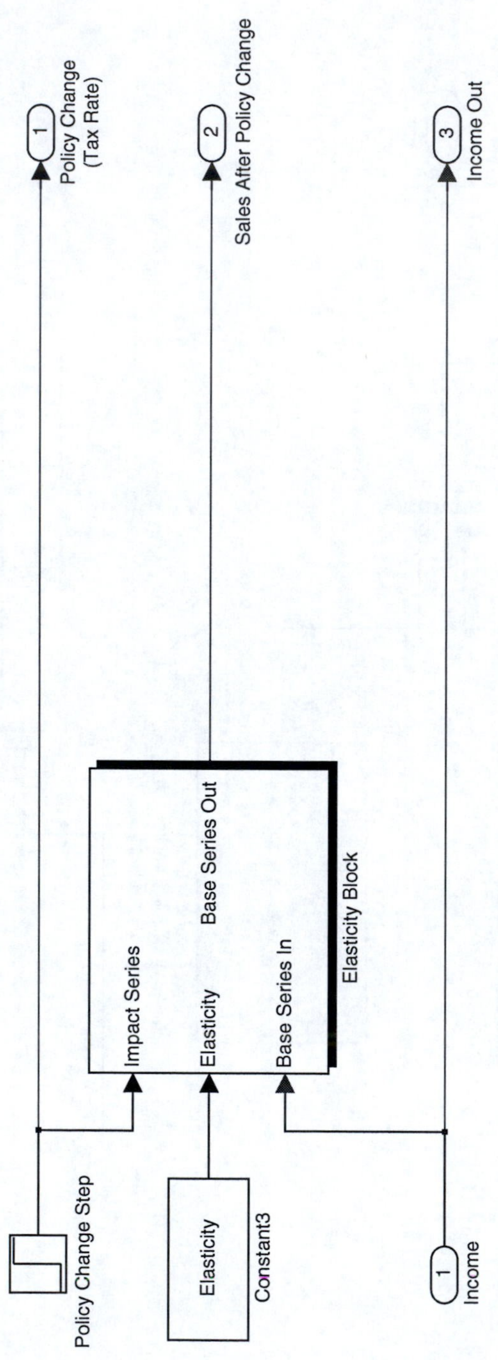

FIGURE 5-8
Policy change subsystem.

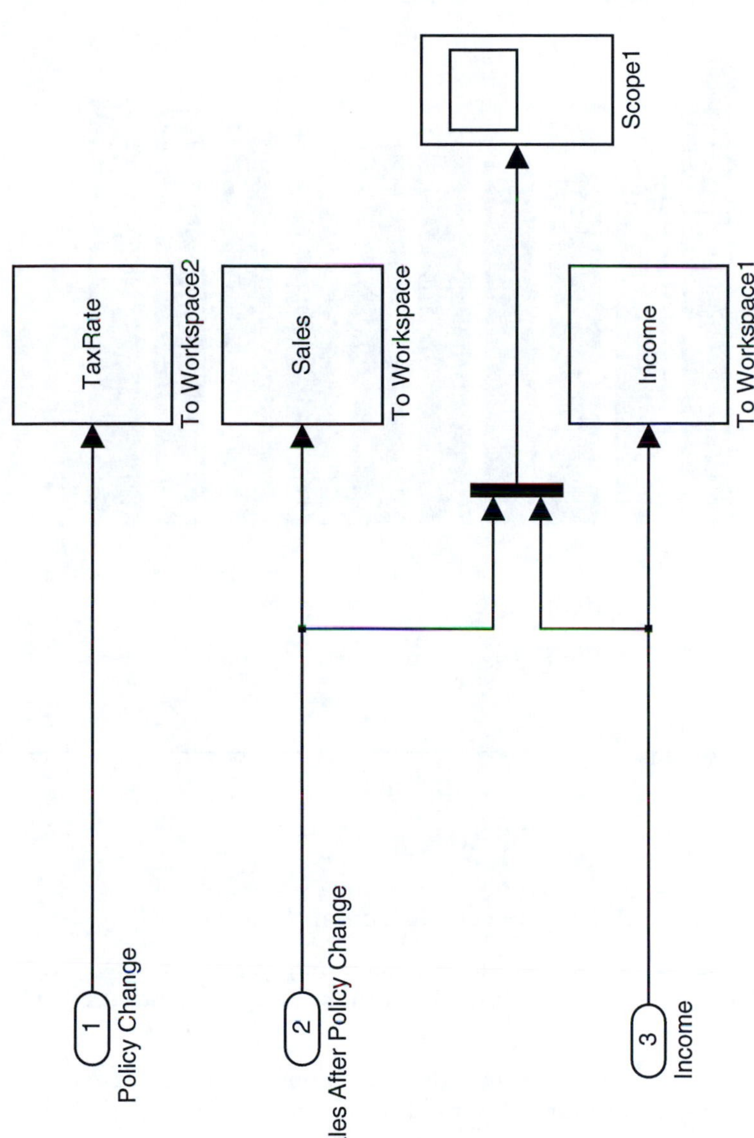

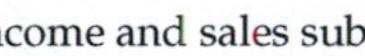

FIGURE 5-9
Income and sales subsystem.

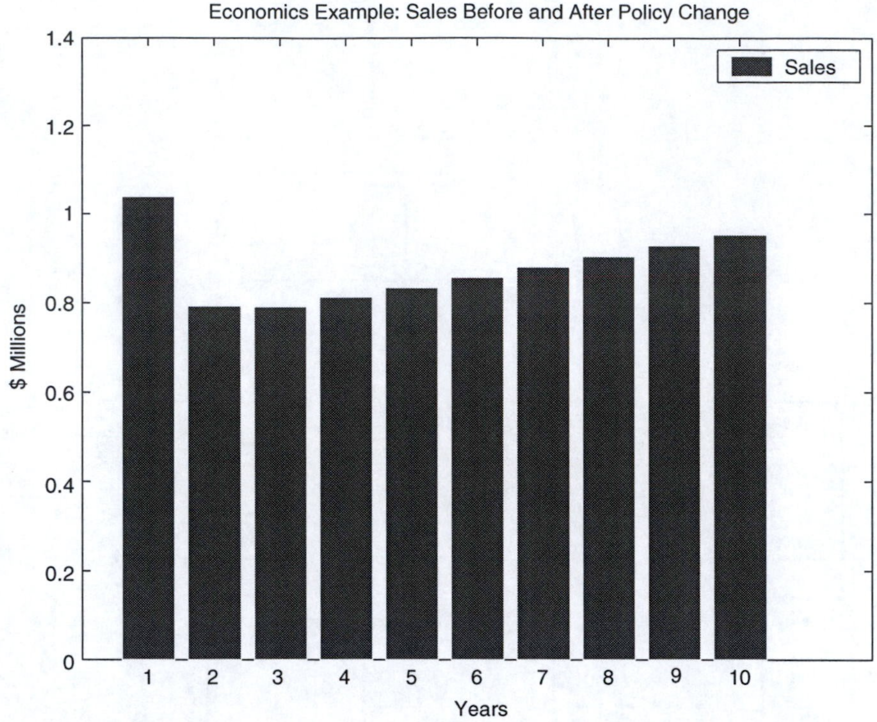

FIGURE 5-10
Sales projection with policy change.

Appendix

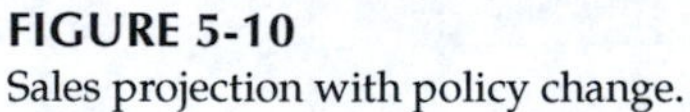

Code Fragment 5-1. Simple Linear Demand Function

```
function [p, q] = simpledemand(price_i, quantity_i,
slope);

%   This simple, linear demand function provides
%   price and quantity demanded,

%   given a slope and one {price, quantity} pair

%   The demand curve is plotted.

%   Example: [p, q] = simpledemand(5, 10, -1);
```

```
%   for a commodity for which 10 units will be
%   demanded at price $5.

%   P.L. Anderson, 15 May 2003; 1 April 2003;

%   (c) 2003 Anderson Economic Group

%   Use by license only.

%   insert default data if insufficient arguments

if nargin <3

    disp('Using default values ...')

    price_i = 25

    quantity_i=100

    slope = -1

end

%   Check Data

if slope > 0

    warning('Price Elasticity of Demand is Positive.
This parameter is normally negative.');

end

if price_i < 0

    warning('Price is negative. This parameter is
normally positive.');

end

%   Call Demand Drawing Subfunction

[p, q] = drawdemand(price_i, quantity_i, slope);

%---------------------------------------------------

function [p, q] = drawdemand(price_i, quantity_i,
slope);

%   create index; don't allow price below 1; rescale
%   if p is entered as a number below 2

if price_i < 2.5

    disp('Rescaling by multiplying price by 100.')

    price_i = price_i*100

end
```

```
inc = min(price_i-1, 10);

p = [(price_i-inc):(price_i + inc)];

%   calculate line from y = mx + b linear form (p =
% mq + b, linear inverse demand; implies q = (p - b)
% /m demand.)

b = price_i  slope*quantity_i;      %y-intercept
(price axis intercept)

q_unlim = (p - b)/slope;

q = max(0, q_unlim);                 %limit quantities
to nonnegative numbers

%----------------Plot-----------------------------

figure;

plot(q, p);

title('Demand Schedule');

elasticity_message = {'elasticity is:' ,
[num2str(slope), ' at starting point.']};

text(.7, .7, elasticity_message, 'units', 'normal');

xlabel('quantity');
ylabel('price');
```

Code Fragment 5-2. Supply Function

```
function [q, sched] = supply(price,
elasticity_param, min_price);

%   This supply function provides quantity produced,
% given information on: price, price elasticity of
% demand, and minimum price for production to start.

%   The supply curve is plotted; "sched" structure
% contains supply schedule and other information.

%   Uses an inverse supply function of nearly
%   constant elasticity:

%     price = quantity.^elasticity + min_price;

%   Example:
```

```
%     [q, sched] = supply(25, 1.2, 10);
%
%     Normally, 0 < elasticity < 2 ; 0 < price < 100;
%                   0 = < min_price = < price;
%
%   PLA, 27 May 2003
%   Business Economics Toolbox, (c) 2003 Anderson
%   Economic Group
%   Use by license only.
%---------check data---------------------------
if nargin < 3
    disp('using default parameters.')
    price = 25;
    elasticity_param = 1.2;
    min_price = 10;
end
if elasticity_param < 0
    warning('Price elasticity of supply is negative.
This parameter is normally positive.');
elseif elasticity_param > 2
    warning('Price elasticity very high. May yield
unusual results with high price input.');
end
if (price * min_price) < 0
    error('Cannot have negative prices.')
elseif price < 1
    disp('...rescaling price by multiplying by
100.')
    price = price*100;
elseif price < min_price
    disp('Price below minimum necessary for
production.')
```

```
end

%--------main program------------------------

%   create supply schedule and graph it

yy = drawsupply(elasticity_param, min_price);

%   calculate (q,p) pair at selected interval

q = find_quantity(yy, price, min_price);

p = price;

%   calculate arc elasticities at each interval;
%   plot it with schedule

elast = check_elast(yy, elasticity_param);

%   save data and notes

notes = {['q and p are price and quantity pairs, yy
is (q, p) supply schedule.'], ...

         ['elasticity_param is defined parameter,
elast is calculated arc elasticities'],...

         ['Reference: Anderson, Business Economics
Using MATLAB, CRC Press, 2003']};

sched = struct('p', p, 'q', q, 'elasticity',
elasticity_param, 'elast', elast, 'min_price',
min_price, ...

    'yy', yy, 'notes', notes);

%--------------------------------------------------

%   Functions

%--------------------------------------------------

function price = supply_inverse(quantity,
elasticity_param, min_price);

%   quadratic form--if m is 2; straight line if m is
%   1.

p1 = quantity.^elasticity_param + min_price;

price = max(p1, 0);     %inequality not necessary if
q is always positive

%--------------------------------------------------
```

```
function yy = drawsupply(elasticity_param,...
min_price);

%   create (q,p) pairs for each row of supply
%   schedule matrix.

Q = [1:50]';

P = supply_inverse(Q, elasticity_param, min_price);

yy = [Q P];

%------------plot for drawsupply-----

figure,

plot( yy(:,1), yy(:,2), 'r-.', 'linewidth', 2);
%Marshall convention: price is on y-axis

title('Supply Schedule', 'fontweight', 'bold');

% elasticity_message = {['Elasticity is close to:
% ' num2str(elasticity_param), '.'], ...

%        ['For most of range.']};      %reserved for
% additional message

% text(0.1, 0.6, elasticity_message, 'units',
% 'normalized');

xlabel('quantity');

ylabel('price');

disp('... plotting supply schedule.')

%-----------------------------------------------------

function q = find_quantity(yy, price, min_price);

%find first q, p pair where price is equal or
%greater to initial argument

if price >= min_price

    [i j] = find(yy(:, 2)>=price);


    %   use to select pair in yy

    q = yy(i(1), 1);

else

    q = 0
```

```
end

%--------------------------------------------------

function elast = check_elast(yy, elasticity_param);

% Calculate arc elasticities (actual changes per
% interval)

% instantaneous "point" elasticities.

[fx fy] = gradient(yy);  % gradient function gets
%differences in both directions; yy is matrix of
%(q,p) pairs in each row.

fy;                        % fy is d(yy)/dy, where dy

%is assumed to be one. Each row of the vector is
%calculated separately.

whos;

prop_change = fy./yy;        %compare with
% TICK2RET.m; return is difference divided by base
% amount;

elast = prop_change(:,1)./
prop_change(:,2);%proportional change in quantity
% (first column)

%                  %divided by proportional change in

%                  %price (second column)

%---------graphically compare arc elasticities with
%supply schedule----------

figure,

subplot(2,1,1);

plot( yy(:,1), yy(:,2), 'r-.');     %Marshall
% convention: price is on y-axis

title('Supply Schedule', 'fontweight', 'bold');

elasticity_message = {['Elasticity is close to: '
,... num2str(elasticity_param), '.'], ...

        ['For most of range.']};     %reserved for
additional message

text(0.1, 0.6, elasticity_message, 'units',
'normalized', 'fontangle', 'italic');
```

```
% xlabel('quantity');
ylabel('price');
subplot(2,1,2);
plot( yy(:,1), elast, 'm-');        %Marshall
convention: price is on y-axis
Ylim = ([-5 5]); hold on;
plot( yy(:,1), elasticity_param, 'c.'); hold off;
title('Arc Elasticity of Schedule', 'fontweight',
'bold');
xlabel('quantity');
ylabel('price elasticity of supply');
%   --------------end------------------------------
```

Code Fragment 5-3. Declaring An Economics Block Library

```
function blkStruct = slblocks
%SLBLOCKS Defines the block library for business
%economics, provided by AEG
%   Copyright (c) 2003 by Anderson Economic Group,
%LLC. All Rights Reserved.
%   Use granted under license from AEG.
%   MATLAB and Simulink licensed by MathWorks, Inc.
%   $Revision: 1.3 $
%   $Date: 2003/04/01 $
% Name of the subsystem which will show up in the
% Simulink Blocksets and Toolboxes subsystem.
blkStruct.Name = sprintf('AEG Business
Economics\nModels');
% The function that will be called when
% the user double-clicks on this icon.
blkStruct.OpenFcn = '';
```

```
blkStruct.MaskInitialization = '';

% The argument to be set as the Mask Display for the
% subsystem.

% You may comment this line out if no specific mask
% is desired.

blkStruct.MaskDisplay = ['plot([1:.1:5],', ...
                                    'sin(2*pi*[1:.1:5])./
[1:.1:5]./[1:.1:5]);'];

% Define the library list for the Simulink Library
% browser.

% Return the name of the library model and the name
% for it

% These are MATLAB structures, so use the format
% "Browser(n).Library

% and then "Browser(n).Name" to identify the file
% name (Library) and

% the screen name (Name).  The "n" is a number
% identifying the order in

% the structure.

Browser(1).Library = 'Economics';

Browser(1).Name    = 'AEG Economics Models';

Browser(2).Library = 'Test2Lib';

Browser(2).Name    = 'Test Libraries';

Browser(3).Library = 'BusEconUser';

Browser(3).Name    = 'Business Economics Blockset';


blkStruct.Browser = Browser;

% End of slblocks
```

6

Economic Impact Models

Economic and Fiscal Impact Models

One of the most common questions faced by state and local governments and their taxpayers is evaluating the economic impact of a project. Many decisions about tax incentives, public financing of roads and infrastructure, changes in laws, and expensive public works and transportation expenditures are made partially on the basis of the economic impact of a proposed project.

Fiscal impact models are also used widely in state and federal governments. Most tax policies are discussed largely in terms of fiscal impact. Fiscal impact includes the direct tax revenue and expenditure effects, while economic impact includes wages, expenditure, employment, and fiscal effects.

Simulation models are a powerful way to address these questions, and we provide detailed instructions on their use later in this chapter. However, because there has been so much poor practice — and some outright charlatanry — in this field, we first discuss the ethics of economic impact analysis and its fundamental mathematics, before turning to the models.

Ethics in Economic Impact

There is no other area of applied economics in which more sloppy work, exaggerated results, and just plain balderdash have been produced than in "economic impact" analysis. Because the claimed economic impact of a proposed development can affect political support for a proposed project — and sometimes taxpayer funding — an incentive often exists to exaggerate the benefits. Sadly, that incentive has proven too much to resist for many practitioners in this field.

Sports Exaggerations

The debates over public funding for new sports stadia is the best documented area in which impact models have been systematically misused to exaggerate benefits. The clear beneficiary in many episodes has been those who own or operate a stadium facility built with public funds.[84] The author of this book has seen multiple economic impact analyses for minor and major leagues' sports stadia and major events that grossly exaggerate the benefits and understate the costs.

A Syndrome: The Missing Report

One of the most pernicious trends in the cottage industry of economic impact analyses in recent years — and one that allows for weak analysis to gain currency — is the syndrome of the missing report. Our efforts to track down the reports that were cited in news stories often revealed that the only documents actually published were press releases.[85] This missing report syndrome is characterized by widely published claims, but no published study. In some of these cases, studies actually did exist, but were not made available for review when the publicity began for the result. In some cases, an understandable and commendable attempt at civic cheerleading results in an enormously attractive claim that, upon inspection, proves to be without firm basis.[86] The host committee, local journalists, and fans want to believe

[84] The seminal work here is by Roger Noll and Andrew Zimbalist, Eds., *Sports, Jobs, and Taxes* (Washington DC: Brookings Institution Press, 1997). See, in particular, the second chapter discussing how *multipliers* for sports stadia are smaller than these for many other developments and should be applied to a smaller (net) figure than is normally used; the conclusions from multiple contributors that municipalities paid a net subsidy to stadium enterprises; and the recounting of the ubiquitous "build the stadium, get the jobs" rhetoric that surrounds campaigns for public financing.

[85] A prime example of this is the economic impact of the Atlanta Super Bowl in 2000. While numerous press releases and news stories cited the $290-million-plus figure, repeated attempts in 2002 and 2003 to actually obtain a copy of the report proved fruitless. Letter from Patrick L. Anderson to NFL Commissioner Tagluabue, August 26, 2003; correspondence with Bruce Seamans, Georgia State University, August 2003. A copy of the June 2000 report was later provided by the NFL in September 2003 — 3 1/2 years after the event.

Other examples include an analysis for a midsize city of a minor league baseball stadium completed by one of the largest accounting firms in the country; and the claims made for the economic impact of the Pontiac Silverdome, first when it was originally constructed in the 1970s and then when it was being abandoned by its only major tenant, the Detroit Lions, after the season ending in 2002. Our (unpublished) 2001 analysis of the Pontiac site indicated that the City of Pontiac would have a larger economic and fiscal impact from bulldozing the stadium. In 2003, after the Lions moved to Detroit's Ford Field, the city accepted proposals to do just that.

[86] For example, in August 2002, a $372 million economic impact was claimed for a Super Bowl in Detroit that would not occur until 2006 and was planned for a stadium that had not yet been completed. The claim was repeated often in the press; see, e.g., R.J. King, "Super Bowl Hotel Rooms Go Fast," *Detroit News*, August 21, 2002, p. 1. As of August, 2003, no report documenting the $372 million economic impact was available. The figure may have been based on

an event will be successful and appear to relax their normal impulse to check whether the claim has a firm grounding.[87]

A Sports Example: The Super Bowl

The Super Bowl is one of the premier sporting events in the U.S. each year. Indeed, if the local economic effects of a Super Bowl were near the $300 million or more claimed in recent years, "super" would be an understatement. Unfortunately, the claimed economic impacts of at least two recent Super Bowls suffer from the missing report syndrome and have resulted in widely published claims that are nearly impossible to document.[88]

A skeptical analysis entitled "Super Bowl or Super (Hyper) Bole," by economists Robert Baade and Victor Matheson, looked systematically at Super Bowls. They concluded that "the economic impact of the Super Bowl is likely on average to be one-tenth or less the magnitude of the most recent NFL estimate."[89] They note that other research based on regression analysis found an impact not significantly different from zero.[90] A review article by Craig Depkin and Dennis Wilson stated that "most economists consider the forecasted impacts offered by advocates inflated," and identified a number of sources of error — practitioner bias, measurement error, and leakage.[91]

We will consider the actual impact of a large sporting event further in "An Example: The Fantasy Game" on page 134.

---

extrapolation from previous reported economic impact figures. Letter to NFL Commissioner (cited above); telephone discussions with Detroit Super Bowl Committee and Detroit Convention and Visitors Bureau, August 2003. A later press announcement — timed to appear on the February 1, 2004, date of the Houston Super Bowl — again stated that a new study would be released showing a $300 million-plus benefit.

[87] C.A. Depkin and D.P. Wilson, cited below, state that: "However, the economic impact studies by local advocates are rarely subject to scholarly review, and inflated estimates can go unquestioned and perhaps unconfirmed after the event."

[88] The two Super Bowls for which we confirmed there were no published reports available at the time of publicity, but widely published claims, are the Atlanta and Detroit events; see citations above. However, an extensive Internet search found thousands of press citations for claimed economic impacts of many recent Super Bowl events — but no published reports or even executive summaries of those reports. In particular, the Houston and San Diego events had widely published claims but no evidence that any actual report was made available at the times the claims were made publicly. In some of these cities, a certain amount of attention was paid by the press to the validity of the claims. For example, the *Atlanta Business Chronicle* published an exchange of letters in February 2000 between Bruce Seamans, a consultant to the NFL–Atlanta Sports Council project that produced the $290 million figure, and Robert Baade and Victor Matheson (cited below) who challenged the NFL estimate. The main topic of the exchange was revealing: whether or not the net effect was zero. Correspondence and manuscript from Bruce Seamans.

[89] R. Baade and V. Matheson, "Super Bowl or Super (Hyper)Bole," found at: http://lanfiles.williams.edu/~vmatheso/research/superbowl2.pdf.

[90] P. Porter, "Mega-Sports Events as Municipal Investments: A Critique of Impact Analysis," Mimeograph 1999; cited in Baade and Matheson, above.

[91] C.A. Depkin and D.P. Wilson, "What is the Economic Impact of Hosting the Super Bowl?" Texas Workforce Commission, *Labor Market Review*, January 2003; reprint found at: http://www.tracer2.com/admin/uploadedPublications/178_tlmrexpert0301.pdf.

Non-Sports Exaggerations

Unfortunately, misuse of economic impact analysis is not confined to sports stadia. Major investments in a state, especially ones that are being considered for tax abatements or direct state investment, routinely have economic impact analyses performed.[92] In many cases, these are performed quite competently under the guidelines of state law or other policy.[93]

Economic Impact of Y2K

Sometimes the economic impact analyses are not performed competently. A spectacular example — hopefully not repeated for another millennium — is the incredible claims about the economic impact of the Y2K bug and the expenditures by government and businesses to control it. We issued a short series of reports in December 1999 indicating the implausibility of such exaggerated claims.[94] During that time, the press and government officials routinely published cost estimates of $300 billion to over $1 trillion with the most common estimate being $424 billion attributed to the Gartner Group.[95]

While there was some uncertainty regarding the actual effects, and government overreaction did trigger unnecessary expenditures in both the public and private sector, there was little real analysis underlying the huge claims.[96] Furthermore, there did not appear to be much journalistic fact-checking going on as the oft-quoted source for one of the estimates turned out to be ephemeral.[97] Our own analysis (which, though not extensive, was at least in writing and based on some actual data) indicated the actual cost to the U.S. economy was about $100 billion, and much of that was caused

---

[92] For example, in Michigan, state law requires an application for a MEGA (Michigan Economic Growth Authority) subsidy to be evaluated after the completion of an independent economic impact analysis.

[93] These policies themselves often create the fundamental assumptions that guide the analysis. For example, a MEGA credit can only be given when a project would not come into the state without the credit. Of course, all applicants for the credit indicate that they will not make the investment without the credit, and the subsequent analyses are based on that assumption.

[94] Patrick L. Anderson, "Y2OverKill," five short articles released on the Anderson Economic Group Web site, December 1999.

[95] See, e.g., P. Despeignes, Economy Ready for Millennium Bug? *The Detroit News*, September 29, 1999, citing estimates or $300 billion to $1 trillion, the latter attributed to J.P. Morgan, and a government report anticipating "severe long- and short-term disruptions to supply chains [that] are likely to occur."

[96] For example, at least one government official acknowledged in testimony before the U.S. Congress that about 50% of Y2K expenditures was upgrading of equipment and software that would have been required anyway. Statement by Robert Albicker, IRS Deputy Chief Information Officer, reported by *Computerworld*, September 1999.

[97] An oft-cited estimate of $424 billion was attributed to the Gartner Group. Repeated efforts by the author to confirm this estimate with the firm resulted in a press release from Gartner and a verbal disclaimer from an employee. No full report, or even summary of a report, justifying the estimate was produced.

by overreaction.[98] Finally, indicating in a comic fashion the underlying exaggeration, the U.S. government set up a toll-free Y2K hotline to deal with emergencies — and scheduled people to answer it only during specified business hours — which stopped just short of midnight on New Year's Eve.[99]

Economic Impact of Broadband

Another analysis was prepared recently for a major industrial state considering the imposition of a tax on telephone lines to support the creation of a development authority. This proposed authority was to help finance the development of broadband Internet technology, which would spur economic growth.

The analysis prepared by the Gartner Group contained an estimate of positive economic impact of "an ubiquitous broadband network." They estimated the positive impact of the authority — which would receive about $420 million over 12 years — would be a Brobdingnagian $440 billion over a 10-year period.[100] This enormous increase in gross state product was specified in some detail, including 770,000 new jobs broken down into categories such as marketing, finance, and information. These new jobs were supposed to rise in a state with the employment of 4.8 million, meaning a 16% increase in statewide employment and $440 billion in GSP from approximately $31 million in additional annual financing.[101] That is a return of over 1000 to 1.[102]

How could such a gargantuan economic impact be projected? The methodology section of the available report indicates that teledensity (phone lines per capita) in third-world countries was compared with GDP, and the same calculations were performed for the U.S. and other industrialized countries. The data show that industrialized countries had more phones per capita. Therefore, putting more broadband connections into Michigan was projected

---

[98] For example, we included in our estimate the cost of cash-hoarding and the U.S. government's efforts to prevent a run on cash. Ironically, the government's own actions, by increasing consumer worries, had the effect of encouraging the behavior it was trying to reduce. Other itemized cost categories included the costs of excess notices, costs to the financial industry for unnecessary preparations, and significant beneficial expenditures on hardware and software. Patrick L. Anderson, "Y2OverKill," press release, December 30, 1999.

[99] The President's Council on Year 2000 Conversion maintained a toll-free phone line (1-888-USA-4-Y2K) that "offers information of interest to consumers in common areas such as power, telephones, banking, government programs, and household product." The Council promised "information specialists will staff the line from 9 a.m. to 8 p.m. (EST), Monday–Friday." New Year's Eve that year was on a Friday. Patrick L. Anderson, "Y2overKill," Press Release, December 29, 1999.

[100] Gartner Consulting, "E3 Ventures for Michigan Economic Development Corporation," November 1, 2001. The ubiquitous network envisioned was defined as a 50% penetration rate, which according to the study was higher than the 20% penetration rate that would have been achieved without state intervention.

[101] Anderson Economic Group, "Fiscal Analysis of Link Michigan Program," December 2001; presentation to Michigan Legislature Committee. This actually overstates the financing effects of the program, as the adopted bills contained no direct taxpayer support for the new broadband development authority.

[102] We projected 12-year revenue of $421 million to the financing authority.

to have the same proportionate impact as industrializing and wiring third-world countries.

Perhaps the authors should have left out any description of their methodology and simply included this statement: "The self-organizing techno-economic paradigm of the Internet cannot be understood by conventional arguments of economic growth."[103]

An Example: The Fantasy Game

How could the net economic impact of a single sporting event exceed $300 million? As an illustration of the vagaries of economic impact analysis, we will analyze a truly fantastic event which we call "the fantasy game." We are interested in the local economic impact, by which we mean the additional income to residents of the local area. Because this is a fantasy, we will throw out all caution in making assumptions about the economics of the event:

- Assume, for a moment, that for the fantasy game, every seat of a 65,000 seat stadium was filled.
- Assume that every single attendee was from out of town and that they all traveled there for just this one event.
- Assume that in addition to the attendees, an extra 1,000 people traveled there to work during the event.
- Assume further that all the fans and all the workers stayed two nights in a local hotel, with each visitor staying in his or her own room, paying $150 per night.
- Assume that every single visitor spent $200 per day on food, tickets, T-shirts, and other local expenditures.
- Since there are some costs — even in our fantasy game scenario — of food, drink, T-shirts, and hotel rooms, assume that such costs are just 50% of what the visitors spent, and that all of the remaining profit goes straight to local owners and employees.
- Now, just for the fun of it, assume further that each of the 65,000 fans and 1,000 workers drop an extra $50 in cash on the sidewalk each day they visit. Despite the hordes of cash-rich/pocket-weak visitors wandering the streets, all of the $3.3 million per day in lost currency is picked up by local residents.

[103] Gartner Consulting, "E3 Ventures." This quote was attributed to University of Texas Broadband Study.

Clearly, this is a fantasy. Whatever heroics might occur on the field would pale in comparison with the supernaturally heroic assumptions made about the fans. Such a game would never be played on earth.

However, what would the local economic impact of such a fantasy be? Table 6-1, "Economic Impact of the Fantasy Game," provides an optimistic calculation.

Sum of Benefits from the Fantasy Game

As the calculations in the table indicate, our fantasy game has a positive economic impact of $13 million for local expenditures other than hotel rooms, another $9.9 million for the hotel expenditures, and $6.6 million for the cash left on the street. The local economic impact totals $29.7 million.

Such an estimate was produced with wildly optimistic assumptions. Yet, the total is still much smaller than the $300 million estimates publicized for recent Super Bowls. It is interesting that the amount is about one-tenth of that amount — exactly the ratio of reality-to-claim found by Baade and Matheson.[104] Indeed, even if we used a multiplier to *double* this amount, the estimate would still be less than one sixth of the $374 million claimed for the Detroit Super Bowl. If we were realistic in our assumptions — noting that some attendees are local residents, that some expenditures would have occurred anyhow, and that most visitors do not drop $50 bills on the sidewalk — the number would go down considerably. There is simply no squaring the circle; you just can't get to $300+ million from 65,000 attendees at a one-day event.[105]

[104] R. Baade and V. Matheson, "Super Bowl or Super (Hyper)Bole", found at: http:// lanfiles.williams.edu/~vmatheso/research/superbowl2.pdf.

As a further check on this, assume that there is a multiplier that should be applied, and that the multiplier is about the same size as the exaggerations embedded in the "throw out all caution" assumptions used above. (i.e., there should be a multiplier of 1.5, but the assumption for spending, etc., were similarly exaggerated by a factor of 1.5.) If this is the case, we would still end up with about a one-tenth ratio of actual effects to claimed effects.

[105] We also considered other possibilities. Broadcast rights are paid by multinational companies to large firms (including the leagues), and therefore both the costs and benefits are spread all around the world. The food consumed by fans outside the park watching on TV is not determined by the site of the fantasy game and has little effect on local income. To get to the figure shown in the fantasy game analysis, we assumed every visitor was from out of town and therefore would not otherwise eat in the area. This unrealistically optimistic assumption is already included in the fantasy calculation. The stadium is not constructed just for one game (and if it were, we would have to deduct the costs). Travel costs to and from the area are spread into multiple states and through multiple companies and government agencies. Taxes (such as sales or excise taxes) paid on expenditures at the event have already been counted and are largely expenditures for services related to activities supporting the event (such as ticket surcharges used to pay for additional police services or gasoline taxes used to pay for roads) or are substantially diffused through various state, local, and federal agencies.

While there may be some direct local benefits missed in the fantasy game analysis, they are small and clearly cannot close the chasm between the likely local impact and the impact that has been claimed for large sporting events.

TABLE 6-1

Economic Impact of the Fantasy Game

	Visitors and Days	Local Food and Other Expenditures		Hotel		Money Left on Sidewalk		Grand Total
		Per Day	Total	Per Day	Total	Per Day	Total	
Workers	1,000							
Plus: Fans	65,000							
Equals: Visitors	66,000							
Times: days	2							
Equals: Visitor-days	132,000	$200	$26,400,000	$150	$19,800,000	$50	$6,600,000	
Less: costs to local economy of expenditures (50%)			$(13,200,000)		($9,900,000)			
			$13,200,000		$9,900,000		$6,600,000	
Grand Total								$29,700,000

There's obviously something going on in these economic impact analyses that are not based on economics. Part of this is an ethical shortcoming among practitioners, as well as promoters eager to show "big" numbers for their projects. The other part is a lack of knowledge among practitioners. These deficiencies motivate the next sections.

A Plea for Ethics in Impact Analysis

There are reasons beside ethical lapses for the prevalence of poor impact estimates. These include:

- Many of these "economic impact" analyses are not produced by economists, but rather by accounting, "futurists," marketing, or other firms.
- There are now a number of commercially available models that can be used as a "black box," in which a few numbers are placed, and out of which a much larger set of numbers comes.
- The journalists who write about these events have proven gullible or have agreed to become secret cheerleaders themselves.

However, we are left with the uncomfortable knowledge that some of these analyses are produced by people who are knowledgeable enough to know that the results are exaggerated but who simply succumb to the temptation to produce the big number.

It is my hope that, after reading this book, you will know enough to avoid the mistakes that have produced erroneous estimates in the past. It is my plea that, regardless of your knowledge, you will not succumb to the temptation to exaggerate your results for an economic impact or any other report. To do so is worse than being unprofessional; it's just plain wrong.

The Economics of Impact Models

The fundamental mathematics and economic reasoning behind impact models are quite rigorous. The conceptual reasoning behind what is known as input–output models was developed by Wassily Leontief in the early part of the 20th century.[106] Leontief's innovation, along with those of the pioneers

[106] Wassily W. Leontief, *The Structure of the American Economy 1919–1939*, 2nd ed. (Fair Lawn, NJ: Oxford University Press, 1951). Earlier work by Leontief was published in 1936 and 1941. See "Wassily Leontief" in *The New Palgrave* (New York: Stockton, 1991). See also Wassily Leontief, *Input–Output Economics*, 2nd ed. (New York: Oxford University Press, 1986).

of econometrics and the creators of the national income and products accounts, paved the way for economics to become an *empirical* science. For his contribution, Leointief was awarded the 1973 Nobel Prize in Economics.

Leontief could build on a few of the pioneers of economics, particularly Francois Quesnay whose *Tableau Économique* (1759) predated Adam Smith and became a founding document for the Physiocrats.[107] Quesnay graphed the interactions among sectors in "zig zag" tables, showing how much of the product of one sector went to another.[108]

History and Current Usage

The history of input–output analysis is an interesting one. The U.S. Army used such an analysis during World War II to select bombing targets in Germany, apparently choosing ball-bearings as the most critical industrial component.[109]

The idea was grasped by the U.S. government which has incorporated the reasoning in input–output models of the U.S. economy and the regions within the country. The best-known products of this research are the RIMS II regional multipliers which are produced by the U.S. Department of Commerce, Bureau of Economic Analysis.[110]

The production of the input–output tables by the U.S. government and the further production of multipliers based on those tables have dramatically improved the ability to assess the economic effects of changes in demand in any one area. Not surprisingly, there arose a cottage industry of economists and others who use these tables or products derived from them to project the economic impact of projects and events.

---

107 An excellent summary is available from the History of Economic Thought Web site, sponsored by the New School University and found at: http://cepa.newschool.edu/het/ essays/youth/tableau.htm. Their presentation is based on Hans Brems, *Pioneering Economic Theory, 1630–1980* (Baltimore, MD: Johns Hopkins University Press, 1986).

108 Quesnay is properly credited with the idea of calculating the flow of goods and services through the economy but not for getting that flow right.

The Physiocrats, while advanced for their day, suffered from some deficiencies in their analysis. In particular, they thought that land was a natural gift for which no return could be expected and that manufacturing simply consumed all of its inputs, leaving no net product to add to the country's wealth. This led them to purse a poorly defined "natural order" that balanced the production of the productive classes with the needs of all classes. See the History of Economic Thought Web site, cited above.

109 Charles W. McArthur, "Operations Analysis in the U.S. Army Eighth Air Force in World War II," dissertation; cited by Cosma Shalizi, Santa Fe Institute, found at http://www.santafe.edu/~shalizi/notebooks/economics.html under input–output models.

110 The RIMS (Regional Input–Output Modeling System) matrices were first developed in the 1970s by the Bureau of Economic Analysis (BEA), and have been updated since. Note that, given the enormous number of computations and data required, a complete revision of the RIMS matrices occurs quite infrequently.

See the U.S. Commerce Department, Bureau of Economic Analysis, *RIMS II User Guide*, 3rd ed. (1997) available at the BEA Web site at: http://www.bea.doc.gov/bea/regional/rims/.

Widespread Lack of Knowledge

The improvement in the availability of input–output tables has not, unfortunately, been matched by a growth in the number of people who know how to use them. In recent years, the study of input–output models in academia has often been relegated to a mathematical topic. This has left many economists bereft of knowledge about the underlying theory and assumptions.[111]

Meanwhile, practitioners of economic impact analysis, who may have never received any economic training, are typically not required to learn anything about the fundamental assumptions on which it is based. The *RIMS II User Guide,* which is the explanatory document prepared to accompany the U.S. government's multipliers, relegates the discussion of input–output analysis to an appendix.[112] While the *User Guide* itself contains numerous straightforward cautions on the proper use, these are apparently quite easily ignored.

This lack of familiarity with the theory, combined with the availability of commercial software and multipliers for every region in the U.S., has made it easy to produce bogus economic impact conclusions using apparently scientific methods, including multipliers provided by a U.S. government agency.

The combination of this knowledge gap and the ethical weaknesses of human beings has resulted in the poor state of the economic impact industry. We plead for ethics in economic impact analysis in the section above. In the following section, we address the knowledge gap.

Input–Output Equations

The underlying concept is that the entire economy can be viewed as a huge production function into which we pour inputs and from which comes output. Economists segment the types of inputs by type, including labor from different areas, industrial goods, agricultural goods, and other sources. They then segment the outputs by type as well, including consumption by various sectors, profits, and other outputs.

The fundamental input–output relations within the economy can be described by a system of n simultaneous equations in n variables. The

[111] See the review of sources in Footnote 113. There is a steady trend in these references; those published in the 1970s described much more of the economics, while those published later distilled it more to mathematics. Of course, the references listed are those we think are *good* sources, and do not represent a random sample of all texts. Furthermore, some of these are designed as "mathematics for economists" books and must be allowed that focus. Nonetheless, from these and other sources, it is clear that input–output analysis is widely used, while the fundamental assumptions are not widely taught.

[112] U.S. BEA, *RIMS II User Guide,* 3rd ed. (March 1977), Appendix A. The word "Leontief" appears only in the appendix, and the fundamental input–output equation in footnote 42.

number of goods or services described determines the number of equations. The household sector (which supplies labor) can be included in these equations as an exogenous variable, in which case the input–output system is "open." If we describe all goods and services, including those provided by the household sector, as intermediate goods that are determined endogenously, then the system is "closed." There are many references on the mathematics of input–output models.[113]

The Technology Matrix

Those familiar with matrix algebra will anticipate that these equations can be represented by simple operations among matrices. Indeed, the fundamental equation is an input matrix multiplied by an input vector to produce a vector of outputs. The relationships among industries are captured by the coefficients in the technology matrix. The technology matrix consists of coefficients a_{ij} which describe the amount of input *i* that is used by industry *j*. (The subscripts can be recalled easily by noting that the first subscript denotes the input industry and the second denotes output.)

In the following three-industry example, the technology matrix would be:

$$A = \begin{bmatrix} a_{11} & a_{12} & a_{13} \\ a_{21} & a_{22} & a_{23} \\ a_{31} & a_{32} & a_{33} \end{bmatrix} \tag{1}$$

113 Good references on the mathematics of input–output analysis include:

Carl Simon and Lawrence Blume, *Mathematics for Economics* (New York: Norton, 1994), Section 6.2, 7.1, 8.5, and 8.7 (including notes). This excellent text by two of my former teachers is designed quite as the title suggests. However, it is careful to include, in succinct mathematical prose, the restrictive assumptions on which input–output models are based and gives an example drawn from the work of Leontief. It uses a small set of input–output equations as a running example for instruction in matrix algebra and other mathematics topics. Simon and Blume cite, as the model for their presentation of Leontief's 1958 study of the U.S. economy, Stanley I. Grossman, *Applied Mathematics for the Management, Life and Social Sciences* (Belmont, CA: Wadsworth, 1985).

Alpha C. Chiang, *Fundamental Methods of Mathematical Economics*, 2nd ed. (New York: McGraw-Hill, 1987), Section 5.7. The Chiang text contains more exposition on the theory of input–output models and some history.

William J. Baumol, *Economic Theory and Operations Analysis*, 4th ed. (Englewood Cliffs, NJ: Prentice Hall, 1977), Chapter 22. Baumol describes the restrictive assumptions embedded in input–output models.

Adi H. Mouhammed, *Quantitative Methods for Business and Economics* (New York: M.E. Sharpe, 1999), Chapter 3. The Mouhammed book departs from the unfortunate trend to reduce this topic to mathematics only and includes an explanation of the full input–output table and the derivation of multipliers.

Finally, Leontief's own survey article "Input-output Analysis" in *The New Palgrave* (New York: Stockton, 1991) is a rich source of historical information.

where a_{12} is the number of output units of industry 1 that are necessary to produce 1 unit of output for industry 2.[114] The output of each industry becomes an input to others and to the household sector if it is included. The coefficients on the diagonals are the shares of the input to each industry which are consumed in producing the output of the same industries. Thus, the technology matrix summarizes the input–output ratios among all industries.

The Fundamental Equation

This matrix is used in a fundamental equation relating the vector of outputs *x*, the input matrix of coefficients *A*, and the final demand vector *d*:

$$(I - A)x = d \tag{2}$$

where $(I–A)$ is called the Leontief matrix.[115] Intuitively, this equation means that the total outputs of all industries *x*, minus the amount consumed in the process of production *Ax*, equals the amount demanded and consumed by the rest of the economy *d*.

A Robinson Crusoe Example

This equation can be illustrated by the following thought experiment.

Assume you are the only worker in a "Robinson Crusoe" economy located on an island of a single human inhabitant and the only job you have is to pick coconuts on the beach. You, however, as the most diligent beachcomber on your island, tend to eat some of the coconuts. The rest you save for other projects, including late-night eating, coconut bocce ball, coconut painting, and building coconut outrigger canoes. The consumption during the production process is commonly called *intermediate demand*, while the consumption by the households that ultimately consume goods and services is known as *final demand*. On this island, however, you fulfill both roles.

If you eat one out of four coconuts while you pick them, then the only coefficient in the input matrix *A* is $a_{11} = 1/4$. If the input matrix *x* is the single entry $x = 10$, representing 10 coconuts, then the fundamental equation for your economy is:

$$(1–[1/4])\cdot[10 \text{ coconuts}] = 7.5 \text{ coconuts} \tag{3}$$

---

[114] In an open system, the third row and column would relate to the household sector, making it technically a two-industry economy.

[115] At this point, we should note that there is some variation in terminology among the references cited earlier. The matrix *A* is sometimes called the input matrix, and *I-A* the technology matrix. There are also some variations in usage for terms such as *open* and the various types of multipliers.

Thus, the final demand for coconuts on this island is seven-and-a-half coconuts, which becomes the payment to the household sector as well as its final consumption. The input matrix explains how the total output of coconuts in the economy, 10, was allocated.

Extending the Example

We can now expand this to economies where there is one open sector that contributes a factor input and receives the resulting production. In most economies, the household sector provides labor and purchases the final product; therefore, most economies are modeled with an open system. In such systems, the sum of the factor inputs for each industry (the column sum in the input matrix A) must total less than one, or there is no incentive to produce it.[116] The difference between the column sum (the sum of all factor inputs other than household) and 1 is the amount given to human labor, including workers and management.

Thus, $(I\text{–}A)x = d$ makes sense for small and large economies.

Restrictive Assumptions

There are important, restrictive assumptions that underlie input–output analysis. The most important are:

1. *Fixed input–output ratios:* This is a much more restrictive assumption than is normally used with production functions. It means that the producers cannot vary the relationship among inputs despite growing or shrinking demands, or fluctuations in prices.[117] If we remember that the matrix A summarizes the technology of production, this assumption boils down to one of fixed technology. This assumption is clearly not fulfilled in most situations. However, it is a reasonable approximation when the

[116] There would be no incentive to produce if the sum of factor inputs was greater than 1 for each unit of production because the industry would then consume more goods and services than it would produce. This would mean (in a capitalistic society) that it would lose money and the investors would shut it down. The removal of this profit motive (and loss penalty) is one of the causes of the failure of nonmarket economies. See A.C. Chiang, *Fundamental Methods of Mathematical Economics*, 2nd ed. (New York: MacGraw-Hill, 1987), Section 5.7.

We return to this assumption later in "Mathematics of Multiplier Analysis" on page 143.

[117] Note that this is more restrictive than the "constant returns to scale" assumption that is frequently used for production functions. In such production functions, one can still substitute capital for labor or make other substitutions. In fixed-technology models (such as input–output models), this is not allowed.

See W.J. Baumol, *Economic Theory and Operations Analysis*, 4th ed. (Englewood Cliffs, NJ: Prentice Hall, 1977), Chapter 22.

overall changes in demand or supply are modest and occur within a short time frame.

2. *Single-industry production of single goods:* The model assumes that each industry produces a single product and that no other industry produces such goods. This, again, is a condition rarely fulfilled. However, the assumption can be somewhat relaxed by redefining the goods in certain industries as composites of multiple goods and services. This assumption becomes more visible when we consider substitution across goods or industries. As the labor force can readily substitute across industries and as consumers (and intermediate producers) can often substitute as well, we see the need to carefully consider the effect of this assumption. In practical terms, we can often conjective that this assumption is fulfilled when the required changes being considered occur during a short period of time and when no other major substitutions are likely.

Mathematics of Multiplier Analysis

An input–output model can be used to predict the change in inputs resulting from a change in final demand and vice versa. By deriving multipliers from the input–output equation, we can directly calculate the change in one sector caused by a change in another.

Any results derived from a model will only be valid when the assumptions underlying the model are valid. For the following derivation, we will assume that this is the case.

Recall the fundamental equation $(I–A)x = d$. If we solve the equation for the vector of outputs from each industry x, we find by inverting the $(I–A)$ matrix:[118]

$$x = (I - A)^{-1} d \tag{4}$$

[118] This is sometimes called the Leontief inversion.

Mathematically trained readers will note that not all matrices have inverses, and we have not formally required that all the entries in the final demand vector Y will be positive. Simon and Blume, cited previously, show a formal proof for the existence of the inverse. See C.P. Simon and L. Blume, *Mathematics for Economists* (New York: Norton, 1994), Section 8.5.

One of the assumptions on which this is based is that the column sums of the technology matrix A are less than one. This is equivalent to saying that the sum of all the costs of all the inputs to these industries is equal, or less, than the sum of its outputs. If we assume industries are, on the whole, profitable, then this assumption is satisfied. While industries can lose money on the whole during one or more periods, they cannot go on losing money indefinitely. Therefore, this assumption is generally satisfied.

where the $(I–A)^{-1}$ matrix is sometimes called the *requirements matrix* as it contains the required quantities of final demand for all sectors, based on new output for each sector.[119] The requirements matrix can be seen as containing the partial differentials of output of each sector x with respect to the final demand d:

$$(I-A)^{-1} = B = [b_{ij}] \text{ where} \tag{5}$$

$$b_{ij} = \frac{\partial x_i}{\partial d_j} \tag{6}$$

Consider the contents of the requirements matrix B for an economy with two industries plus the household sector:

$$B = \begin{bmatrix} b_{11} & b_{12} & b_{13} \\ b_{21} & b_{22} & b_{23} \\ b_{31} & b_{32} & b_{33} \end{bmatrix} \tag{7}$$

In this economy, b_{11} and b_{22} are the coefficients indicating the same-industry (direct) multipliers for industry 1 and 2, respectively. The indirect effects are indicated by the sum of the other coefficients in the same column (other than the row for the household sector), meaning the effect of a change in the final demand of sector i on the output of each sector j. As this is an open economy model, the induced effect (caused by spending and respending of the household sector) is denoted by coefficients b_{3j} for sectors j = 1, 2, 3. The last coefficient, b_{33}, would be change in output from the household sectors due to a change in demand of the same amount.

Total Output Multipliers

In this structure, the columns of matrix B summarize the impact of changing total demand for one industry's goods on the output of all industries. If we include the household sector, the induced effect caused by the respending of earnings is included in the multiplier.[120] The total output multipliers that include induced effects are usually greater than one because a new dollar spent in a community would be used to buy factor inputs (including wages

[119] For open systems, this matrix is subdivided into the direct, indirect, and induced requirements matrices as it contains the coefficients for the household sector. See A.H. Mouhammed, *Quantitative Methods for Business and Economics* (New York: M.E. Sharpe, 1999), Chapter 3.

[120] This is sometimes called a Type II multiplier, whereas the multiplier confined to intermediate goods and industries is called a Type I multiplier.

paid to local workers), and a portion of those wages and purchase funds would be respent in the same community on additional purchases.[121]

Uses of Input–Output Analysis

The discussion on economic input models in this chapter should not imply that this is the only use for input–output analysis. Other — often better — uses include planning for economic growth; examining the impact of strikes, work stoppages, and wars; and understanding the effects of tax policy and location changes.

Rules for the Proper Use of Multipliers

Now that we have reviewed the fundamental theory and mathematics, we can present the proper use of multipliers in economic impact analysis. The key rules for doing so, with references to the theory underlying the rules and the consequences for violating them, are:

1. Only *bona fide* changes in final demand or factor outputs should be multiplied. Recall that the multiplier matrix B consists of partial differentials. A differential is, by definition, a change caused by a change. If there is no change, there is no differential. Therefore, there is nothing to multiply. Violating this rule is one of the most common reasons for exaggerated impact projections. If a project or event causes $1 million in spending to shift from one entertainment venue to another, and entertainment is an industry in your economic impact model, then the change in final demand for the entertainment industry is zero. It does not matter whether the multiplier is high or low.

   The *RIMS II User Guide* contains multiple examples in which this rule is followed.[122] It states the rule explicitly: "These multipliers measure the economic impact of a *change in final demand*, in earnings, or in employment *on a region's economy*."[123] As the emphasized words indicate, a multiplier should only be applied to a change in final demand that occurs within a region's economy.

---

[121] To see how this is true, assume that 60% of each dollar spent is respent in the same community. Of the first dollar spent, 60 cents remains; of this, 0.6×0.6 remains and so on. After seven rounds, the amount of additional expenditures in the community is $1.43.

[122] Example events in the *User Guide* in which BEA discusses substitution effects, including the opening of a new shopping mall and a factory shutdown.

[123] U.S. BEA, *Regional Multipliers: A User Handbook for Regional Input–output Modeling System (RIMS II)*, 3rd ed., p. 8; available at http://www.bea.gov.

2. Select and use the appropriate region consistently. The input–output tables for the entire U.S. economy are quite different from those for any one state or region. The U.S. economy as a whole produces just about everything. Most regions do not. The households in the U.S. can buy most goods and services from producers within the U.S. For states, especially regions or counties within the states, this is often not true.[124]

   If you are looking at the economic impact on a county, then all expenditures and income from outside the county should not be considered. This will considerably reduce the multiplier effect of expenditures inside the community, as relatively little respending will occur in the community. However, it will mean that it is more realistic to consider additional expenditures by outside entities as new expenditures. The choice of region is often determined by others. However, once it is chosen, it must be consistently used.
3. Subtract all substitution effects before using a multiplier. Many projects result in the substitution of one form of production or consumption for another. A new sports stadium will probably be associated with either the demolition, or a sharp reduction in the usage, of an existing stadium.[125] A new shopping mall will certainly reduce the sales at nearby existing retailers. The opening of a new gaming casino will probably cannibalize existing casinos (if there are others nearby) and will certainly cannibalize other types of entertainment. All these projects can be analyzed properly, and in some cases the economic impact will be positive.[126]

[124] This is particularly the case if you do not include services, government, and shelter. For example, food is largely produced elsewhere, even in agricultural states, which tend to specialize in certain crops.

[125] The construction of the new Comerica Park in Detroit directly resulted in the abandonment of Tiger Stadium known as the "queen of diamonds" for its illustrious history and close-to-the-field fan experience as a venue for professional baseball. The construction of Ford Field nearby caused the City of Pontiac to seek proposals for the reuse (and possible demolition) of the Silverdome. Chicago's new Comiskey Park opened in 1991 across the street from the old Comiskey Park where the White Sox had played since 1910. It was then demolished. The great Detroit Red Wings teams of the 1950s played at the Olympia. When the modern Joe Louis Arena was built, the Olympia was demolished. The "Joe," however, has since hosted more championship teams. As of this writing, Comerica Park, Ford Field, and Comiskey Park are still waiting.

For an extensive discussion of sports stadia and economic impact analysis, see R. Noll and A. Zimbalist, Eds., *Sports, Jobs, and Taxes* (Washington D.C.: Brooking Institution, 1997).

[126] We have performed economic impact analyses for all these types of projects and in each case accounted for substitution effects. In the case of sports stadiums, our analyses of Tiger Stadium (after the Tigers had left for Comerica Park) compared retail activity in a planned conversion to no activity at all, which was the alternative. For the Pontiac Silverdome, we found the alternative usage of the stadium site (rather than hosting a small number of professional football games and other stadium events) would actually result in more economic benefits to the community. For a retail development in Arizona, we accounted for displacement effects on existing retailers. Finally, for a proposed casino in western Michigan, we accounted for cannibalization of the casino and other expenditures.

In other cases, after adjusting for substitutions, there will be little or no net economic impact.[127]

4. Use the *bill of goods* method when available. Often, information will be available on the likely allocation of expenditures by sector for a new project. For example, a publicly funded facility plan will normally include some accounting for the amount of money to be spent on land, construction, operations, and employment. This information should not be ignored in favor of lumping the whole project into an industry category and hoping the averages on which the input–output tables are based are close. They are probably not.

   The *RIMS II User Guide* specifically recommends that practitioners use a bill of goods method for allocating expenditures to different industries, along with the appropriate multipliers (after deducting any substitution effects) for those industries.[128]

   For example, consider the operation of a new facility. (We deal with construction in the next rule.) Assume that the new facility will be a manufacturing facility employing 1000 workers of which 250 will work in a distribution center, and that 500 of those workers are transferring from other facilities in the area. Assume further that almost all the facility's factor inputs will be shipped in from out of state and that little of the remaining production will remain in the state.

   This information allows you to separate out substitution effects (such as the transferring employees and operating expenditures) for about half the plant. It also allows you to use the multiplier information for a wholesale trade or similar distribution category for a portion of the plant's operations. Finally, you can also make adjustments for the shipping in and out of most of the plant's materials and output, including accounting for the significant

---

This last study and a critique of a separate economic impact study is publicly available; see P. L. Anderson et al., *Critical Review of Gun Lake Band of Potawatomi Indians Environmental Impact Study*, Lansing, MI, Anderson Economic Group, February 2003 (also submitted to the Bureau of Indian Affairs, and part of the public record there); and *Market and Economic Impacts of a Tribal Casino in Wayland Township*, March 2003; available at http://www.andersoneconomicgroup.com; also available from the Grand Rapids (MI) Chamber of Commerce.

127 Similarly, we have analyzed the economic impact of new businesses in a tax-free zone in Detroit and concluded that the net effect of some businesses (primarily fast-food restaurants and service businesses such as laundromats) in the zone were zero. This is because the new businesses did not represent *bona fide* new investment or employment in the city but rather existing businesses moving down the street and bringing their customers with them. It seemed unlikely that the tax-free zone would result in people eating more or people driving in from outside the city to get their laundry done.

In contrast to the net zero effect of these business, we found a positive effect due to new industrial development that would otherwise have not located in the city.

128 BEA, *RIMS II User Guide*, 3rd ed., p. 8.

transportation and logistics operations that it implies. This will be far more accurate than simply using a manufacturing multiplier for the plant's annual expenditures in the area.

The Simulink models we present in this chapter and the next frequently rely upon separating the economic impact of multiple categories of expenditures, consistent with the bill of goods method.

5. Separate construction from operation. Most new facilities have a construction phase and then go into operation. The construction phase will normally be qualitatively and quantitatively different, in terms of its economic impact, from the operation phase. These should be handled separately. For the construction phase, carefully consider how much of the materials will come from within the region. In some cases, the regional averages for construction will approximate the effects. In others, a bill of goods method will be far more accurate.[129] Construction phases may also include demolition and other activities that should not be ignored. For example, major projects often involve upgrades for transportation networks (such as roads, bridges, highway ramps, port facilities, etc.); when these are *bona fide* new investments, they should be considered.[130]

   For operations, be sure to carefully consider employment and wage earnings. If workers are likely to be absorbed from the existing labor pool (especially if one effect of the new project will be a decline in employment at other businesses in the area), then the net economic impact of their wages will be small and may be close to zero.

6. Remember the assumption of constant technology. Recall that one of the limiting assumptions in input-output models is that of unchanging technology. The multipliers that derive from these models are based on this assumption. If your project involves a deviation from the average use of factor inputs for specific outputs, then you must adjust any multipliers or other uses of the input–output tables to account for this. This happens much more often than is expected.[131] New facilities arise because something is new. That newness may just be new population, in which case

[129] For example, we once analyzed a half-billion dollar power plant construction project on the west coast of Michigan. As we knew that a large portion of the expenditure would go for giant turbines to be shipped from Europe, we deducted this from the expenditures in the local economy. We also adjusted for transportation costs and for the share of the construction payroll that would remain in Michigan.

[130] Do not also assume that all the benefits of the new transportation facilities accrue to the new project. Do not forget the costs to the taxpayers of those facilities.

[131] Given the dearth of knowledge about the fundamental theory and mathematics of this field, perhaps this is to be expected. However, of the many studies we have reviewed over the past several years, not a single one explicitly addressed this.

the assumption is probably correct.[132] However, a new type of plant, a new way of distributing goods and services, a cheaper production process, significant changes in prices, changes in law, changes in environmental regulations, and other changes that occur quite frequently would violate this assumption.

Violating the assumption does not mean you cannot perform an economic impact analysis. It means you must use the standard multipliers only on the portion of the project that is standard.

Conclusion: Rules for the Use of Multipliers

The preceding rules are not terribly difficult to follow. If they are followed, economic impact analysis can be a rich tool, allowing customized, region-specific analyses of many events.

In the following section, we will look at an example analysis that properly uses multipliers to determine the economic and fiscal impacts of a change in policy. In the succeeding chapter, we will look at more models that examine different phenomena.

Economic Impact Simulation Models

Below we describe an actual economic impact simulation model and explain briefly how it was used. This approach can be used as a guide for other economic impact analyses.

Economic and Fiscal Impacts: City of Detroit Model

We developed a sophisticated economic model for the City of Detroit to estimate the effects of a change in a state law that would eliminate the city's long-standing residency requirement for city workers. This required a particularly complex analysis, as all of the following were direct results of the policy change:

- The number of residents in the city would change.
- The economic contribution of those residents would change.

[132] For example, putting in enough retail service establishments to service a growing population probably does not require rethinking the "constant technology" assumption.

- The taxes paid (and city services used) by those residents would change.

In addition to these direct results, we anticipated some indirect results as well.

Key Assumptions

The key assumptions we used included:

- Detroit is one of the largest cities in the U.S. with a population of approximately 1 million, and has a well-defined border and separate taxing authority. Local effects were only those within the city.
- The new policy would go into effect on a specified date. The alternative was current law.

Modeling Resident Behavior

The fundamental microeconomic dynamics of the model matched that of the residents. Should the law change, those residents that were formerly constrained in their residence choices would become free to move to other locations. Some of these individuals would move outside the city.

We anticipated that most residents who wanted to migrate outside the city would not do so immediately, given the large transaction costs associated with selling and buying a home, as well as the other disruptions that occur when a family moves. Therefore, we built into the simulation model a partial adjustment feature and projected the results out five years or more.[133]

Scenarios: GUI for Model

The model allowed for one change-of-residency to affect the migration of workers in and outside the city. This change, in turn, affected five different tax bases and the tax revenue.

We created a custom graphical user interface (GUI) to allow us to input different scenarios. Each scenario assumed different speed of adjustment, number of migrants, and other factors. Thus, we were able to confidently project the economic and fiscal impact of the change in law after reviewing the likely impact under numerous scenarios.

---

133 For an extended discussion of this important behavioral dynamic, see "Modeling Dynamic Behavior: Adjustment" on page 156.

Use of the Model and Output

The resulting report was used in a number of arbitration proceedings on labor contracts, budget preparations for the city, as well as in testimony to the state's legislature.[134]

Implementing the Model

We took the following approach:

1. Tax, economic, and geographic data were analyzed to provide key assumptions for the model. These assumptions included the baseline tax revenue, tax rates, tax bases, collection factors, population, income, and other variables for the model.

   The demographic and geographic data were analyzed in a geographic information systems (GIS) facility which allowed for current residence locations of city employees to be used as predictors of the likelihood of further migration.[135]

2. The key behavioral assumption — the effect of the change in migration given a change in the residency law — was based on survey data, census data, and a variety of other sources.

3. A Simulink model was created. It had subsystems that calculated the key variables at each step in the cause-and-effect chain. The Simulink model allowed others, through visual inspection, to see these cause-and-effect relationships.

4. We designed the model with explicit control variables that governed the speed and magnitude of adjustment. Different control variables were used to generate multiple scenarios, allowing us to test the sensitivity of the model to changes in assumptions.

5. We created a GUI that allowed the user to directly input or change the key behavioral and economic assumptions.

6. A callback simulated the model under two scenarios: a baseline scenario (no change in law) and a change-in-law scenario. The callback then calculated the difference in the output variables over a multiple-year period.

---

[134] P.L. Anderson, *Economic and Financial Impact Assessment of a Change in Residency Requirements in the City of Detroit, Michigan,* Lansing, MI: Anderson Economic Group, December 1999; available at: http://www.andersoneconomicgroup.com; Report and testimony given at State of Michigan Act 312 arbitration hearings, 1999, 2000; testimony to the Michigan House of Representatives, local government committee, 2000.

[135] In this case, high concentrations of employees living right on the border were viewed as an indication of likely further migration.

7. Output variables were imported into Excel for presentation. The Simulink report generator was also used to describe the model. Certain figures were printed for use in the report.
8. In addition to a written report, we constructed a Web page that provided the executive summary of the report, along with selected exhibits.

Schematics of the Simulink Model

The following figures show schematic diagrams of the various subsystems within the fiscal and economic impact model, starting with the top-level view and going through the major subsystems.

1. Top-Level View: Figure 6-1,"Fiscal and economic impact model," shows the top-level view. The key behavioral variables are shown on the left side. These include resident and nonresident population, overall population and economic growth, and a switch to turn the residency requirement off and on.
2. Population Migration: These variables flow into a population module which calculates the population (as affected by migration in and out of the city) under different scenarios. This module is illustrated in Figure 6-2, "Population migration subsystem."
3. Economic Growth: The base variables also feed into an economic growth module which calculates income and property tax base growth factors. These are used in the remainder of the model.[136]
4. Tax Base and Tax Revenue: From the outputs of these subsystems, the resulting tax bases for city taxes are calculated for multiple years in the tax base module. Using these tax bases, tax revenue is calculated in the tax revenue subsystem which is illustrated in Figure 6-3, "Tax revenue subsystem."

[136] Typically, we use a constant for the value of a variables in base year and then multiply by growth factors to project their values in the future. This is done explicitly in this model, although a simpler way to implement the same growth pattern is by using the variable or compound growth blocks described in Chapter 5, "Economic Impact Models" under "Custom Business Economics Blocks" on page 113.

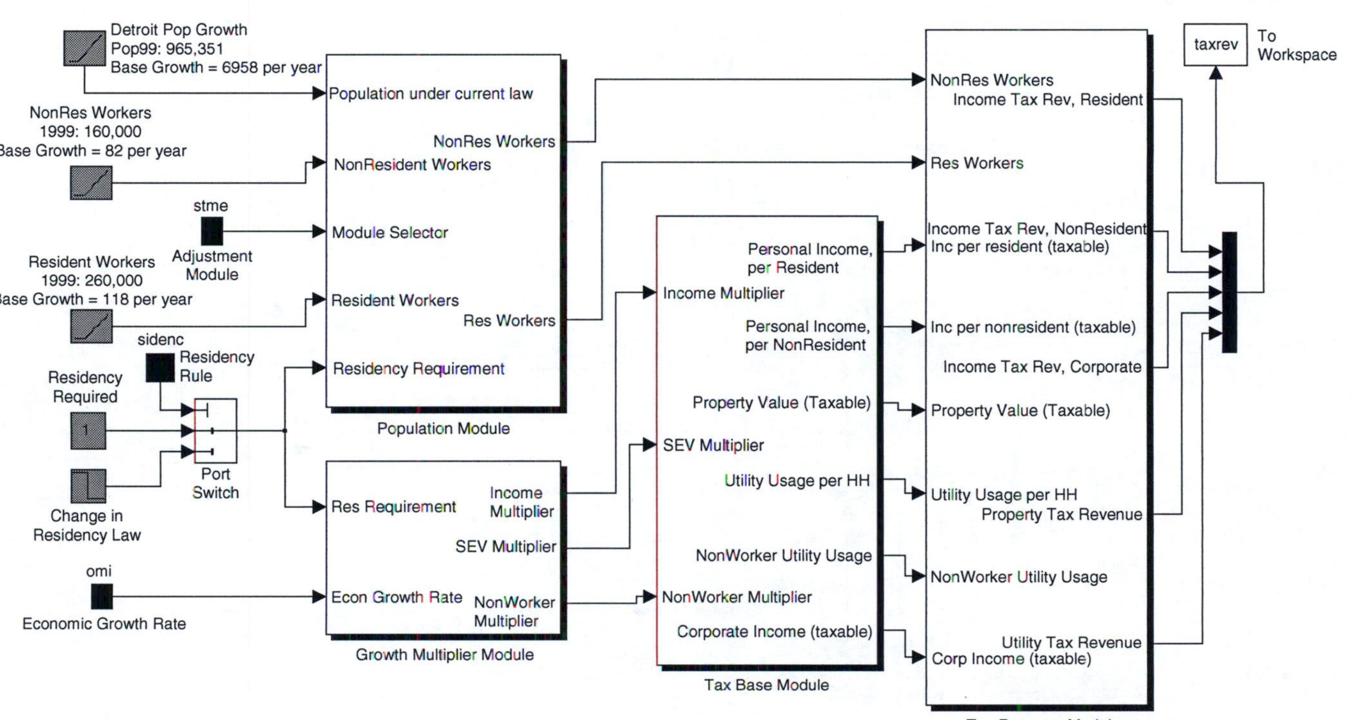

FIGURE 6-1
Fiscal and economic impact model.

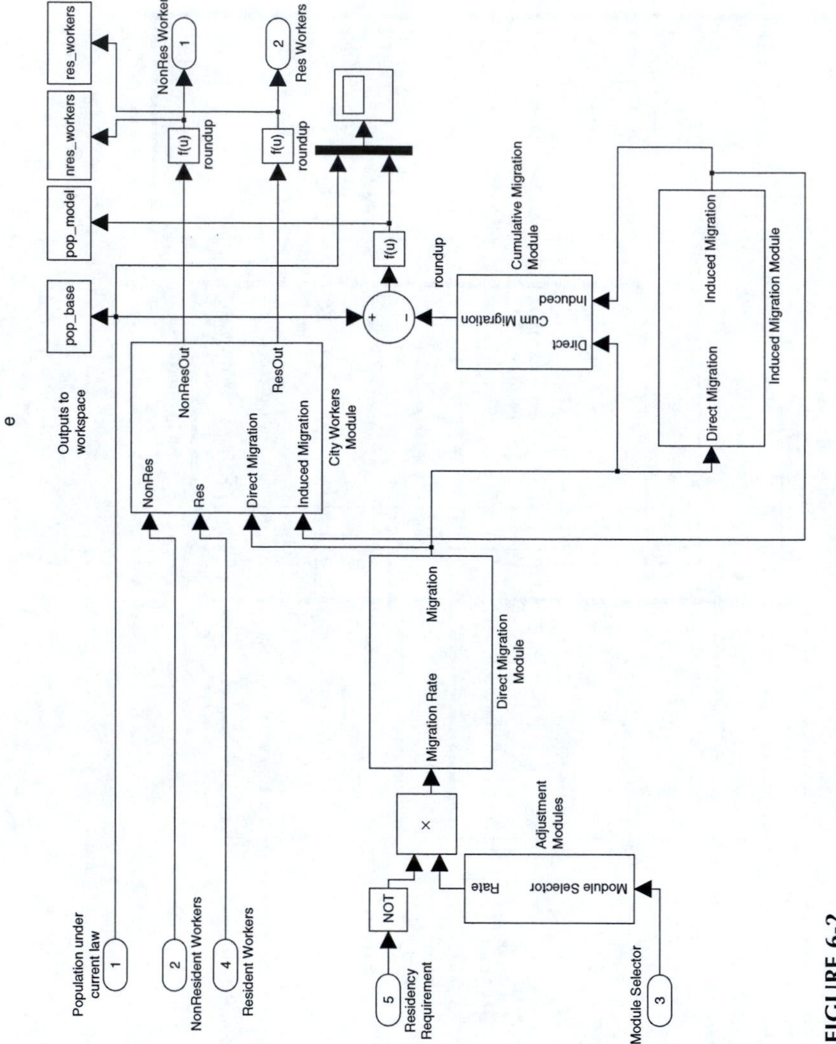

FIGURE 6-2
Population migration subsystem.

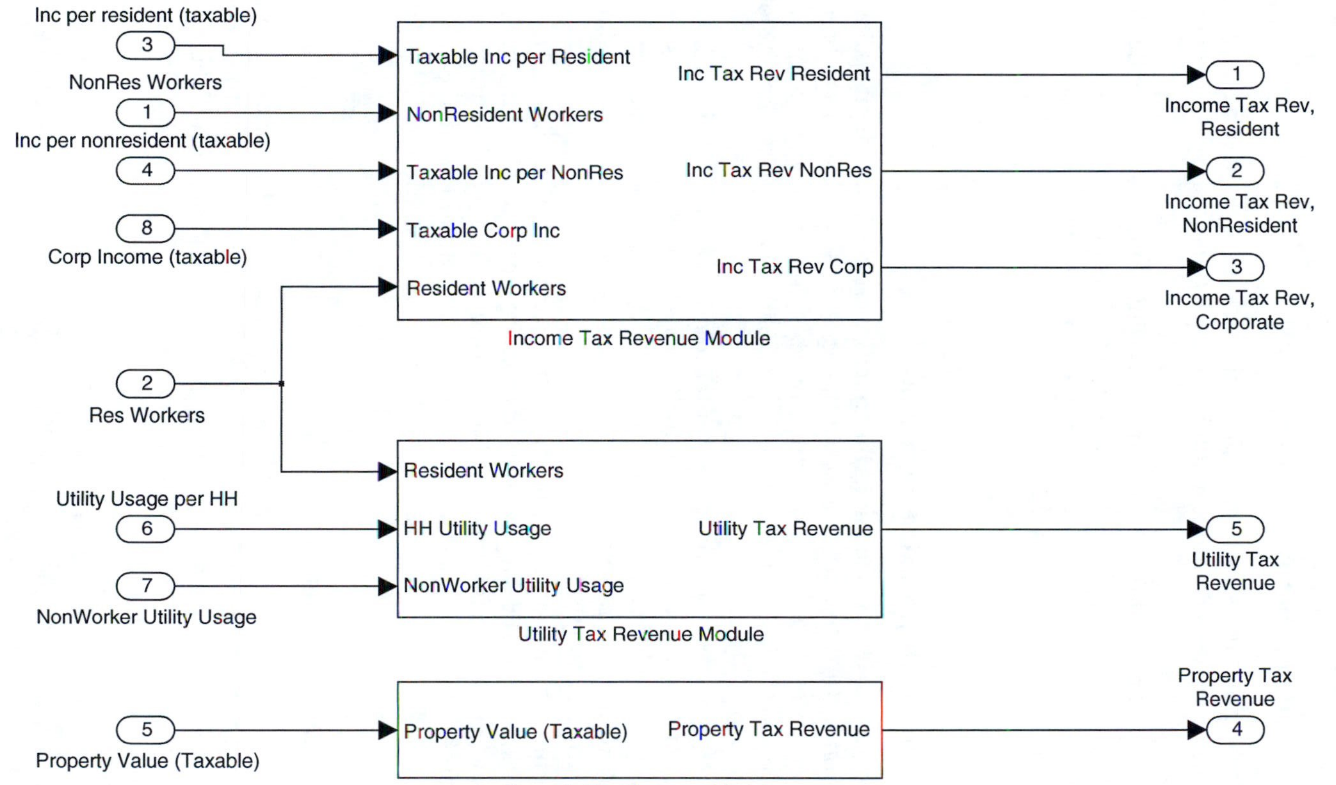

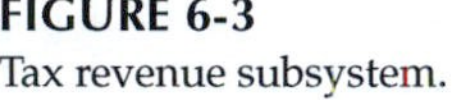

FIGURE 6-3
Tax revenue subsystem.

Modeling Dynamic Behavior: Adjustment

One key behavioral question is the speed of adjustment. If we assume that 20% of the employees will move, that tells only part of the story. How many will move each year?

If this is an important question in its own right, and data are available, the best approach is to directly estimate the adjustment path. However, this is often impossible. Therefore, we offer two practical approaches for cases in which the speed of adjustment is not the most critical aspect of the model.

Option I: Borrow Depreciation Algorithms

One approach is to borrow functions that have been defined for depreciation. Depreciation in finance and taxation is the scheduled decline in the book value of an asset.[137] Automobiles are probably the most well-known of depreciable assets. Numerous guidebooks and services provide estimates of their resale value.[138]

To project migration, we could use depreciation algorithms. For example, a simple method would be to assume that if 20% of the people wish to migrate over the next 5 years, one fifth would move each year. This is analogous to straight-line depreciation, as plotting the migration over time would produce a straight line.

Most human behavior, however, does not follow straight lines. Thus, a more accurate adjustment path would often be closer to "double declining balance" or other methods in which earlier years have faster adjustment. Using MATLAB functions, numerous adjustment paths can be programmed, but in many cases these depreciation functions will do the trick.

Implementating this Model

To allow us to vary the adjustment method, we inserted a switch in the subsystem. This switch was controlled by a drop-down box on the GUI. Thus, scenarios differing only in the speed of adjustment could be run sequentially, revealing the sensitivity of the model to changes in this assumption.

---

[137] For a discussion of depreciation, including example MATLAB functions, see Chapter 12, under "Auto Depreciation," on page 301.

[138] Auto resale guides are much more accurate in predicting the actual market value of assets than most depreciation schedules, as they are based on actual sales information. Most depreciation schedules are based on accrual accounting principles or on the Internal Revenue Code, neither of which is grounded in actual market value.

Option II: Specify Partial Adjustment Model

Another option is to specify a partial adjustment model that governs the path of adjustment toward a desired level of a variable.[139] Such models can have strong theoretical grounding based on the costs of adjustment.[140]

For example, investment can be considered movement toward a desired level of capital stock. Such models are sometimes called stock adjustment models. Specifications for partial adjustment models include those based on polynomials, ratios, and geometric series.[141] Such models can be readily implemented in MATLAB.

139 Depreciation schedules set a path, but we do not consider them partial adjustment models because they are conventions. In addition, partial adjustment models may move toward a dynamic objective (a desired level that may change over time), while depreciation is used to drop the book value of an asset from one static level (historical cost) to another (salvage value or zero).

140 Indeed, the transaction costs involved in selling a home, plus the additional disruption and costs of moving, provide excellent practical and theoretical basis for expecting partial adjustment in the City of Detroit model.

141 An older but excellent survey of the use of partial adjustment models in empirical work is M.D. Intrilligator's, *Econometric Models, Techniques, and Applications* (Englewood Cliffs, NJ: Prentice-Hall, 1978), Chapter 6. Note that the topic is often discussed in the econometric literature as "distributed lag" models.

7

Fiscal Impact Models

Fiscal Impact Methodology

We describe in this chapter fiscal impact analysis, which we define as the impact on revenue and expenditures of an organization from some outside event. In this chapter we assume the organization in question is a government, but the term *fiscal* could also apply to a private company or even a family.

Definition of Fiscal Impact

As discussed in the chapter on economic impact models, we insist on a specific, conservative, and realistic definition of impact. We define fiscal impact by including only *bona fide* new tax revenue or reduced government expenditures. Therefore, to arrive at the net fiscal impact of a proposed event, we must subtract any lost tax revenue or increased expenditures from the increased tax revenue that would be realized from the event.[142]

Relationship between Economic and Fiscal Impacts

Tax revenue is normally generated by economic activity. Income, sales, and value added taxes are directly related to economic activity, whereas property taxes and other taxes are indirectly related. Certain excise taxes based on usage (notably gasoline taxes in most states) are also directly related to economic activity.

Therefore, most fiscal impact analyses are based on assumptions about economic activity. A common error in fiscal impact analysis is to assume the

[142] See Chapter 6, under "A Plea for Ethics in Impact Analysis" on page 137, for the related rule for economic impact analysis.

level of economic activity without considering how the change in policy affects the activity. The most obvious illustration of this is to estimate the effect of a change in tax policy on tax revenue. Elementary economic theory reveals that tax rates affect the after-tax earnings of an activity or the after-tax cost of a service or good. Therefore, tax policy must affect economic activity, taxable earnings, and taxable sales of goods and services.

While economic theory indicates that tax rates affect economic activity, it does not tell us how much activity is affected. That is the realm of applied economics.

In some analyses, the level of economic activity involved will be a straightforward assumption. In such cases, make this assumption explicit and support it. Remember, theory states that tax policy will affect economic activity. If you are assuming that it does not, you must explain why you are assuming that human beings do not behave as they normally do.[143] If you are constrained by institutional policy to performing *static* analysis, then state this prominently and note how actual human behavior is not static. For further discussion, see Chapter 8, "Tax Revenue and Tax Policy."

On the Political Economy of Taxes

The question of tax policy is connected with expenditure policy through the government's budget constraint. The two are sometimes discussed separately and sometimes artfully combined.[144] Make sure you are explicit in limiting the explanation of your analyses to the topic you actually studied.[145] In particular, if you analyze taxes, do not confuse the analysis of taxation with one about how the revenue is spent.

[143] One approach which, though honest, is incomplete, is to acknowledge that you have ignored the impact on behavior caused by a change in tax policy. In such cases, state that your estimated tax revenue is likely to be higher (or lower) than the estimate you have made ignoring human nature.

[144] The classic hidden pairing is to support (or oppose) tax increases when you are really advocating increased (or decreased) government expenditures.

[145] For example, if you analyzed property taxes and their effect on business location decisions or economic development, you should ensure your report describes the effect of such taxes on business and economic development. Even if the tax is used to pay for public schools, do not use the analysis of tax policy to form a pretense for a discussion of education policy unless that policy is directly connected to tax policy.

Some discussion of the uses of the tax, however, is useful in explaining the entire issue. When doing so, be sure you do not fall into the related trap of presuming that more tax revenue will invariably result in better services.

Fiscal Impact of Industrial Development

Methodology Summary

Fiscal impact methodology includes categorizing and modeling both the costs and benefits to a local government and its taxpayers due to the subject development. Consider the analysis of a proposed industrial development. The analysis should start with describing the scale of the project in terms of employment, tax base, and other factors for each year over the next decade or two. These variables are then used as a basis for calculating both the *benefits* of development (including those associated with new jobs, new employees, additional taxes, revenue sharing, payments from the state, and permit fees) and *costs* to the community (including additional police and fire burdens, capital expenditures, general financial administration, and other burdens).

The analyst should also identify key assumptions, describe the methodology, and identify any important factors that cannot or were not quantified.

Description of Each Part of the Model

We illustrate a fiscal impact analysis performed on a proposed industrial development in a rural township in southeastern Michigan.

We describe each part of a Simulink fiscal impact model using a schematic for each system and subsystem in the model.

Along with the data and some model-specific parameters, the schematic describes the mathematics of the model.

The following are brief descriptions of each subsystem in the model.

Overview: Fiscal Impact Model

The overview of the fiscal impact model shows the big picture of the model. You will note the Project subsystem on the left (See Figure 7-3, "Fiscal model schematic — I," on page 167). Within this system we project new residences, employment, and property tax base variables. The output from this system — these variables — flows into two other subsystems: cost and benefit.

This overview, the most important view of the model, illustrates how the same projections about the project facility drive analyzes the related costs and benefits. The following paragraphs describe the portions of the model that are within each subsystem. In each subsystem, variables have the initial values described in the data tables.

Project Subsystem

This subsystem is composed of residential growth, employment, and property tax base subsystems.

Residential Growth Subsystem: This uses a number of relocated workers, spin-off residences as a share of relocated workers, spin-off residences as a share of new workers, portion of new workers resident in the township, and portion of relocated workers resident in the township to calculate the number of new residences from employment, spin-off residences, and total new residences.

Employment Subsystem: This separates construction and operations workers. The look-up tables for construction, transfer, and new workers specify the schedule of employment during the time period for each category.

Property Tax Base, Development Subsystem: This summarizes the property tax base, including change in personal property value, initial personal property value, annual growth rate of personal property, change in real property value, initial real property value, and annual growth rate of real property. A compound growth module is used to calculate the values of these variables as they grow over time.

Cost Module

The cost module includes calculations for the costs of new residences, which include fire, police, and government administrative costs per new residence, plus other direct costs per new residence; and employment cost which includes fire, police, and government administrative costs per job, plus other direct costs per job.

Benefit Module

The benefit module calculates the benefits to the community. These benefits include state revenue sharing, permit fees, and property taxes. A portion of these benefits is estimated within the property tax revenue and permit fee subsystems. Other benefits are estimated from variables for average residents per new household, state revenue sharing and other revenue per capita, and portion of average state revenue per capita to the township.

Property Tax Revenue Subsystem: This uses variables of average house value, property growth rate, assessment ratio, and property tax rate to generate property tax revenue. A zero property tax rate for the township means that property tax revenue will be zero.

Permit Fee Subsystem: This summarizes permit fee revenue flow to the local government.

Notes on Model Structure

The following are additional notes on the model structure:

1. This is a fiscal impact analysis. It does not define or estimate the larger economic impact, economic substitution effects, or other variables, including concerns about amenities, character, quality of life, access to jobs, tax base, and other issues.
2. There are generally four sources of revenue generated by this kind of development: permit fees (usually a one-time fee), property tax revenue (which normally grows and arises from both direct investment and any indirect or induced development), revenue sharing (provided on a per capita basis by the state to the township), and fee-based revenue for specific services. We have modeled each effect.
3. Other costs are largely offset by fees paid for services, such as utility costs, as explained in the report. For this category of services, we have excluded both the fees and associated costs.
4. Fiscal costs are allocated separately for residential and employer taxpayers. These include fire, police, and administrative costs. U.S. Census data were used to estimate the per-capita costs of this category of municipal costs for small and medium-sized municipalities in the state of Michigan. These total per-capita costs were further divided into per-resident and per-employee costs.
5. We also projected separately the increase in employment and residents created by the new plant.

Source Data

The spreadsheets summarizing input data we used in our model and data sources were included in the data tables in our report. Though the spreadsheets are omitted in this book, the method used to import the data and export the output is described in Chapter 4, "Importing and Reporting Your Data," on page 63.

Recapitulation of Exhibits

The following exhibits also summarize the methodology and results of the analysis.

1. The fiscal impact projection graphs are in Figure 7-1, "Municipal costs," and Figure 7-2, "Cost and benefit."
2. The fiscal impact model is illustrated by schematic diagrams in Figure 7-3, "Fiscal model schematic — I," Figure 7-4, "Fiscal model schematic — II," and Figure 7-5, "Fiscal model schematic — III."

These diagrams illustrate, in flowchart fashion, how variables describing the project itself (including employment, tax base, and additional residences) are used to estimate costs as well as benefits to the community.

Additional Technical Notes

The following are additional technical notes.

1. We used a script program to run the simulation model. This "callback" program:
    a. gets the input data in .mat file format
    b. runs the simulation model
    c. gets the results
    d. gets new input data and runs the model a second time in situations where a second set of assumptions is compared with the first[146]
    e. creates graphs

For reference, the key portions of the m-file in the fiscal impact callback we created for the project are given in the code fragment that follows.

[146] In this case where the alternative to the development under consideration is no development, we need to run the simulation only once.

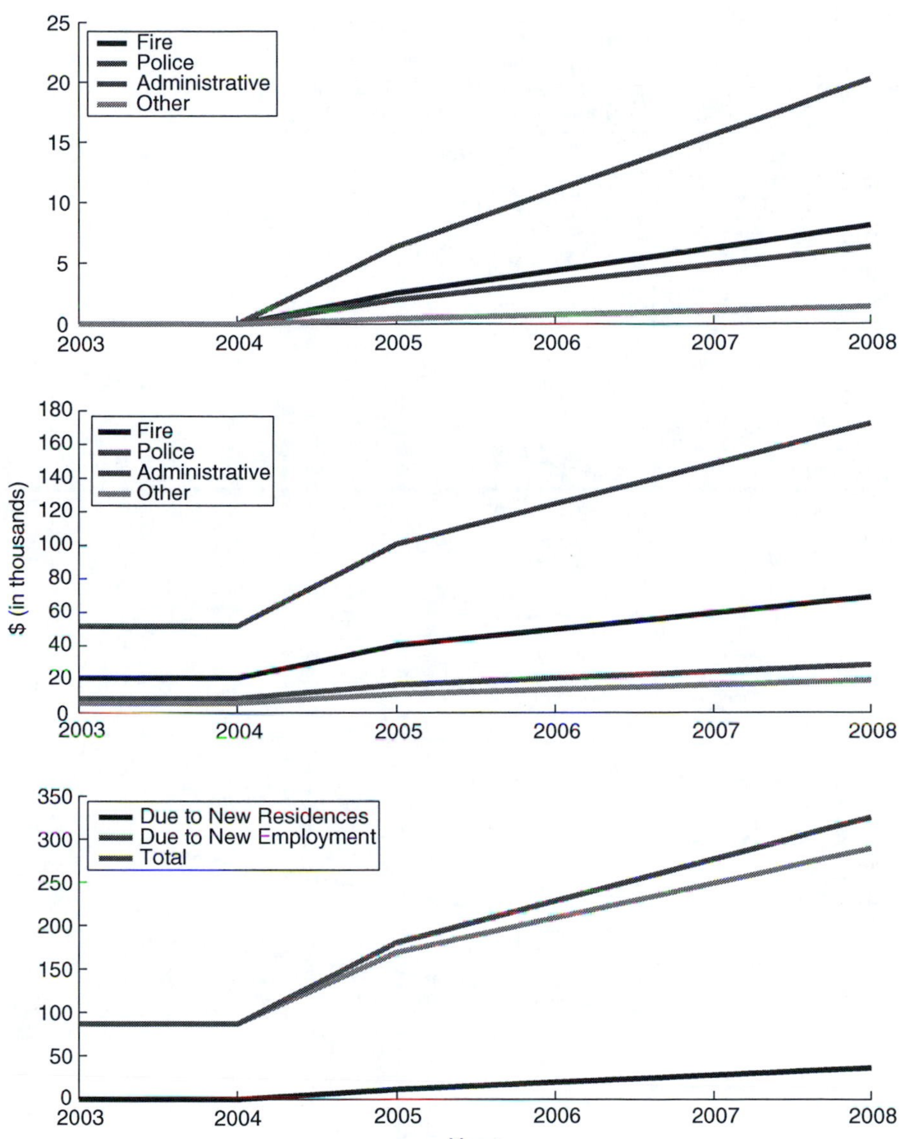

FIGURE 7-1
Municipal costs.

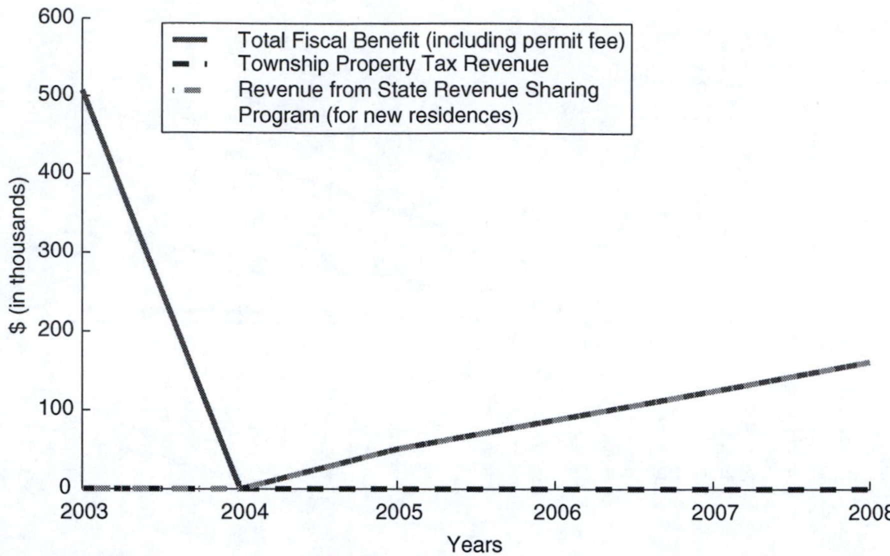

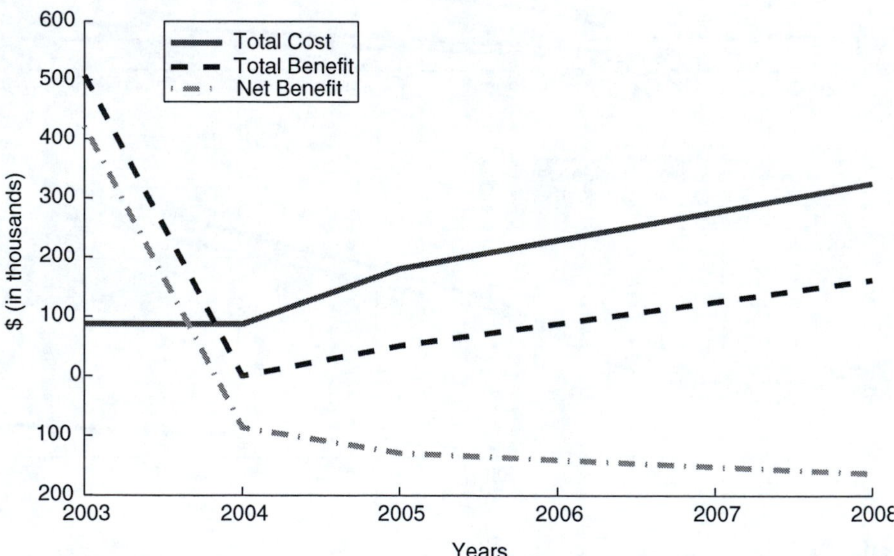

FIGURE 7-2
Cost and benefit.

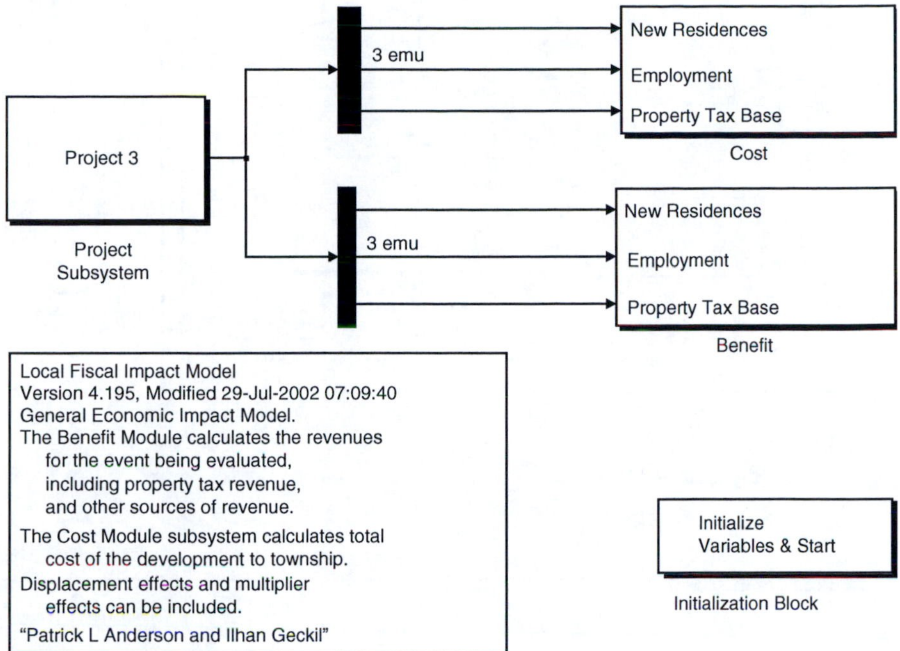

FIGURE 7-3
Fiscal model schematic — I.

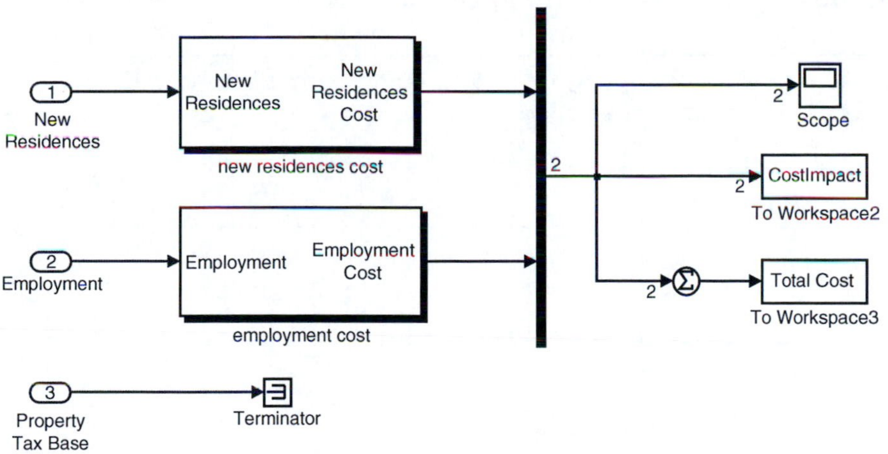

FIGURE 7-4
Fiscal model schematic — II.

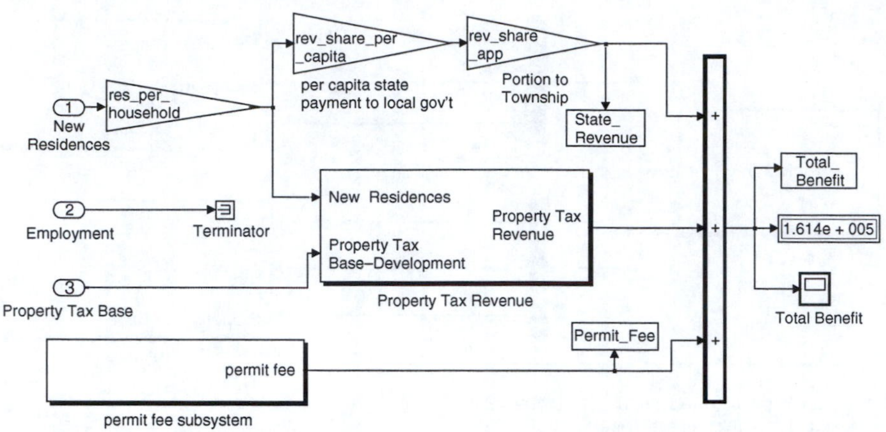

FIGURE 7-5
Fiscal model schematic — III.

Appendix

Code Fragment 7-1. Fiscal Impact Callback

```
%--code is copyright 2002, Anderson Economic Group.
% All rights reserved.--
% Adapted for fiscal impact model.

%-----------------------------------------------------

% Load workspace (.mat) file, if necessary
% (make sure file is located in reachable
% directory.)
load X_AutoSupplier_input.mat;

% Get data from GUI (workstation) or from HTML (Web
% server) or from .mat
% file.
%-----------------------------------------------------
```

```
% Time Parameters

Tstart = 2003;

Tstop = 2008;

Tstep = 1;

Years = [Tstart:Tstep:Tstop]';

sset = simset('SrcWorkspace','base');

% --------------------------------------------------

% Sim command takes model, timespan, options, and
% external input arguments.

% Options structure and Timespan should be set
% above.

sim('X_AutoSupplier',[Tstart:Tstep:Tstop],sset,[]);

display('Model Simulated.')

% Show output variables

display('Echoing Key Variables and Parameters.');

echo on;

Years;

initial_real_property_value;

change_real_property_value;

initial_personal_property_value;

change_personal_property_value;

real_growth;

personal_growth;

construction_workers;

new_workers;

transfer_workers;

prty_tax_rate;

permit_fee;
```

```
rev_share_per_capita;
rev_share_app;
resident_worker_app_new;
resident_worker_app_relocated;
spin_off_residences_new;
spin_off_residences_relocated;
change_property_value;
avg_hous_value;
fire_cost_per_job;
police_cost_per_job;
govt_adm_cost_per_job;
other_direct_cost_per_job;
fire_cost_per_residence;
police_cost_per_residence;
govt_adm_cost_per_residence;
other_direct_cost_per_residence;
echo off
% --------------------------------------------------
% Graph Results
display('Graphing Results.');
% Define variables used for graphing
cost_labels = {'Fire','Police','Administrative',
'Other'};
revenues_labels = {'Payroll Revenue','Local
Purchases Revenue',...
'Property Tax Revenue','Imports Revenue','Tourism
Revenue'};
% Y = [Var1./millions, Var2./millions];
```

```
% Call specific graphing commands or call graphing
% callback here.

figure(1),

subplot(3,1,1);

plot(Years, Res_Cost./1000,'linewidth',2);

legend([cost_labels],2);

subplot(3,1,2);

plot(Years, Emp_Cost./1000,'linewidth',2);

legend([cost_labels],2);

subplot(3,1,3);

plot(Years, CostImpact./1000,'linewidth',2);

hold;

plot(Years, (CostImpact(:,1)+CostImpact(:,2))./
1000,'r','linewidth',2);

legend('Due to New Residences','Due to New
Employment','Total',2);

orient tall;

figure(2),

subplot(2,1,1);

plot(Years, Total_Benefit./1000,'r-','linewidth',2);

hold;

plot(Years, Property_Tax_Revenue./1000,'b--',
'linewidth',2);

plot(Years, State_Revenue./1000,'g-.',
'linewidth',2);

legend('Total Fiscal Benefit (including permit
fee)','Township Property Tax... Revenue','Revenue
from State Revenue Sharing Program (for new...
residences)');

subplot(2,1,2);

plot(Years, TotalCost./1000,'r-','linewidth',2);
```

```
hold;
plot(Years, Total_Benefit./1000,'b--
','linewidth',2);
plot(Years, (Total_Benefit-TotalCost)./1000,'g-
.','linewidth',2);
legend('Total Cost','Total Benefit','Net Benefit');
orient tall;
```

8

Tax Revenue and Tax Policy

Introduction

One of the constants of public debate is tax policy. Indeed, British tax policy was one of the irritants that caused the outbreak of the American Revolution. In this section, we discuss methods of analyzing government taxes.

We are particularly interested in:

- Tax rates
- Tax revenue
- The incidence of the tax (who pays it)
- The effect of the tax on work effort (supply side effects) and output or sales (demand side effects)

Tax policy is often modeled as a naive function. The common static approach to reviewing tax policy is to assume that behavior stays exactly the same even though the government imposes a different tax regime. Such an assumption is thorough nonsense. However, it is currently the standard assumption for most analyses of tax policy, as was discussed in Chapter 7, "Fiscal Impact Models."

One reason for the stubborn persistence of a patently false assumption is the politics of most tax policy discussions. Another is the difficulty in properly modeling the dynamic supply and demand effects of tax policy changes. While this book cannot affect the former, it can demonstrate techniques that make the latter's incentive effects tractable.

Advantages of Simulation Models

Spreadsheets provide an ideal environment to calculate the excise tax due on a particular transaction or the income tax due on a particular firm in a particular year. In such cases, we simply assume that the transaction has already occurred (or the income has already been earned), and we are simply calculating the tax due.

However, this simple 2-D environment is not well-suited to analyze tax policy. At least as far back as Adam Smith, economists have recognized that taxes affect behavior. Part of that behavior is the earning of income and the purchasing of goods. Thus, changing tax policies does not just change the rate of the tax, but *also changes the tax base*.

Simulation models provide an ideal environment to model this complex phenomenon. Using this approach, we can adjust the base and the rate, run multiple scenarios, and compare the results.

Sections in This Chapter

In this chapter, we briefly discuss a variety of taxes levied by state, local, and federal governments in the U.S. Similar taxes are levied in other countries.

We then perform an extensive analysis of the tax burden on workers, which includes a number of taxes. We consider the earned income tax credit, which is a tax policy that is designed to offset one set of taxes by a credit against another. Finally, we present a model of state taxes that simulates the effect of a state earned income tax credit, including both the static and dynamic effects.

Income and Payroll Taxes

We discuss below the different types of payroll and income taxes levied on workers in the U.S. These taxes, especially federal income taxes, change constantly. In general, we present the laws as they were in effect for the year 2002.[147]

Federal Income Taxes

The Federal Revenue Code establishes personal exemptions and standard deductions that allow a certain amount of income to be earned without tax liability. These amounts are indexed annually for inflation.

The personal exemption of $3,000 plus the standard deduction of $4,700 for a single person made the first $7,700 of earnings tax-free for a single

[147] This discussion draws extensively from Patrick L. Anderson, *A Hand Up: Creating A Michigan EITC*, Lansing, MI, Michigan Catholic Conference, 2002; available at: http://www.micatholicconference.org and http://www.andersoneconomicgroup.com.

person in 2002.[148] A married couple with one child paid no federal income taxes on the first $16,850 of earnings.

Above the threshold amount, taxpayers paid marginal rates of 10%, 15%, 27%, 30%, 35%, and 38.6% as their income increased.[149] For example, in 2002, married individuals filing jointly paid no federal income tax on the initial exemption and standard deduction amounts, and then paid tax at 10% on the first $12,000 in taxable income; single individuals paid tax at 10% on taxable income of up to $6,000.

There are other, more complicated aspects of federal taxes for individuals, especially individuals whose income is in excess of $100,000, and who, therefore, qualify for phase-outs of deductions as well as may be subject to Alternative Minimum Tax.

State and Local Income Taxes

Many states and cities also levy income taxes. In Michigan, taxpayers faced a flat rate state income tax of 4.1% in 2002.[150] Many cities also levy an income tax on wages earned within the city.[151]

Taxable income for the Michigan income tax is based on federal adjusted gross income (AGI) which is calculated before deductions and exemptions. With the state personal exemption at only $3000 and no deductions, almost every dollar of wage earnings is taxed in Michigan. Taxable income in other states is based on different definitions of income which vary widely.[152]

OASDI (Social Security and Medicare) Taxes

The formal name for Social Security taxes is Old Age, Survivors, and Disability Insurance (OASDI). Both employees and employers must withhold 6.2% of wages for the Social Security portion of the tax and an additional 1.45% for the Medicare portion of the tax.

---

148 The 2003 federal tax law reduced these marginal rates.
Amounts shown for 2002 were estimated by Commerce Clearing House of Chicago, IL, as of November 2001; See *2002 Master Tax Guide*, Paragraph 127.

149 Readers are cautioned that actual tax returns, especially federal income tax returns, are much more complicated and may have actual marginal rates that are greatly distorted by the existence of alternative minimum taxes (AMT), phase-outs of exemptions and deductions, loss carry forwards, and other fiendishly complicated aspects of the Internal Revenue Code.

150 Public Act 6 of 1999 requires the state income tax to be reduced in 2002 to 4.1%.

151 The Michigan City Income Tax is generally levied at a rate of 1%, with nonresidents paying half that amount. Some cities levy higher or lower rates. The City of Detroit, through a special allowance in state law, levies a much larger rate, which was 2.65% in 2002 but is scheduled to decline over the next decade. Public Act 500 of 1998 specifies a declining rate structure, although it allows for certain conditions under which the city could retain higher rates. The decline in rates began in July 1999. For more information see the Citizens Research Council organization Web site at www.crcmich.org.

152 One of the best sources of state tax rate and tax policies is the Tax Foundation. Their Website is at http://www.taxfoundation.org

When combined, this payroll tax burden is 15.3% of wages, up to the phase-out amount for the Social Security portion of the wage base, which was $84,900 in 2002. There is no phase-out amount for the Medicare portion of this payroll tax. In effect, it is an additional income tax on wage earnings.

FUTA (Federal Unemployment Tax Act)

The federal government also imposes a 6.2% federal unemployment tax assessment on wages, up to a phase-out amount. However, most states have their own unemployment insurance systems, and for these states a substantial credit is offered that reduces the net tax to 0.8%. This tax is subject to a phase-out at $7000 in wages per employee.

Unemployment Insurance

States normally require government-sponsored unemployment insurance contributions which are another payroll tax. For example, the State of Michigan requires employers (even households) to pay unemployment insurance taxes. For new employers, the rate had been 2.7% for several years. Large employers and employers with several years of experience pay a rate partially based on their own unemployment benefit experience, which could be higher or lower. Construction workers, for example, were charged at an initial rate of 8%. The average rate across the state has ranged from 4.5% to 2.7% over the 10 years prior to 2002.[153]

This tax is subject to a phase-out at $9500 in wages per employee in the State of Michigan. The phase-out limit decreased to $9000 in 2003, due to the passage of Public Act 192 of 2002.[154] It is likely that the net effect of these changes will be to increase the average unemployment insurance tax rate in the future.[155]

Who Pays Payroll Taxes?

It is clear that income taxes are borne by employees, as the tax is withheld from their paychecks and the employees file annual returns. However, many managers are unaware of the burden payroll taxes place on employees.

Social Security taxes are an example. Although half of this tax appears to be charged to employers, in reality the worker bears the entire cost of such

[153] State of Michigan Bureau of Workers' and Unemployment Compensation Web site at: http://www.michigan.gov/bwuc., Table "Average Contribution Rate Based on Total and Taxable Payrolls, 1936–2000."

[154] The same Act increased maximum benefits from $300 per week to $362. See fact sheets 97 and 98, available on the State of Michigan Bureau of Workers' and Unemployment Compensation Web site at: http://www.michigan.gov/bwuc.

[155] Because the tax rate is set by calculating the tax revenue necessary to pay benefits, increasing the benefits has the effect of increasing the rate. However, the rate is adjusted over time, so the full effect of PA 192 of 2002 will not be felt for several years.

payroll taxes. When a worker is hired, the employer makes an economic decision whether or not the product the worker provides — his or her output of goods and services — is worth more than the cost of employing the worker. The cost of employing that worker includes his gross wages and any payroll taxes. If this total cost is more than the output of the worker, the company will lay off the employee or eventually go out of business.

From economic reasoning alone, it is clear that the worker must bear the burden of producing enough goods and services to pay for his or her wages and all payroll taxes. Past studies that have attempted to disentangle statistically the share of burden that falls on employees have commonly found at least 70% of the payroll tax burden is borne by employees.[156] However, even this understates the degree to which employees actually bear the burden.[157] There is a broad consensus among economists today that employees (not employers) shoulder the burden of payroll taxes.[158]

Modeling Income and Payroll Taxes

The taxes described above, especially the federal income tax, are complicated burdens. None, with the exception of the Medicare tax, is a flat share of wage earnings.

Fortunately, we have programmed functions in MATLAB that estimate tax liabilities for all these taxes, given a level of wage earnings. Of course, these programs estimate liabilities, given certain assumptions, and are not suitable for preparing individual returns. However, they are especially well-suited to examining tax policy.

We have collected the following functions at the end of this chapter:

Social Security Taxes: These ubiquitous taxes can be modeled in a straightforward fashion. A function to do so is in Code Fragment 8-1, "Social Security Tax," on page 183.

Federal Income Taxes: By far the most complicated tax discussed in this chapter, we model this tax as reprinted in Code Fragment 8-2, Federal Income

---

156 See, for example, P.M. Anderson and B. Meyer, The effects of firm specific taxes, *J. Pub. Econ.*, August 1997; and J. Gruber and A.B. Krueger, The Incidence of Mandated Employer-Provided Insurance, *Tax Policy and Economy* (Cambridge, MA: MIT Press, 1991); cited in M.D. Wilson, How Congress Can Lower Federal Taxes on American Jobs, Heritage Foundation *Backgrounder* no. 1287, May 27, 1999.

157 Employees also benefit from the profitability of a company, so lost employer earnings results indirectly in lost wages. However, for the purposes of this study, the difference between the employee bearing 70% and 100% of the payroll tax burden is not material.

158 As a further example, liberal economist Paul Krugman writes in a recent book, "Again, there is generally universal agreement that the real burden of the [payroll] tax falls almost entirely on the worker. Basically, an employer will only hire a worker if the cost *to the employer* of hiring that worker is nor more than the value that worker can add." P. Krugman, *Fuzzy Math* (New York: Norton, 2001), p. 43. [Emphasis in original.]

Tax Calculation, on page 184. In looking at this function, note that the filing status is an argument, along with the wage income.

Earned Income Tax Credit

Introduction

Payroll taxes (including unemployment insurance, federal unemployment taxes, OASDI taxes on both the employee and the employer, and Medicare taxes) provide a disincentive to working for low-skill workers whose wages are unlikely to climb over the Social Security maximum.

Because the effects of these taxes are largely unexamined, many people are unaware of the negative effects they have on workers. A previous report by this author indicated that for a typical single worker in Michigan making just $20,000 per year, the wage earner received only 76% of the labor costs incurred by his or her employer.[159] The largest share of that burden is payroll taxes, not income taxes.

Total payroll taxes on a worker earning just $5000 per year in Michigan often exceed 20.1%. In contrast, workers earning above the Social Security maximum ($84,900 per year in 2002) incurred a payroll tax burden that drops to 2.9%.

The Earned Income Tax Credit (EITC)

One powerful way of addressing these disincentives is through an earned income tax credit (EITC). An EITC is a credit for income earned by workers at the lower end of the income scale. An EITC offsets most or all of the payroll taxes borne by a worker earning a wage around the poverty level, and then phases out as his or her income increases.

Tax Policy Analysis Methodology

In order to bring together the various local, state, and federal tax structures, we used a variety of m-files to calculate each tax separately. We then created a master file, bringing together these taxes to calculate and graph the effective tax rates. This examines tax rates and tax incidence.

A fiscal estimation model was then created in Simulink to estimate the fiscal cost of a Michigan Earned Income Tax Credit (EITC).

---

[159] See Patrick L. Anderson, *A Hand Up: Creating A Michigan EITC,* Lansing, MI, Michigan Catholic Conference, 2002; available at: http://www.micatholicconference.org and http://www.andersoneconomicgroup.com.

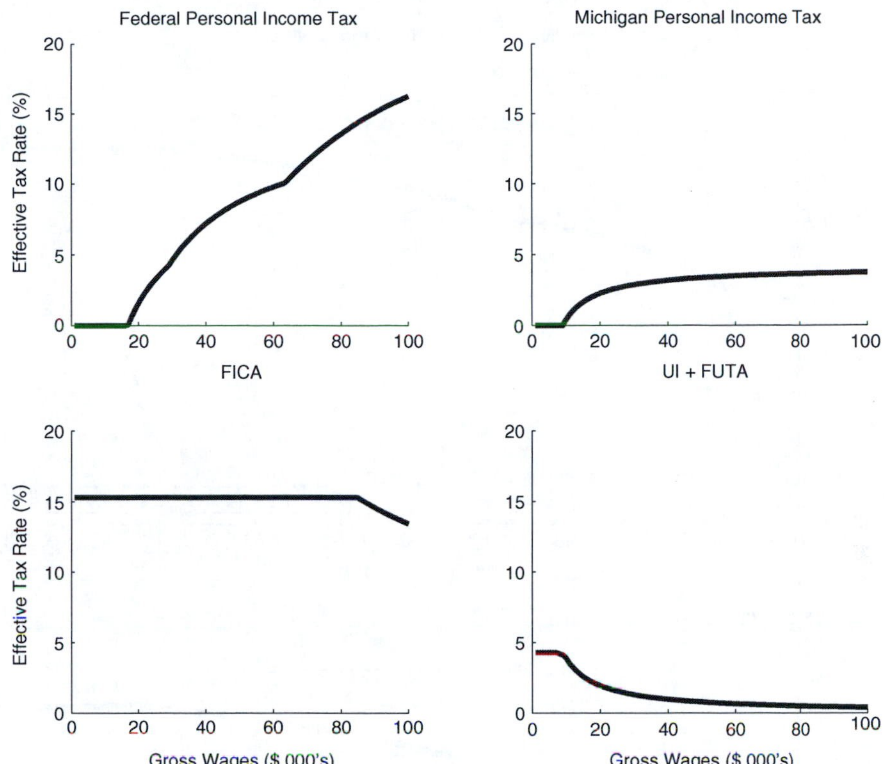

FIGURE 8-1
Effective tax rates, selected income, and payroll taxes

Payroll Taxes As a Share of Taxes on Wages

Using this methodology, we were able to add up the various taxes on earners in different income groups and model the overall incidence of the various taxes, again by income group.

Results of Tax Incidence Analysis

Figure 8-1, "Effective tax rates, selected income, and payroll taxes," graphically represents four groups of taxes and their impact on wages to a married couple with one child, earning from $1,000 per year to $100,000. Figure 8-2, "Total tax rates on wage income," illustrates the effect of income and payroll taxes on workers with different earnings, both before and after the application of an EITC. You will note that, consistent with the goals of the program, the credit offsets much or all of the payroll tax burdens on low-income workers.

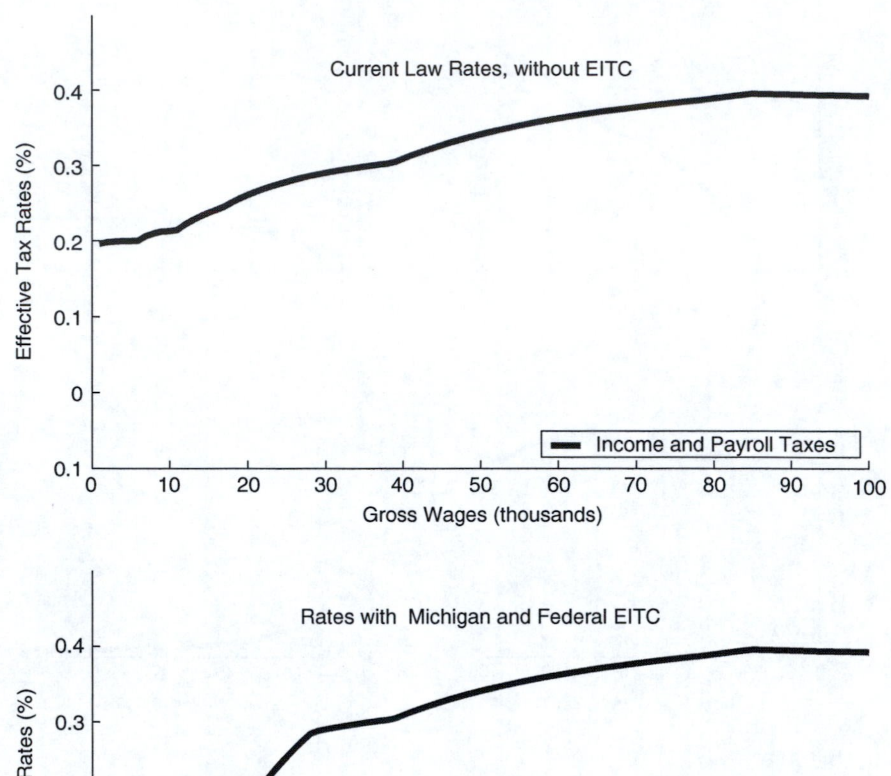

Source: Anderson Economic Group

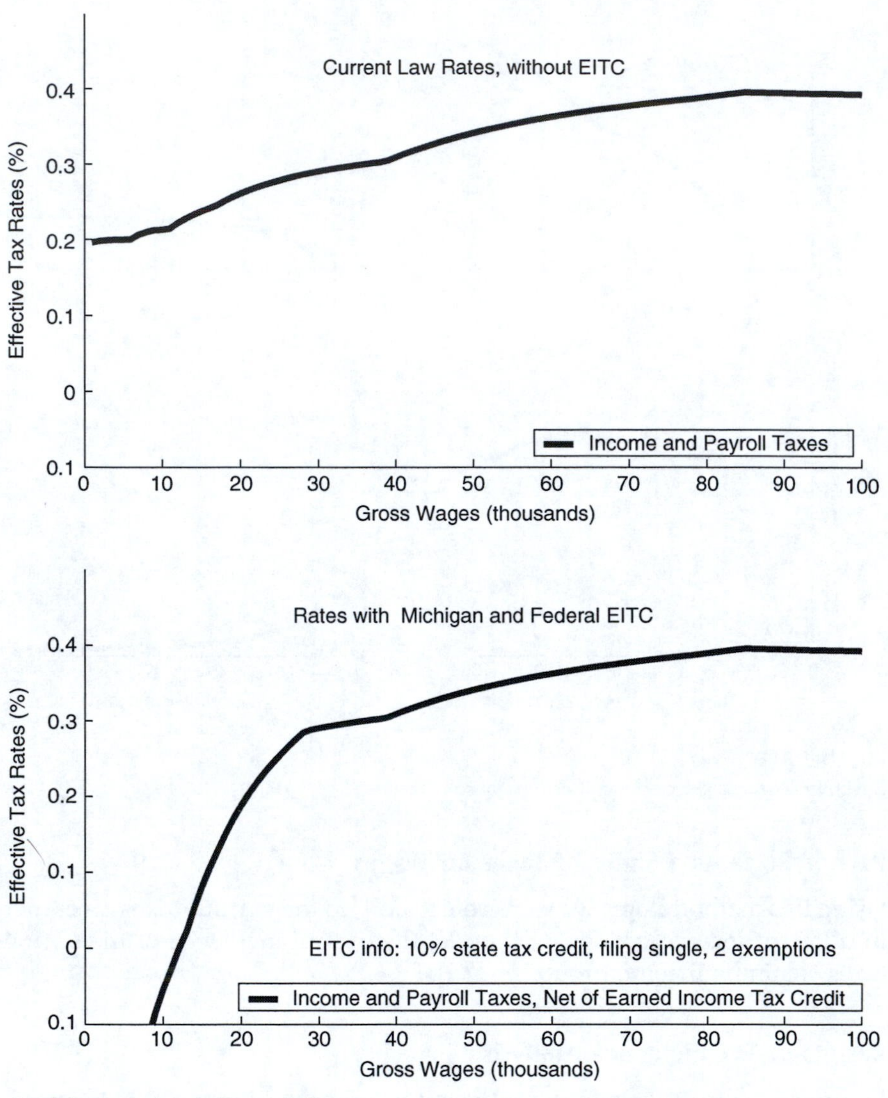

FIGURE 8-2
Total tax rates on wage income.

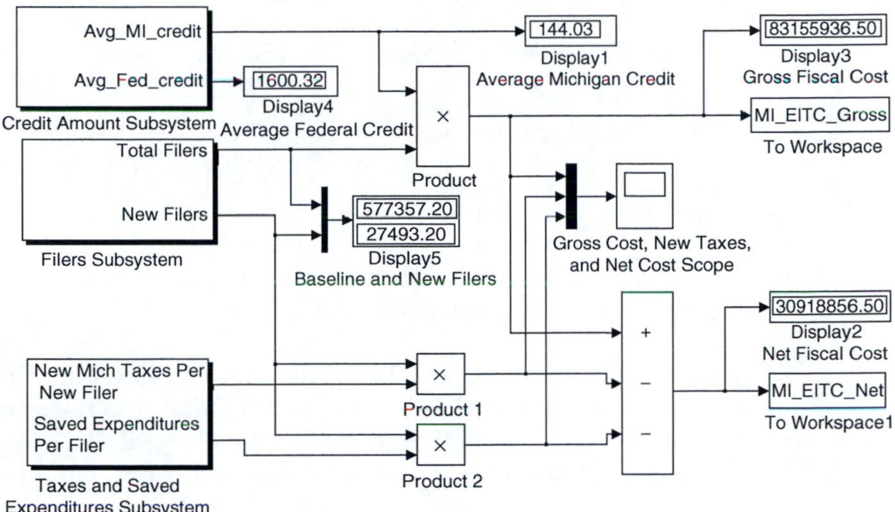

FIGURE 8-3
Fiscal estimation model.

Fiscal Analysis

To illustrate the effects of income and payroll taxes, we created a simulation model that calculates tax liabilities depending on gross wages.[160] The model incorporates exemptions, the standard deduction, and tax rates, but does not include itemized deductions, special credits, or the alternative minimum tax.[161] Figure 8-3, "Fiscal estimation model," illustrates the simulation model.

We used this model to calculate total taxes paid, including both income and payroll taxes, on workers at various income levels. We also included the effects of the federal EITC for those workers that qualify. The model includes dynamic effects on labor supply, allowing the increased after-tax wage for low-income workers to increase their participation in the labor force.

Fiscal Impact of Tax Policy Change

Table 8-1, "Gross and Net Costs of a 25% Michigan State EITC," projects the net and gross costs of a Michigan EITC starting in calendar year 2004,

---

160 The model uses certain simplifying assumptions about federal income taxes. We also assumed that wage and salary earnings are the dominant form of income for the taxpayer. This will be an accurate reflection of most taxpayers with earnings of $100,000 or less, and almost all working taxpayers with earnings under $25,000.

161 While the model accurately projects the approximate tax burden on wage and salary income at various earning levels, it is not intended to calculate the specific taxes of any individual taxpayer.

TABLE 8-1

Gross and Net Costs of a 25% Michigan State EITC

Years	MI EITC Gross Costs	MI EITC Net Costs
2004	$212,839,599	$82,246,899
2005	$215,314,478	$84,721,778
2006	$217,789,358	$87,196,658
2007	$220,264,237	$89,671,537
2008	$222,739,116	$92,146,416
2009	$222,739,116	$92,146,416
2010	$222,739,116	$92,146,416

Source: Anderson Economic Group fiscal simulation model.

set at a 25% share of the federal EITC. The gross cost includes only the revenue foregone from the tax credit. The net cost includes the revenue foregone, expenditures saved, and other state tax revenue. Both the gros-sand net cost figures assume some increase in labor force participation

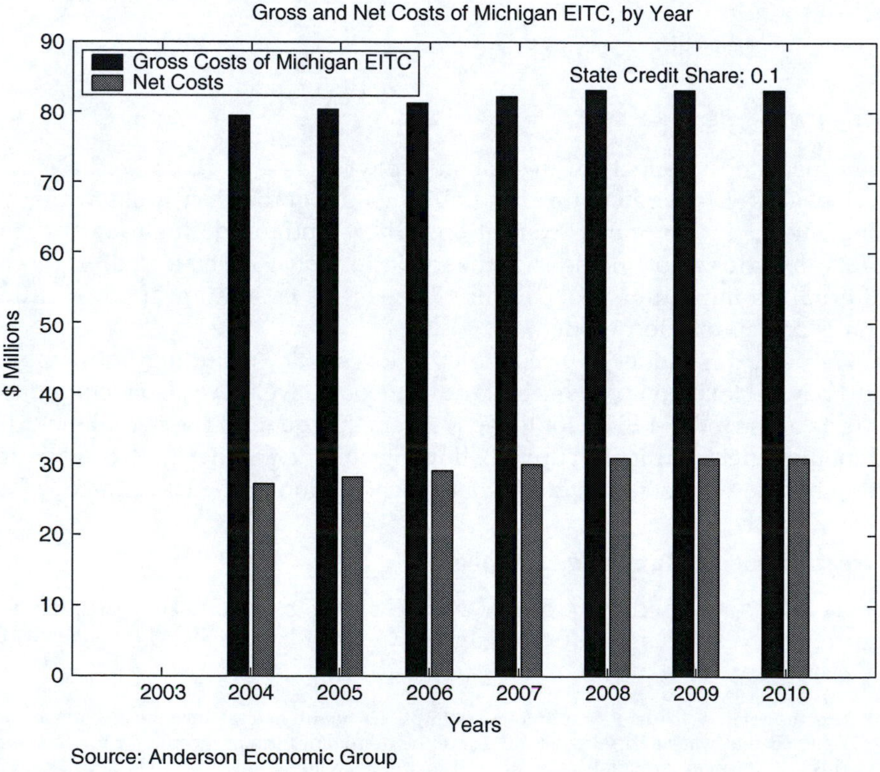

Gross and Net Costs of Michigan EITC, by Year

Gross Costs of Michigan EITC
Net Costs
State Credit Share: 0.1
$ Millions
0
10
20
30
40
50
60
70
80
90
2003
2004
2005
2006
2007
2008
2009
2010
Years

Source: Anderson Economic Group

FIGURE 8-4
Gross and net costs of Michigan EITC by year.

(and hence increases in the number of EITC filers) due to the increased incentives to work.

Figure 8-4, "Gross and net costs of Michigan EITC by year," compares the gross and net costs graphically.

Master File for Tax Impacts and Graphics

We used a master callback file to initalize and run the simulation model, and generate the graphics. The callback is shown in Code Fragment 8-3 on page 187.

Appendix

Code Fragment 8-1. Social Security Tax

```
function [liability, marginal_rate]=fed_OASDI(wages)

%FED_OASDI Federal Old Age, Survivors, and
%Disability Insurance (OASDI) Tax

%Liability and marginal rate, as a function of
%wages.

%[liability, rate] = fed_OASDI(wages)

%   Note that medicare tax is in addition to this
%   tax.

%   Tax is levied on min($maximum base wages), paid
%   by the employer

%    for each employee, during the calendar year.

%------------current wage base and rate-------------

rate=[.062];

wagebase=84900; % Taxed on first amount of wages,
for each employee


% Calculate liability

if wages < 0,

  liability=0;

else

  liability=min(wagebase,wages)*rate;
```

```
end

%  Reporting marginal rate

if wages <= wagebase,

    marginal_rate = rate;

else

    marginal_rate = 0;

end
```

Code Fragment 8-2. Federal Income Tax Calculator

```
Function [liability,
rate]=fed_ind_inc(AGI,status,exemptions)

%FED_IND_INC Tax Liability Federal Personal Income
%Tax Liability, 2001

% [TaxLiability, TaxRate]=Fed_Ind_Inc(AGI,
% FilingStatus, Exemptions)

% Where

% AGI = Federal Adjusted Gross Income (Income before
% deductions & exemptions)

% FilingStatus = single, jointly or separately

% Exemptions = Number of exemptions

% rate = marginal rate on earnings

%----------------------------------------------------

% Status & Standard Deduction amount

if nargin<2

   st=1; % Default filing status is single

   std_deduct=4700;

exempt_threshold=137300;

elseif strncmpi('single',status,3),

   st=1;

   std_deduct=4700;
```

```
exempt_threshold=137300;
elseif strncmpi('jointly',status,3),
   st=2;
   std_deduct=7850;
exempt_threshold=206000;
elseif strncmpi('separately',status,3),
   st=3;
   std_deduct=3925;
exempt_threshold=103000;
else
   error('Filing status was not entered correctly.
Try "jointly" or "single" or "separately".');
   return
end
%----------------------------------------------------
%Computation of Exemption amount and Phase-outs
exempt_amount = 3000;
if nargin < 3, exemptions = 2; end % Default number
of exemptions
if AGI <= exempt_threshold,
   exempt_amount=exempt_amount*exemptions;
elseif AGI > exempt_threshold,
   exempt_reduction=3000; % Exemption amount reduced
by 2% for every $3000 over Exemption Threshold
amount
   exempt_amount=3000*exemptions*max(0,1-
.02*round((AGI- exempt_threshold)/3000));
end
%----------------------------------------------------
%Taxable Income Calculation
t_inc=max(0,AGI - std_deduct - exempt_amount); % Do
not allow negative Taxable Income
```

```
if t_inc >10000000
    error('Function not defined for income above
$10,000,000.');
end
inflection=[0   6000  27950  67700 141250 307050
10000000          % Single Individuals
            0   12000 46700 112850 171950 307050
10000000          % Married, jointly
            0   6000  23350 56425  85975 153525
10000000];        % Married, separately
%rates=[ .10 .15 .27 .30 .35 .386]; % All filers use
same rate scale, for 2001.
%---------------------------------------------------
diffmatrix = diff(inflection')';
%   ratematrix = repmat(rates(1:5),3,1)---former
ratematrix = repmat(rates(1:6),3,1);
pay_amounta=diffmatrix.*ratematrix;
pay_amount = [ [0 0 0]' pay_amounta];
display(['taxable income is: ', num2str(t_inc)]);
%   Rate calculations--based on IKC matrix logic
if  t_inc <= inflection(st,1+1),
      rate=rates(1);
      liability=sum(pay_amount(st,1:1))+(t_inc-
inflection(st,1))*rate;
elseif (t_inc > inflection(st,2) & t_inc <=
inflection(st,1+2)),
      rate=rates(2);
      liability=sum(pay_amount(st,1:2))+(t_inc-
inflection(st,2))*rate;
elseif (t_inc > inflection(st,3) & t_inc <=
inflection(st,1+3)),
      rate=rates(3);
```

```
      liability=sum(pay_amount(st,1:3))+(t_inc-
inflection(st,3))*rate;
      elseif (t_inc > inflection(st,4) & t_inc <=
inflection(st,1+4)),
      rate=rates(4);
      liability=sum(pay_amount(st,1:4))+(t_inc-
inflection(st,4))*rate;
elseif (t_inc > inflection(st,5) & t_inc <=
inflection(st,1+5)),
      rate=rates(5);
      liability=sum(pay_amount(st,1:5))+(t_inc-
inflection(st,5))*rate;
% added this bracket 10/29/02
elseif t_inc > inflection(st,6),
      rate=rates(6);
      liability=sum(pay_amount(st,1:6))+(t_inc-
inflection(st,6))*rate;
end
```

Code Fragment 8-3. EITC Master M-file Code

```
line = '-------------------------------';
echo off;
estimate_eitc;
% display 'EITC Model Simulated'
eitc_model
disp('...estimated EITC Fiscal Model...')
disp(line)
%   Tax Plots---------------------------------------
% Note: calls figure 3 and 4; single plus one
% dependent
% PayrollPlots3;                %(this replaces
% PayrollPlots1 and 2)
```

```
% display('                    ...PayrollPlots3...')

% display(line)

% Note: calls figure 5, 2 subplots for rates with
% and without eitc; single

% plus one dependent

PayrollPlots4a;

display('                    ...PayrollPlots4a...')

display(line)

% Note: creates single plot showing eitc credit
% amounts for 3 types of

% filers. Calls figure

eitc_plots;

display('                         ...eitc_plots.')

display(line)

% Show 'Tax Rates on Wage Income' graphs (Selected
% Income & Payroll Taxes),

% 4 subplots; joint plus 1 additonal dependents (3
% exemptions)

figure(9),

TaxPlots;

display(' ...plotting tax rates...')

% Show "Payroll Taxes as a Share of Total Tax
% Burden" Graph; joint plus 1

% additional dependent

TaxSharePlot;

%   Note: calls figure 10;

display('     ... plotting tax shares...')

display(line)

display(line)
```

9

Regional Economics

The Primacy of Regional and Local Economies

Most academic economics texts, particularly macroeconomics texts, use an entire country as the typical unit of analysis. For financial markets and for the analysis of government fiscal and mo netary policy, such an assumption is approximately correct. However, for the market for most goods and services, including the market for labor, it is local and regional markets that are the relevant areas of analysis.

In this chapter, we focus on the tools of understanding local economies (the economies of a city, county, or area) and regional economies (regions which would typically stretch across a number of cities and sometimes cross state boundaries).

Diversification and Recession Risk

Most discussions of economic diversification are based on one goal — the reduction of economic risk. By spreading the use of its resources and particularly employment into various sectors (so it would not have all its eggs in one basket), a region could lower its risk of losing significant employment in a future recession. For our analysis, let us pick a county in the state of Michigan. As Michigan's economy has historically been vulnerable to downturns a risk is a significant cause for concern for Michigan counties. At the same time, this and other former "Rust Belt" states lost jobs in manufacturing during the 1980s and again in the beginning of the new century. Therefore the "reward" of new jobs is also important.

However, a thorough discussion of diversification should recognize another objective besides reducing risk — increasing reward. A regional economy can increase its reward by utilizing its comparative advantages so that the region's workers can benefit from the unique qualities of the workforce, institutions, and natural resources of the area. A useful measure of diversification will take into account both risk and reward.

Traditional Measures of Diversification

A careful review of economic diversification measures shows that traditional approaches for measuring diversification are fraught with difficulty. In particular, the standard approaches are not rooted in a firm concept of lower risk, comparative advantage, or independence. Instead, they are based on a notion of conforming to a standard of dispersion of employment across industries. The standard approaches then calculate a measure of how far the regional economy is from that standard.

These approaches typically take one of the following two distributions as the standard.

1. Equal shares, i.e., an equal number of employees in each division
2. The current U.S. or a specific state's employment distribution

Practioners of these approaches then calculate, using a weighting scheme, an index of diversification. This index can then be compared across states or counties. We present these measures in two categories below.

Equal-Share Basis Measures: Measures such as the Ogive Index take equal share in each industry sector as the norm.[162] Other measures, developed for intraindustry comparison, such as the Herfindahl–Hirschman Index and the Top 4 Market Share, have also been adapted to measure interindustry diversification.

Such statistics are based on the implicit assumption that the most diversified economy is the one that has an equal number of employees in the various industrial sectors. This is an absurd proposition. There is no reason to believe that an equal-share economy would have low risk, make good use of comparative advantage, or be independent of other influences. The measure would depend critically on the classification scheme, ignoring the reality that some sectors, when compared to others, should employ more people in any healthy economy. Thus, we find no reason for a regional economy to seek to perform well on such measures.

State-Share Basis Measures: A somewhat better standard is the state-share distribution, which reflects the reality of the marketplace today rather than some artificial equality across sectors. This standard for computing measures of diversification is now given preference among most economists writing in this field. Their assumption is that the economies of most states, or of regions covering multiple states, are diversified and therefore approximating these proportions would be beneficial.

However, the key weakness of this approach is that comparative advantage is disregarded, as is the desire to have some independence from outside influences. If a Michigan county, for example, has an advantage in manufacturing and can earn higher wages in those industries, why should it

---

[162] See M.J. Waslenko and R.A. Erickson, On measuring economic diversification: comment, *Land Econ.*, 54, 1, February 1978.

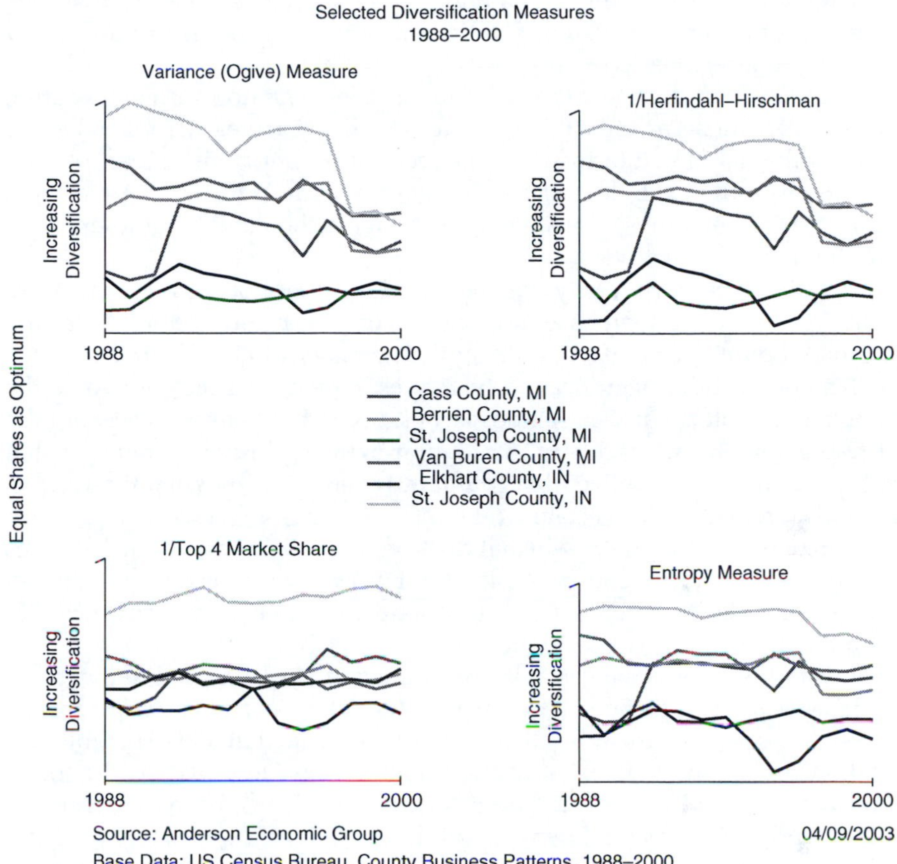

FIGURE 9-1
Selected diversification measures.

be penalized for specializing in it? Conversely, if the U.S. as a whole is more dependent on foreign oil imports than the Great Lakes states, and a Michigan county seeks to reduce the chance that a disturbance in OPEC countries will crimp its economy, should not that diversification goal be measured in some way? Unfortunately, there is no perfect measure of diversification, and the state-share measures suffer from the weaknesses just mentioned.

We plot these measures for a selection of Michigan counties in Figure 9-1, "Select diversification measures."

Recession-Severity Measure of Diversification

As all the standard measures of diversification have serious weaknesses and are not rooted in the fundamental motivations for diversification, we developed a new measure in 1999. This unique approach is motivated by the desire to measure the risk to a region's economy of a broad recession.[163] This is the primary motivation for diversification in general and a primary concern for a regional economy.

Measuring Recession Severity: As we are unsure about how or when a future recession will arise, an effective recession-severity risk measure will take into account the risks in all industries. The recession-severity risk measure used here assesses the likely loss of employment in each industry in a recession year. It then sums those risks, using as weights the actual employment in each industry for each year.

Fifth, the recession-severity risk measure here does not presume the cause of the recession and then base the measure on that presumption.[164] Instead, it is based on historical data showing the volatility of all industries.

This approach has significant advantages. First, it completely avoids the basic flaws of all equal-weight approaches. Second, it takes into account the historical risk in each industry in a region by using historical data for that industry. Third, it recognizes that comparative advantages will push a region towards production in certain areas and does not penalize a region for specializing in industries in which it does have an advantage. Fourth, it does not confuse a secular trend in employment growth which can be anticipated by most people in the industry with unanticipated economic changes that normally cause the most dislocation.[165]

This approach, however, is not without some weaknesses. First, it relies on historical data to predict future risk (as do all statistical approaches). Second, it does not include all the information in that data, notably the relationship among industries. Some industries tend to move *counter* to the business cycle and others *with* the business cycle. This approach presumes that all industries have a down year and uses that as the measure of risk.

The formula for the recession-severity statistic is presented in Appendix II to this chapter, which follows and relies upon the next section.

Portfolio Theory Analysis

Modern financial portfolio theory provides a new approach to viewing industrial diversification. We view a local economy, as rooted in the desire

[163] Patrick L. Anderson, Ian Clemens, and Robert Kleiman, *Michigan Economic Diversification Study*, Lansing, MI, Anderson Economic Group, August 1999; available at: http://www.andersoneconomicgroup.com. Also available from the Michigan Economic Development Corporation at: http://www.michigan.org.

[164] Measuring recession-severity risk by counting the number of manufacturing workers, for example, presumes that a recession would be caused by a manufacturing slump. The statistic we propose — while it has some weaknesses — is not based on such a presumption.

[165] Much of the recent research on the business cycle focuses on this point, namely, that unanticipated changes are more likely to cause recessions than anticipated economic shifts.

of the citizens to use their comparative advantages to earn a good living, on a consistent basis, from their labor.

Risk and Return: Portfolio theory looks at investments in terms of risk and reward. Risk includes all types of losses and gains. Rewards include earnings and jobs.

States and regions should view their land, buildings, equipment, and most importantly their human capital as investments in their economy. These investments produce a return. This return can be measured in terms of personal income or employment.[166]

Each investment also carries with it a certain amount of risk. In the portfolio theory approach, risk generally applies to the variation in returns. If an investment tends to have big swings in returns — big negative and positive returns and lots of volatility in between — we say it is risky. Typically, an investor who holds such investments will demand a higher average return to compensate for the added risk.[167]

Limitations of Portfolio Theory: While portfolio theory offers a powerful approach, it does have some limitations when applied to this field, including:

- Portfolio theory assumes that the return on a financial asset does not depend on the asset holder. This is an appropriate assumption for most investors choosing among many different securities. However, states and regions do affect the profitability of industries that exist in their jurisdictions. Therefore, there is some interplay between the investor (the workers in the region) and the investment (the industries themselves).
- Portfolio theory assumes that investors can choose any asset. However, the returns for some industries are so region-dependent that they would be hard to reproduce elsewhere. For instance, Michigan's established base of auto manufacturing and related enterprises makes the return on a venture in this field to be greater in that state than in other areas. California's weather and rich soil mean that agricultural investments have a higher chance of success there than in most other states. Different industries have different barriers to entry, making the cost of entry into an industry unequal across all states.
- States and regions cannot trade for different industries with the ease an investor can trade securities.
- Like any investment, future results may be different from the results of the recent past.

---

166 Other measures of that return, including the benefits these industries and their employers and employees provide to our society, cannot be easily measured. They are normally assumed to be highly correlated with employment and income. There may also be environmental and social benefits and costs which are not correlated with output.

167 There is no negative moral connotation in the term *risk* in this context; many prudent investors deliberately choose more risky investments because they provide better returns over long periods.

Even with these limitations, the approach offers a very powerful way of analyzing state economies.

Methodology

We take the following steps in applying portfolio theory to the diversification question:

1. Using the data on employment for various industries, calculate the average compounded growth rate for employment in each industry. This describes the average return on each industry.

   Note that, for many areas in the U.S., this will require SIC to NAICS conversion, as employees formerly considered to be in one industry are now, quite often, considered to be in another.[168] The NAICS system is much more precise, but the bridging of the classifications requires some work. This is especially the case with high-tech and manufacturing workers.
2. Calculate the risk in each industry so as to determine which is the typical variation in the return. Some industries have larger swings than others and therefore will show a higher risk. (See "Appendix II: Risk and Return Measures for Diversification Measures" on page 204 in this chapter for a discussion.)
3. Using the periods at the beginning and the end of our time frame, create model investment portfolios of industries.
4. Using the financial and statistical techniques standard for financial portfolios, answer the following question: "As the concentration of employment in these industries has changed over time, how have the expected returns and risks changed"?
5. Graph the results for different time periods on a risk–return chart. An example for a county in Michigan is shown in Figure 9-2, "Employment risk and reward."

Results of This Methodology

Our first analysis of this type, for the state of Michigan, was completed in 1999. Since then, we have completed similar analyses for a number of counties. Some of these reports are publicly available.[169]

For the State of Michigan, formerly a "rust belt" state, the results indicated that the state's economy had diversified during the past decade. In our

---

[168] SIC and NAICS are two classification systems for industries in the U.S.

[169] For example, P.L. Anderson, Ian Clemens, and Chris Cotton, Economic Diversification and High-Tech Employment in Oakland County, Lansing, MI, Anderson Economic Group, 2001, available at: http://www.andersoneconomicgroup.com; also available from Oakland County, MI. See also the example cited below.

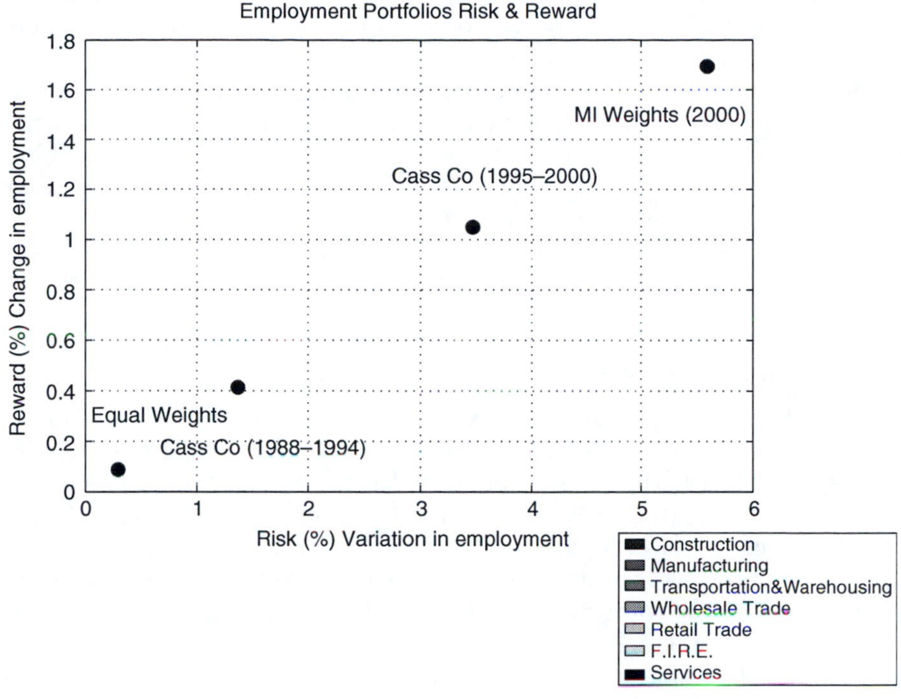

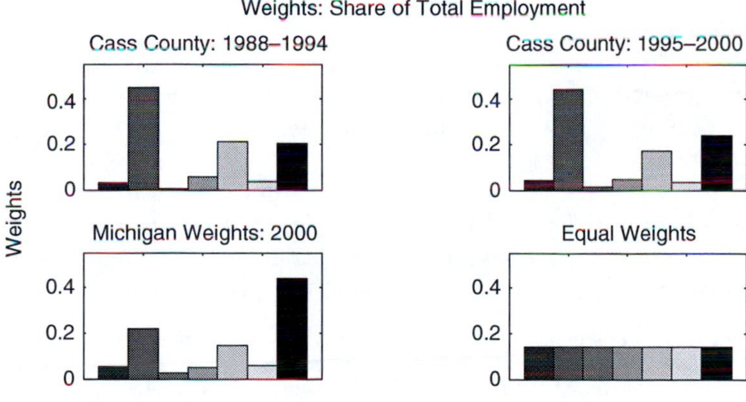

Source: Anderson Economic Group
Base Data: US Census Bureau, County Business Patterns, 1988–2000

FIGURE 9-2
Employment risk and reward.

risk renewal paradigm, "diversification" is not a goal — the reduction of risk is a goal. In Michigan's case, the states reduced concentration in manufacturing, along with other changes, did result in lower risk. However, the state actually improved the trend growth rate for employment for the weighted average of workers in various industries. This was a double winner — the equivalent of shifting to investments that had both lower risk and higher returns.

The results of our analysis received a major test 2 years later. Based on the diversification analysis, we projected that in the next major recession Michigan would lose significantly fewer workers than it did in the 1980–1982 recession, and less proportionately than in the in 1990–1991 recession. Indeed, while unemployment had skyrocketed to well over 10% in the 1980–1982 recession in that state, its unemployment rate in the 2001–2003 recession moved to only about 1% higher than the national average.[170]

Example Analysis of Cass County, MI

We completed a similar analysis of Cass County, MI, in 2003. Cass County is a rural county on the border with Indiana. Although rural, it has a significant manufacturing base. The county had lost a number of employers to other states during the 1980s, largely due to substantially higher taxes and other business costs in Michigan at the time. During the 1990s, however, Michigan substantially reduced its tax burdens, making the state's business climate much more competitive.[171] How did these economic changes in the state as a whole affect this county?

We summarize the analysis in Figure 9-2, "Employment risk and reward." The chart at the top plots risk-and-reward characteristics of this particular county at present and at a time in the past. It also compares this county with the state as a whole.

As indicated on the chart, both the expected risk and reward increased for Cass County over the past decade. This means that while the county's growth prospects have improved due to the changes over the past 10 years, the amount of risk also increased.

In the lower portion of the graphic, we provide summary information on concentrations of employees in industrial categories, both at present and in the past, to show which industries have been hiring and which have been reducing employees.

---

[170] This analysis is based on preliminary data and a preliminary dating of the recession period. Because the Michigan unemployment rate rose *after* the recession was over, our statement here is actually conservative. At the November 2001 trough of the recession, Michigan's unemployment rate was within one point of the national rate and under 6%. By contrast, in January 1982 it was 16%, and it averaged over 12% for all of 1981.

For dating of business cycles, see the NBER site at: http://www.nber.org. The dating used here is from the July 2003 report of the NBER Business Cycle Dating Committee. For labor market information for Michigan, see the Michigan Department of Career Development Web site at http://www.michlmi.org or the U.S. BLS at: http://www.bls.gov.

[171] For a systematic comparison of business costs among major business centers in Michigan and several other competing states, as well as in the Canadian province of Ontario, see *Michigan Business Climate Benchmarking Update Study*, Patrick L. Anderson, Lansing, MI, Anderson Economic Group, 2000; available at: http://www.andersoneconomicgroup.com; also available from the Michigan Economic Development Corporation Web site at: http://www.michigan.org.

One implication of this approach is that there is no one prefect position; we illustrate multiple possibilities in Figure 9-3, "Frontiers of risk and return."

Urban Sprawl[172]

Urban sprawl has emerged as one of the foremost issues concerning communities across the country and is destined to be a focal point in the next decade. Like most emerging issues, the topic is subject to considerable confusion. Much of that confusion is created by the lack of definition of "sprawl."

In this section, we define sprawl rigorously. We then create an index which can be used to compare the degree of sprawl among metropolitan areas across the country as well as in single metropolitan areas over time. This new sprawl index properly adjusts for natural population changes and the varied metropolitan governing structures of different cities. It measures only those changes in location that conform to the general notion of sprawl.

After defining sprawl, we then measure the degree of sprawl among major Michigan cities as well as a number of other national cities. The results are quite surprising.

Defining Sprawl

Sprawl is generally known as the movement of individuals from cities to inner suburbs and then from inner suburbs to outer suburbs. These suburbs sometimes form "edge cities," with many of the accoutrements of urban core cities, including offices, complexes, shopping, entertainment, and medium-density residential areas. A survey of the extensive literature discussing the issue reveals no clear definition of sprawl. Indeed, most discussions of the issue merely describe the maladies associated with sprawl, often confusing cause and effect. Obviously, we cannot understand and discuss a problem without defining it.

Sprawl can be defined as a *change in density preference*, in which individuals make deliberate choices to move from high-density cities to medium-density or low-density suburbs and rural areas. This definition of sprawl also explains much of its perceived negative attributes. As new residential areas gobble up old farmland, new residents bring with them a desire to shop and work, thereby pulling office complexes, strip malls, and traffic congestion into formerly rural or low-density suburban areas. At the same time, much of the infrastructure in urban core cities — extensive highway

[172] This section is drawn from P.L. Anderson and I. Clemens, What Cities Sprawl the Most? A Comparison Among Cities, unpublished manuscript, March 2000.

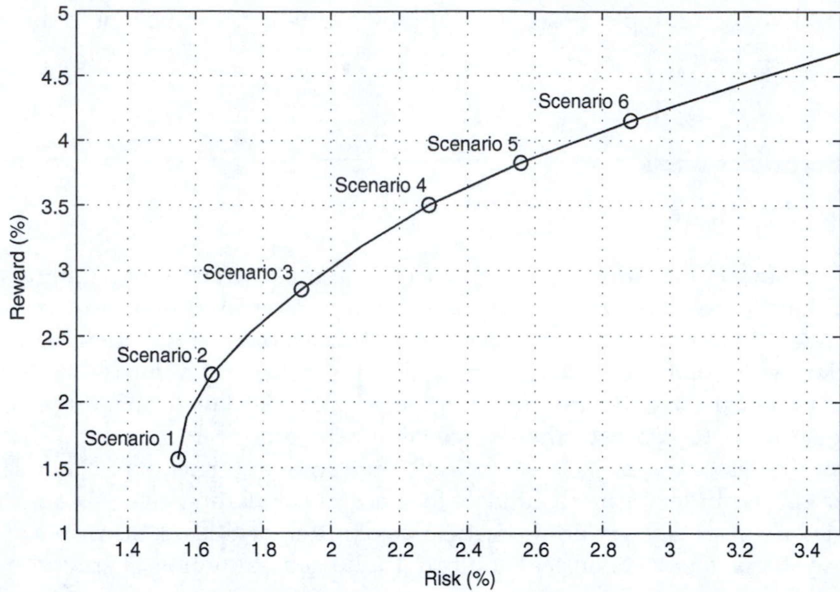

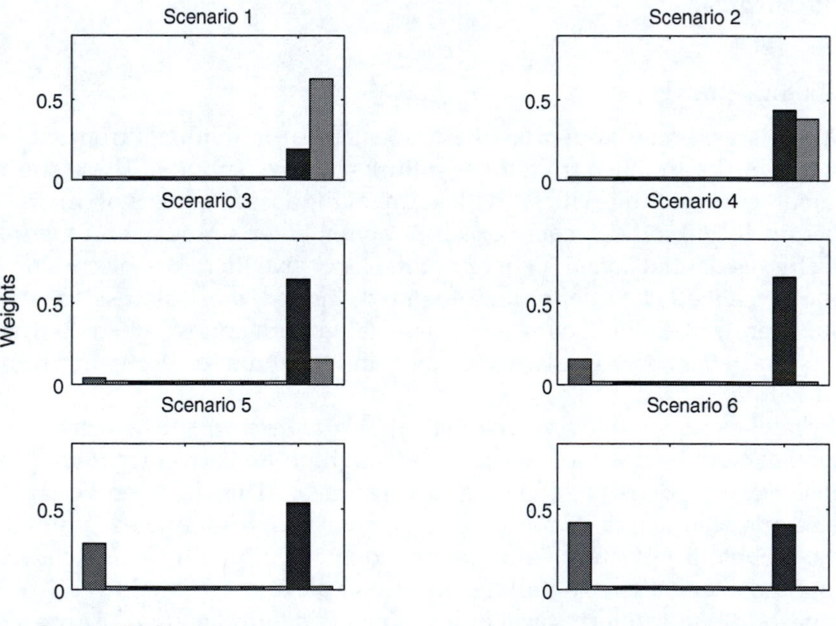

FIGURE 9-3
Frontier of risk and return.

and other transportation networks, industrial facilities, large surface streets, and high-rise office buildings — may go underutilized.

Calculating Sprawl

We calculate the sprawl index by measuring the density preference of individuals in a region, both within a core city and in the surrounding metropolitan area. We first do this in a base year. The selection of a base year index for each city is important because there are reasons to believe that density of population should vary across different regions. One reason for a different density is simple geography. Cities that are located on rivers would naturally have more development near the river if the river was used for trade or recreation. Communities with limited water supplies, those hemmed in by mountains, or with other natural features that enhance or limit their options, would adjust to those facts of their existence. Thus, the fact that two different cities in any selected year had different densities would not indicate anything about their desires to sprawl.

In this analysis, we chose the year 1980 as the base year. We then calculated density ratios for both core urban cities and their surrounding areas. The ratios of these two quantities — the relative density of a city and its suburbs — were then set to 100, establishing an index variable.

We then recalculated these density ratios for a more recent year. The ratio of the relative density ratios of core cities and their metropolitan areas in two different periods (their ratio multiplied by 100) was defined as the sprawl index for the city. The mathematics of the Sprawl Index are shown in Appendix I to this chapter.

Rural Sprawl

While the major emphasis in recent years has been on urban sprawl (the change in relative density of cities as compared to their suburbs), we should also consider what might be called rural sprawl. We define rural sprawl as a change in relative density preference between suburbs and rural areas. If sprawl is consuming our nation's farmland, it is probably rural sprawl rather than urban sprawl. The majority of farmland is located in rural areas. However, the definition of "rural" is imprecise.[173]

We currently have conflicting evidence about rural sprawl. In many rural areas, population has been declining as residents move to urban areas to

[173] This presents a practical problem that, for urban sprawl, is addressed by using city boundaries and surrounding metropolitan areas. For suburbs, the "surrounding" rural areas are hard to define.

find jobs. In others, population has been exploding as farmland is plowed up to create new suburban residential and commercial areas.

Rigorous analysis of rural sprawl will help us determine whether there really is a crisis in farmland, and if so, where it is occurring.

Limitations of the Sprawl Index

As in any statistic, the sprawl index has certain limitations and drawbacks. The first limitation is the selection of the base year. Residential patterns are always in flux, and the selection of any base year inevitably adds some noise to the system.

The second major limitation is defining the metropolitan and core city areas. As part of their metamorphoses, cities often annex other areas. Boundaries for metropolitan areas, such as those defined as MSAs, PMSAs, and CMSAs by the Census Bureau, also change over time. In order to remain consistent, we use the same boundary areas when calculating the density ratios for different years even if the city has annexed land or the metropolitan area has changed.

The final major limitation is the fact that density, measured as the number of residents per square mile, does not fully measure all the maladies generally associated with sprawl. In some communities, a change in density preference may be accompanied by well-planned development with all the sensitivity that residents often desire. In others, haphazard development may result in the destruction of beautiful areas, the fouling of waters, and congestion of roads without any efforts to rationalize the development.

Even with these drawbacks, the sprawl index, based on a rigorous measurement of changes and density preference, is an accurate, reproducible, and well-motivated measure of sprawl.

Preliminary Findings

We computed density measures and calculated the sprawl indices for a number of cities in Michigan and, for comparison, cities in other states. The sprawl index for these cities is contained in Figure 9-4, "Sprawl indices for selected cities."

The figure shows that the top sprawl cities for the 1980–1990 period were Detroit, MI, and Gary, IN, both with sprawl indices exceeding 120. Battle

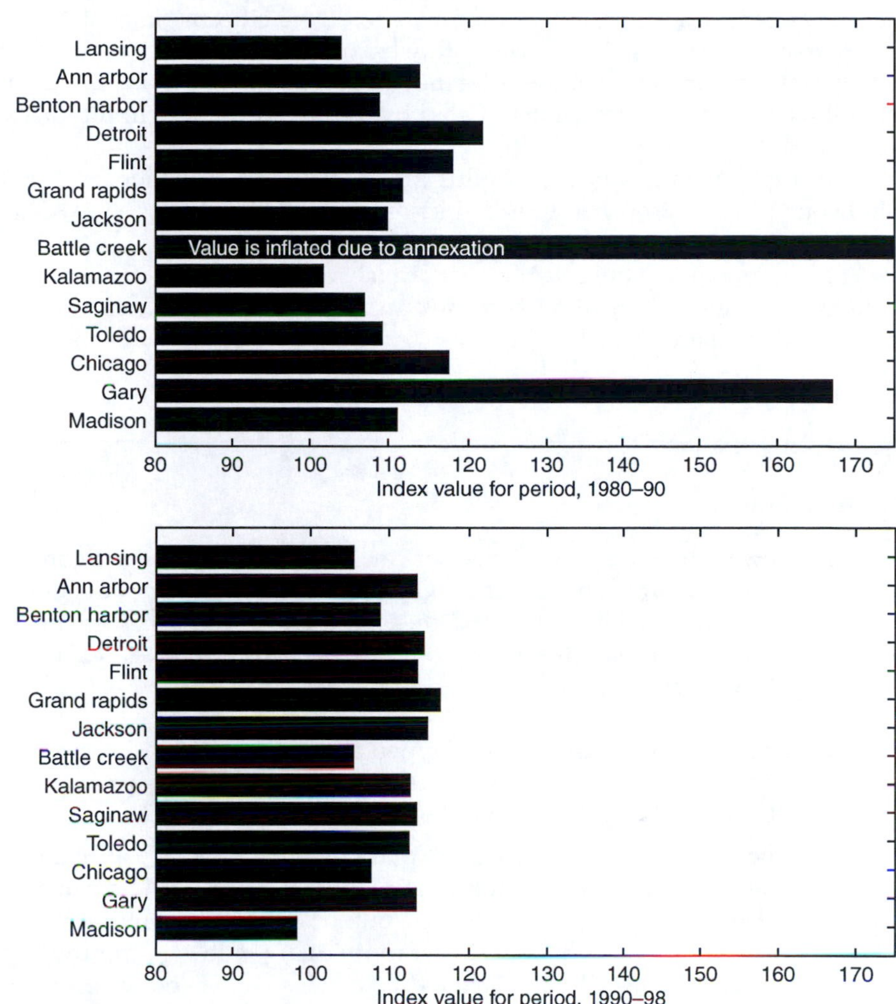

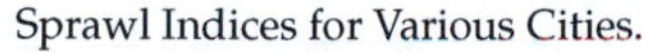

FIGURE 9-4
Sprawl Indices for Various Cities.

Creek, MI, is noted on the chart as having annexed property, thus distorting its (uncorrected) sprawl index.

A comparison of the 1980–1990 and 1990–1998 periods for the same group of cities seems to indicate a slowdown in urban sprawl. While at the time of this analysis the data were not yet available for an entire decade, it appears that these cities were no longer sprawling as fast as they once were. Gary, IN, Detroit, MI, and Grand Rapids, MI, all had 8-year sprawl indices in the 110–120 range. Lansing, MI, seems to have more sprawl recently than in the past decade.

There are a surprising number of cities that have not sprawled. These cities may represent a postsprawl environment for their citizens, in which the desire to migrate out of the city is roughly balanced by the desire to migrate into the city. In some cases, these cities may have experienced changes earlier than the large cities which were buffeted by industrial changes in the 1970s and 1980s.

Cities that have not sprawled in this group include Kalamazoo, MI, and Madison, WI. Madison has the distinction of having negative sprawl in recent years.

While more work should be done rigorously analyzing the actual sprawl patterns of cities, this method provides an objective, quantifiable way to define and measure sprawl.

Conclusions

Our assessment of the sprawl in American cities focused on a rigorous definition and measurement of sprawl. Once sprawl is properly understood as a change in density preference and measured as such, it becomes much easier to rationally discuss the policy options that are available. Based on our analysis, we have the following recommendations:

1. Sprawl should be rigorously defined as a change in density preference so that the debate over the causes and effects of sprawl is not confused due to a vague definition of sprawl itself.
2. Policy makers must recognize that the fundamental cause of sprawl is individual's choices about where they wish to live. Such choices can hardly be considered immoral or destructive when they reflect the genuine desire of individual families to improve their conditions of living. While the overall effect of sprawl, especially when communities are not prepared for it, can be negative, nothing is gained by attributing malevolent intentions to people who choose to relocate.
3. To the extent sprawl is caused by rational location decisions on the part of individual citizens, policy makers should look first to see whether their policies are encouraging sprawl. The policies they should examine include tax policies, environmental regulations, and transportation networks. For example, levying heavy taxes in central core cities encourages people to move to suburbs. Making the development of industrial facilities in "greenfields" less risky than in urban "brownfields" practically guarantees sprawl.[173a]

[173a] Indeed, all policies that create cost premiums in urban areas encourage sprawl.

4. Policy makers should consider whether those who use public goods pay properly for those goods. This does not mean simply punishing suburban residents with awkwardly defined fees, but it does mean recognizing and properly charging those who make new demands on an area's government and its taxpayers.

Appendix I: The Sprawl Index in Mathematical Terms

We define a sprawl index C for city i over time period t_0 to t_1 as follows:

$$C_{i01} = R_{i1} / R_{i0} \tag{1}$$

where R indicates the ratio of densities between a city and its suburbs at specific times.

$$R_{ij} = D_{ijcity} / D_{ijsuburb} \tag{2}$$

where D_{ij} is the population density of city or suburb i at time j, defined as:

$$D_{ijk} = Resident\ Population_{ijk} / Area_{ik} \tag{3}$$

where k stands for the city or its suburbs.

Example Calculations

Consider a city with a density of 7000 people per square mile (close to that of Detroit) with a surrounding suburb of density 3000 in the year 1990. Suppose that, over the next decade, the density of the city dropped to 6700 people per square mile, while the density of the suburbs increased to 3200.

The relative density (R) between the city and its suburbs in 1990 would be:

$$R_{Detroit,\ 1990} = 7000 / 3000 = 2.33 \tag{4}$$

The relative density in 2000 would be:

$$R_{Detroit,\ 2000} = 6700 / 3200 = 2.09 \tag{5}$$

The sprawl index (C) for the city, between 1990 and 2000, would be:

$$C_{Detroit,\ 1990:2000} = R_{1990} / R_{2000} * 100$$

$$= 111.44 \tag{6}$$

Thus, we have a sprawl index of 111, indicating 11% change in relative density between the city and the suburbs over the time period.

Appendix II: Risk and Return Measures for Diversification

Risk and Return

For each major employment division, risk and return statistics were calculated by the following formulas.

The return for each industry was calculated as:

$$R_i = \log(X_{it}) - \log(X_{it-1}) \tag{7}$$

where X_{it} indicates the employment in sector i in year t and log indicates the natural log. The historical volatility for each sector was calculated as:

$$V_i = std[(\log(X_{it})) / (\log(X_{it-1}))] \tag{8}$$

where *std* indicates the standard deviation calculation.

This calculation, widely used in financial theory, is superior to volatility calculated as the standard deviation of the return, for it accounts for any growth trend that may skew the results. The calculation also expresses volatility in percentage terms.

Thus, an industry that tended to grow about 1% each year, but with a typical variation of around 3% each year, would have 3% volatility and 1% return.

Standard Diversification Measures

The following statistics are based on differences in each industrial classification between one state's employment and a number calculated as a standard employment in all industries. They differ mainly in the standard used and the weights applied to each industry.

For example, an *equal-share* standard for 11 industries would be 9.9% for each industry. Using an equal-share statistic, we would take a service sector share of, say, 30% and subtract that number from the standard of 9.9%. We would then weight (multiply) this difference and combine it into an index number.

TABLE 9-1

Standard Measures of Diversification

Measure	Standard	Formula
Variance/Ogive measure	Equal shares	1/(Variance about the mean)
Herfindahl–Hirschman index	Equal shares	1/(Sum of shares squared)
Top 4 market share	Equal shares	1/(Sum of Top 4 shares)
Entropy measure	Equal shares	Logarithmic measure of dispersion
Variation	U.S. shares	1/(Variance about the U.S. shares)
Hachman index	U.S. shares	1/(Sum of shares squared over U.S. shares)

Table 9-1, "Standard Measures of Diversification," describes the six standard tests referenced in the text. As indicated above, we place little faith in these statistics (especially the equal-share-based statistics) as they are not firmly rooted in the theoretical or practical motivations for diversification.[174]

Recession-Severity Risk Statistic

This statistic provides a quantitative indication of the likely loss of employment in an economic downturn for a specific region.

We defined above volatility V_i of employment X_i in sector *i*. The recession-severity risk statistic is the weighted sum of the volatility measures of each sector, where the weights are the share of total employment in each industry. Note that this statistic changes over time, reflecting a changing severity risk.

As indicated in the text, this statistic does not capture the covariance among the industries. Furthermore, it is an indication of severity and not likelihood of a recession.

$$\mathrm{RR} = \sum_i \mathrm{W}_i \mathrm{V}_{ij} \qquad \Sigma w_i = 1. \tag{9}$$

[174] In order to make all measures comparable, some are inverted.

10

Applications for Business

Introduction

This chapter begins the focus on the economics of the firm. There is substantial accounting literature on the proper way to *account* for the past actions of a firm. There are also numerous finance books covering specific topics in the *capitalization* of a firm. In this book, we touch on relevant accounting concepts and analyze briefly the capitalization of a firm. However, our focus is something often treated as an assumption in many of these texts: *how a firm earns money.*

The fundamental economics of a firm revolve around earning a profit. Accounting and capitalization are important, but there can be well-capitalized, properly accounted firms that go out of business and sloppily managed firms that stay in business for years. Therefore, an understanding and modeling of the process of earning a profit should be the core of business economics.

Contents of this Chapter

We discuss in this chapter:

- Methods for projecting the future values of business variables (such as earnings or revenue), first with certainty, and then allowing for uncertainty
- How to model different types of uncertainty, including random deviation around a well-determined path; Brownian motion (or "random walk") processes; and "jump" processes, which simulate the effects of unexpected defaults, terminations, and other adverse events
- The derivation of net present value formulas for various types of cash flows
- How earnings from individual lines of business produce the gross earnings affect the firm

- How earnings — or losses — affect the firm's equity investors and lenders
- An introduction to the use of simulation models to represent the revenue, losses, and earnings of a firm over time
- An introduction to iterative methods, and an example method to estimate the value of a real estate enterprise when its earnings are partially based on the market value of property
- Advanced topics such as the expected present value of an enterprise (or security) subject to business termination (or default) risk.

The material in this chapter is essential for understanding the following chapter on business valuation (Chapter 11). The methods for projecting future values are used in other applications in this book.

Projecting Trend Growth: Simple Uncertainty

One of the most elemental tasks in projecting the future earnings of a company is projecting other variables into the future. Typically, these projections are phrased as statements such as, "We project 6% growth in GDP next year," or "Experts anticipate sales of fancy new technology gadgets will grow 10% next year."

These statements imply a kind of certainty that simply does not exist. *Whenever* we project the future, we project with uncertainty.[175]

While we cannot avoid uncertainty, we can anticipate it. Therefore, even when we forecast trends that seem well established, it is best to anticipate some type of uncertainty. Therefore, we deal first with forecasting trend growth with simple uncertainty. Later, we will introduce more complex forms of uncertainty.

Simple Uncertainty: Baseline, High, and Low Scenarios

Most knowledgeable people routinely forecast important variables with some confidence. By "some confidence," we mean approaching the statistical definition of confidence intervals, although not reaching them.

[175] Wise men and women would do best to remember the following aphorisms: "Prediction is very difficult, especially if it's about the future." (Niels Bohr, Nobel laureate in Physics); "To expect the unexpected shows a thoroughly modern intellect." (Oscar Wilde); "A good forecaster is not smarter than everyone else, he merely has his ignorance better organised." (Anonymous). These and other remarks on forecasting were collected by D.B. Stephenson and found at http://www.met.rdg.ac.uk/cag/forecasting/quotes.html.

For example, the "experts" cited in the statement above would probably be able to recommend high- and low-bound estimates to accompany their baseline estimates of sales growth. The spread (interval) between those high- and low-bound estimates implies a type of confidence interval, even though some elements of a statistical confidence interval will be lacking. A true statistical confidence interval will have *both* a "confidence level" (say, 90%) and an interval.[175a]

A Trend Projection Function

For most tasks in business economics, important variables should be forecast using some allowance for uncertainty. We describe a utility program TrendProject.m that takes a base level of activity and forecasts it into the future incorporating the simple uncertainty of a high, medium, and low forecast trend.

This function, a part of the Business Economics Toolbox, uses the following steps:

- Get the key assumptions about growth.
- Get the starting period and amount.
- Run the growth calculations three times, creating a high- and low-range estimate, in addition to a best-guess estimate. This is a true compounded growth calculation.
- Plot the results.

Uses for this Function

We will use growth projections in most applications. This specific function is used to forecast income in Chapter 11, under "Estimating Business Values" on page 277. Similar growth projections are used in most simulation models described in this book.[176]

Trend Projection Code, Graphics, and Output

The graphical output from this function is in Figure 10-1, "Trend projection graphic." The program code is reprinted in Code Fragment 10-1, "Trend Projection Function," in the Appendix to this chapter on page 235. The command window output shows the high, low, and middle projections for the default input assumptions.

[175a] Despite the ubiquitous statements in the general press about the confidence intervals from survey research ("Voters favor Smith by 10 points with a confidence interval of plus or minus 3 points"), such statements are meaningless without an accompanying confidence level.

[176] See "Other Methods for Compound Growth Projections" in this chapter on page 211.

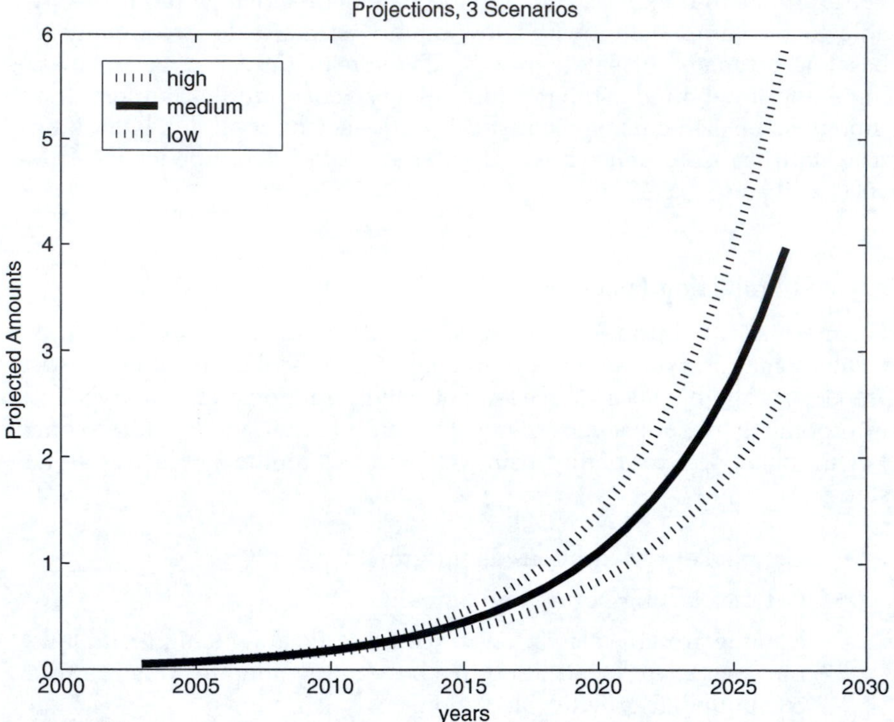

FIGURE 10-1
Trend projection graphic.

Function Details

The function is organized as follows:

1. It checks for the proper assumptions. If none is specified, it opens a dialogue box. This also sets a flag to graph the results. Organizing the function in this fashion allows for it to be used interactively with graphical information provided, or as part of a larger program in which no interaction or graphics are desired.
2. It finds the current year and uses this as the default starting date.
3. Using a subfunction, it calculates the projected future values, assuming the baseline growth rate and a high and low deviation from that rate. The algorithm used is an iterative calculation[177] in

[177] For ease of understanding, we present a number of iterative algorithms (using "for loops") that could be more efficiently programmed as "vectorized" code. See the section on "Optimizing for Speed" in the MATLAB help documentation.

which a variable for revenue in each period would be calculated as:

$$rev_{t+1} = rev_t \times (1+g) \tag{1}$$

which, generalized to all periods, means the amount in period i can be directly calculated from the base amount rev_1:

$$rev_i = rev_1 \times (1+g)^{i-1} \tag{2}$$

The subscript in these equations indicates the time period of the variable, and the variable g indicates the growth rate. In the first period, $i-1 = 0$, meaning that the growth factor term reduces to one.

4. It determines whether the graphing flag has been set, using the `switch ... case` syntax. If the flag has been set, then each growth scenario is plotted on the same x-y axis plot.
5. The key variables used in the function are saved in a structure `inputs`, and the resulting projections are produced in an array including the low, high, and baseline forecasts. Any or all of these forecasts can be selected easily in other routines.

Other Methods for Compound Growth Projections

There are many other methods for calculating variables growing at compounded growth rates. In particular:

- We include, for use in Simulink simulation models, two growth blocks in the Business Economics Toolbox. See Chapter 5, "Library Functions for Business Economics," under "Custom Business Economics Blocks" on page 113.
- MATLAB has numerous other functions, including logarithmic and power functions, which can be used to project variables.
- A spreadsheet can be used to calculate a growth projection with variable or constant growth if the formula and inputs are copied across a number of cells. However, it is difficult in such an environment to incorporate uncertainty, especially complex uncertainty.

We present more complex types of uncertainty in "Projection: Complex Uncertainty" on page 212.

Projection: Complex Uncertainty

In this section, we describe mathematical models that have been found to represent the risk characteristics of important business variables.[178] We will use the term "process" to describe the generation of random variables using a specific statistical distribution and mathematical model.

Markov Processes, States, and Uncertainty

When we use the terms "risk" or "uncertainty," we often describe a randomness in variables that follows a certain statistical distribution. This does not mean a complete lack of foreknowledge, but rather a considerable knowledge about the factors that will govern the future. That knowledge is usually encapsulated in a description of the underlying process by which the future value of variables unfolds. This process may involve both a determinist component and a random component.

In particular, we will often speak of a *Markov process*, in which the value of variables at one time provides all the information necessary to make the best possible prediction of their values at the next period.[179] This is formalized as a probability statement about the random sequence $\{X_t\}$ and a subset of possible values A:

$$Pr\{X_{t+1} \in A | X_t\} = Pr\{X_{t+1} \in A | X_t, X_{t-1}, X_{t-2}, \ldots\} \tag{3}$$

meaning that the entire usable information set (the variables shown after the conditioning operator) for predicting the next period is contained in the present period. This "memoryless" restriction is less confining than it might appear; depending on how the periods are defined and the amount of information in the state at time *t*, most economic and business variables of interest can be approximated in this manner.

The information available at a certain time can be formalized as a *state*. In the equation above, we encapsulate the state in the variable X itself. In later

[178] To be more precise, we will identify specific statistical distributions and methods and related parameters to generate random variables using them. Many common and important business variables appear to behave like random variables generated in this manner.

[179] A *Markov process* can have many values; a *Markov chain* can only take specific integer values. In recent years, the study of *Hidden Markov Models* has grown rapidly, particularly in Bayesian statistics. MATLAB contains a number of functions to estimate hidden Markov models. Other MATLAB utilities for use with Markov processes are described in M. Miranda and P. Fackler, *Applied Computational Economics and Finance* (Cambridge, MA: MIT Press, 2002); and Lars Ljundqvist and Thomas J. Sargent, *Recursive Macroeconomic Theory* (Cambridge, MA: MIT Press), 2000.

discussions, we will use a state variable, or state vector of variables, to capture more information.[180]

Simple Uncertainty: Deviation Around a Trend

We already demonstrated a simply type of uncertainty: random deviation around a trend. Certain economic and business variables appear to be generated by such a process. A very important example is the real output of the economy, often represented by real gross domestic product, or "real GDP." Extensive economic research, going back over a century, analyzes the trends in economic growth, and in particular, the causes and characteristics of the "business cycle."[181]

GDP illustrates a characteristic of a large class of business variables: they represent the sum of many, many other independent variables. Thus, this approach is often useful for setting the backdrop of long-term trends in a major industry or economy.

We have described a numerical method for generating random deviation around a determined path in "Projecting Trend Growth: Simple Uncertainty" on page 208.

Problems with Simple Uncertainty

Despite its common appearance in both the academic and professional literature, simple random deviation around a trend growth rate is normally *not* a good model for the behavior of individual businesses or even individual sectors of an industry or economy. An illustration of why this is the case follows.

Many times, managers will forecast variables using a shorthand that appears to fit this approach. For example, consider the statement "We expect revenue to be up 6% next year, ..." followed by further statement such as "plus or minus 2%." This seems to be a reasonable approach for an industry that, on average, has grown 6% a year for the past decade. However, the manager's statement is not the same as assuming that the trend growth rate is 6%, with random variation around that trend. Re-reading the statement, note that the baseline is not a trend; it is the actual value last period. That value probably was not 6% higher than that of the previous year, and almost certainly not $(1.06)^2-1$ higher than that of two years ago. Thus, constantly following the manager's approach to forecasting ("up 6% next year") is not equivalent to expecting a 6% trend over the next several years.[182]

---

180 For an excellent discussion of the use of state variables, co-authored by one of the pioneers in the use of state variables in economics, see "Brief history of notion of the state," in L. Ljungqvist and T.J. Sargent, *Recursive Macroeconomic Theory* (Cambridge, MA: MIT Press), 2000.

181 Signal examples of this include Milton Friedman and Anna Schwarz's *Monetary History of the United States*, 1867–1960 (Princeton, NJ: Princeton University Press, 1963); John Maynard Keynes, *The General Theory of Employment, Interest and Money* (1936); and the more recent "real business cycle" literature, including the rational expectations models.

182 A well-known example is the targeting of certain economic variables by governments. The

Another Simple Uncertainty: Random Walk

A well-known metaphor for stock prices (and other natural and manmade phenomena) is known as a "random walk." This process generates numbers that are either up a certain amount or down a certain amount each period. The baseline for the next period is the last period's value. A random walk is a very simple Markov process and one that appears often in applied work.

Betting on coin flips is a good example. Say, you capitalize a gambling business with $10. If you bet $1 on a series of fair coin flips, the net assets of your business follow a simple random walk. For each win, you gain $1; for each loss, you lose $1. On average, over a large number of coin flips, you break even. However, your unbiased next-period-ahead forecast of the value of your business is not $10, except on the first coin flip (and any other times when your cumulative winnings are zero.) Your next-period-ahead forecast (expressed as a confidence interval) is the same amount as you have now, plus or minus $1.

Such a process is known as *Brownian motion*, or a *Weiner process*.[183] Increments in the process are independent and follow a standard normal distribution. Therefore, the sum of those increments also has a normal distribution, with an expectation of zero and a variance of $n(\Delta t)$, which is the product of the time increment size (Δt) and the number of increments n.[184] Note that this is an explosive variance, which grows linearly with the time period. Over indefinitely long periods the process cannot be expected to remain within a defined range.

---

182 U.S. Federal Reserve Board has, at various times, established growth targets for M1 and other measures of the money supply. These aggregates usually trended outside their target intervals after some time. The Fed would then retarget them, starting at a new base point; this new base point would rarely be near the original baseline.

For a discussion of targeting prices vs. monetary aggregates — and an example of aggregates trending outside their target intervals — see William Gavin, The FOMC in 1996: A step closer to inflation targeting, *Review*, Federal Reserve Bank of St. Louis; found at: http://research.stlouisfed.Org/publications/review/96/09/9609wg.pdf. A multicountry analysis is in George A. Khan and Klara Parrish, Conducting monetary policy with inflation targets, *Economic Review*, 1998-3 , Federal Reserve Bank of Kansas City; found at: http://www.kc.frb.org/publicat/econrev/PDF/3q98khan.pdf.

183 The term comes from the botanist Robert Brown, who in 1827 first described random, successive motion of small particles suspended in liquid. Later, Albert Einstein proposed a mathematical theory of Brownian motion, which was further developed by Norbert Weiner in 1923. The phrase "random walk" became popularized when the process was later used to describe stock prices. Ironically, this brought the idea full circle, as Louis Bachelier had first suggested such behavior in a stock market in 1900.

This historical note and much of our presentation of this topic and generalized Ito processes generally follow Avinash K. Dixit and Robert S. Pindyck, *Investment Under Uncertainty* (Princeton, NJ: Princeton University Press, 1994), Chapter 3.

184 More formally, a Weiner process z is defined as sum of a series of independent random variables distributed identically:

$z_{t+1} = z_t + e_{t+1}$;
z_0 = a constant (such as zero); and
$e_t \sim N(0, 1)$.

Difference Between Trend and Random Walk Forecast

Note the important difference between forecasting a trend and forecasting a random walk. The trend, if your assumption are correct, will probably stay within a defined band. A randon walk, however, will probably "walk" outside of a narrow band — even if your assumptions are correct.

Complex Uncertainty: Geometric Brownian Motion

The random walk behavior of the coin flip may be combined with other systematic behavior to present complex forms of uncertainty. Returning to the previous example, the manager forecast revenue using a "random walk with drift" model, meaning that he or she took last period's level and added a 6% "drift." Even if the drift was consistent over time, the path of a variable generated by a random walk with drift will not be the same (except by coincidence) with one generated by a deviation-around-a-trend process.

The forecasting behavior of the manager over a number of years probably distilled the information about the past into a small information set — the value of the current year's sales, the trend growth rate, and the historic variation around that trend. Using such information, the manager simply took this year's sales, added 6% for the trend, and assumed a variance about that trend. This is a Markov process and a very useful one for modeling business variables. We define it more formally as *geometric Brownian motion*, a Markov process for the sequence $\{X_t\}$ in which:

1. The change in the value of the variable at each period consists of two components: a *drift* component that consistently moves in one direction, and a *diffusion* component that will randomly move up or down.
2. The diffusion process is a random walk or "Brownian motion" process multiplied by the constant σ and the variable X.
3. The drift process is a simple geometric growth, meaning a compound growth like that described under "Projecting Trend Growth: Simple Uncertainty" on page 208.

---

[184] Increments in a Weiner process (dz) have the following properties, where $E_t(.)$ denotes the mathematical expectation operator conditional on information available at time t.

$E_t(dz) = 0$;

$E_t(dz_t dz_x) = 0$ (increments are uncorrelated);

$Var(dz) = E(dz^2) = 1*dt$ (variance for each increment is 1, which means its standard deviation is also one.)

This is, strictly speaking, an abuse of notation since dz is not a random variable, and hence has no expectation. However, this is an intuitive way of understanding the process. See, for example, D. Duffie, *Dynamic Asset Pricing Theory*, 3rd ed. (Princeton, NJ: Princeton University Press, 2001) Section 5c.

We can describe geometric Brownian motion for the series X with the following equation, where dX is the differential in X, dt is the differential of the time index, and dz is an increment in a standard Weiner process:

$$dX = \alpha X dt + \sigma X dz \tag{4}$$

Here, the α and σ parameters are the drift and diffusion parameters. Note that, over time, the drift parameter will systematically move the series in one direction at a growth rate that is compounded. The diffusion parameter, by comparison, causes the extent of the likely diffusion of the series. Because the random walk is truly random, we do not know whether it will "walk" up or down.

Ito Processes

The complex uncertainty above can be generalized to the family of Ito Processes, which take the form of the stochastic difference equation:

$$dx = a(x, t)dt + b(x, t)dz \tag{5}$$

Here, rather than the linear functions in the geometric Brownian motion process above, the drift and diffusion functions can take any form and involve both x and t. Evaluating the dynamics of Ito Processes involves the use of the celebrated *Ito's Lemma*, which uses stochastic calculus to differentiate Equation (5).[185] We will simply note here that the forms of complex uncertainty we described belong to the larger family of Ito Processes.

A Projection Utility for Complex Uncertainty

We presented above a projection utility for simple uncertainty. For analysis of the more complex uncertainty that prevails in many business and economic situations, we present a similar utility for a subset of Ito Processes.

---

185 Ito's Lemma can be stated as: for a function $F(x,t)$ that is twice differentiable in x and once in t, where x follows an Ito process of the type in Equation (5), the differential of F is:

$$dF = \frac{\partial F}{\partial t} dt + \frac{\partial}{\partial x} F dx + \frac{1}{2}\left(\frac{\partial^2 F}{\partial x^2} \cdot (dx)^2\right)$$

A careful review will reveal that the differentiation of the equation using a Taylor expansion under standard calculus would leave one less term. However, for this stochastic process, dx behaves like $\sqrt{dt}$, and dx^2 like dt, so the extra term captures movement of the same magnitude as dt and cannot be ignored. For a more careful derivation, see one of the references cited in this section, including A. Dixit and R. Pindyck, *Investment Under Uncertainty* (chapter 3), and M. Miranda and P. Fackler, *Applied Computational Economics and Finance* (Appendix A).

The ItoProject.m function, which is part of the Business Economics Toolbox that accompanies this book, will project variables using three types of processes: simple Brownian motion (a random walk), Brownian motion with drift, and geometric Brownian motion. You supply the parameters, including the starting point, the drift and diffusion parameters (if applicable), and the length of the time period. The utility program generates the path and will also generate a graph showing the results.

Code Fragment 10-3, "Ito Process Projection," on page 237 in the Appendix to this chapter contains a fragment of the code for this utility.

Complex Uncertainty: Jump Processes

For many business and economic variables, even the complex uncertainty modeled above will not be sufficient. In particular, there is a certain class of variables subject to risks that are not well modeled by random deviation around a trend, random walks, or by geometric Brownian motion. In particular, some variables are subject to sudden, unexpected changes that may dramatically change the prospects of a business.

The risks borne by insurance companies are an obvious example, but nonfinancial companies also bear such risks. For example, companies that manufacture or distribute brand name products or products under a franchise agreement bear the risk that the brand itself will decay, the products will no longer be produced, or the franchisee will lose its franchise rights. Investors bear a default risk that may be small but is not negligible for many classes of bonds. Almost all parties to contracts bear a risk of nonperformance by other parties.[186]

We consider in this section events that have a small probability of occurring in any one year. The normal distribution, or at least a nice symmetrical distribution of events, cannot be used.[187]

Instead, we suggest the *Poisson* distribution as an appropriate model for such risks. This statistical distribution is close to that of a binomial distribution in which the number of trials is very high and the probability of success in each trial is low. The Poisson is typically used in studies of errors, breakdowns, queuing behavior, and other phenomena where the chance of any one subject facing a specific event is small, but where the number of subjects is large.

A Poisson process is governed by the following probability density function for discrete variables:

---

[186] See Patrick L. Anderson, "Valuation and Damages for Franchised Businesses," AEG Working Paper 2003-12, presented at the Allied Social Science Association convention, January 2004; and available at: http://www.andersoneconomicgroup.com.

[187] This has implications in many fields, including finance. In particular, if there are not a wide variety of securities that create a risk-reward frontier, or that could, under additional assumptions, be expected to have normally distributed returns or risk characteristics, many of the nice, standard conclusions of modern portfolio theory are undermined.

$$P(x) = \frac{e^{-\lambda}}{x!}\lambda^x \qquad (6)$$

$$x = 0, 1, 2, \ldots,$$

Note that the Poisson is a *discrete* probability distribution; it provides positive probabilities only for nonnegative integers.[188]

Uses for Jump Processes

Such a process is useful in modeling business variables that change infrequently, but when changed, cause either catastrophic losses or large gains. We will describe later a present value rule for assets whose returns are subject to Poisson risk.

A Utility for Modeling Poisson Uncertainty

We describe a utility for projecting jump processes, using a Poisson distribution, in the Appendix. The JumpProject.m utility function is part of the Business Economics Toolbox accompanying this book and projects variables subject to Poisson event risk, using a mean arrival rate parameter supplied by the user. Like the other utility functions, it generates data and can produce a graphical display. See Code Fragment 10-4, "Projecting Jump Processes," on page 239 in the Appendix to this chapter.

The Value of a Financial Asset

The source of value for a commodity is a matter of deep importance to economists. However, there is a surprisingly weak bridge between the theoretical texts on microeconomics and the applications common in finance. In this section, we review with some rigor the concept of the value of a financial asset. From this we will introduce applications to model business operations and estimate their value.

188 The PDF shown in Equation 6 can produce non-integer numbers, but these are probabilities of certain integer values, not the values themselves. Random Poisson numbers always produce integers. For example, a Monte Carlo run of random Poisson numbers with *lambda* = *.25* produced 100 numbers: mostly zeroes, some ones, and one two. The sum of all 100 numbers was 21; 21/100 is close to the mean arrival rate *lambda* = *.25*. For Monte Carlo tests of Poisson processes, see the Appendix to the working paper cited above.

Earnings or Discount Rate?

The value of a financial asset is largely determined by two factors: expected future earnings and the discount rate investors apply to those expected earnings. (We ignore, for the moment, the role of risk.) Most references in finance and valuation leave future earnings growth to a short discussion, generally implying that recent trends are good bases on which to forecast future growth.[189] By contrast, these same references will often devote multiple chapters to the determination of the correct discount rate to use for these future earnings.[190]

Both factors are of vital importance. The growth of earnings is at least of equal importance in determining the value of a firm, and in practical terms is more important. To illustrate this, and to prepare for later treatment of valuation, we consider next a rigorous treatment of present value.

Deriving the Present Value of a Financial Asset

The price of a financial asset in a well-traded market, which provides a known dividend *d* at time *t* and can be resold at a known (ex-dividend) price p_{t+1} in the next time period, must be:

$$p_t = (d_{t+1} + p_{t+1}) / (1 + r) \tag{7}$$

where *r* is the one-period discount rate. We say "must" because, if it was not, arbitrage possibilities would exist that would be exploited by other investors.

Note that we used the term *price*. We are confident the asset can be sold for this amount to a willing buyer. This price is enforced by arbitrage possibilities in real markets and not by theory or textbooks. Some of our confidence is born out of the assumptions of no transaction costs, default risks, or imperfect information. However, these assumptions will be relaxed soon and for some financial assets (such as U.S. Treasury Bills) have small effect.

In Equation (7), we relate the price at a certain time period to a ratio in which the numerator is a future set of priced assets and the denominator is a discount rate. This ratio is the *present value* of the future stream of returns

---

[189] Obviously, recent prices should always be given great weight in forecasting future prices. In most finance applications, however, recent prices characterize the capitalization of the firm better than its performance over the past year. As will be discussed in Chapter 11, "Business Valuation and Damages Estimation," our guidance on information necessary to properly value a firm and the IRS guidance on valuing firms place greater weight on the future earnings of the firm than on its cost of capital. See Chapter 11, under "Checklist for Information" on page 280.

[190] The better references caution against forecasting long-term growth at rates above that of the general economy and even provide variations on the constant growth assumption in which growth is expected to return to a standard growth rate in two or three stages. However, the discussion of capital pricing models (e.g., CAPM [Capital Asset Pricing Model] and Arbitrage Pricing Theory) typically occupies much more space than does a systematic treatment of earnings growth.

from that asset. Returns include both earnings (dividends) and any return of capital (capital gains or losses). From this relation, we deduce that financial assets (under our restrictive assumptions) will be priced at the present value of their future returns.

Now extend the number of periods forward, assuming that the interest rate stays constant. The price becomes:

$$p_t = \sum_{i=1}^{\infty} d_{t+i} \Bigg/ \left(\prod_{j=0}^{i-1} (1 + r_{t+j}) \right) \tag{8}$$

which simplifies to:

$$p_t = \sum_{i=1}^{\infty} (1 + r)^{-i} d_{t+i} \tag{9}$$

where the negative exponent indicates the reciprocal.[190a]

Dividends and discount determine price: Note that the entire derivation of prices of financial assets reduces to dividends and discount rates. Terms like cash flow, EBITDA, CAPM, beta, and depreciation have not been mentioned. Firms pay dividends out of earnings, and EBITDA is one definition of earnings.[191] Getting caught up in these terms at the expense of understanding thoroughly the source of earnings is missing the forest for the trees. In the market, every day, it is expected future dividends, and the discount rate we apply to those expectations, that determine price.

Prices under strong assumptions: If the financial asset is an equity security in a firm (such as shares of common stock) and we assume that we can project the dividends from future earnings, what is the present value of the asset?

Consider the question using the following (dramatically) simplifying assumptions:

- The firm's sales will grow at a constant rate.
- Its operating margin on those sales will remain the same.
- Taxes, working capital needs, capital expenditures, and resulting depreciation will all occupy the same ratios to sales as they do now.
- The firm's payout ratio (the portion of earnings it pays in dividends) will remain the same.

This is clearly a special case, especially when these assumptions are added to our previous ones. However, the results are very instructive.

[190a] Note that the discount rates are inside the summation. See also Darrell Duffie, *Dynamic Asset Pricing Theory*, 3rd ed. (Princeton, NJ: Princeton University Press, 2001), section 2.6 for a rigorous extension.

[191] EBITDA means Earnings Before Interest, (income) Taxes, Depreciation, and Amortization. Cash flow is also an indicator of income.

Special case with simplifying assumptions: In this special case, the net present value of the firm's future earnings in Equation (9) simplifies to the following:

$$p_t = \frac{(1+g)d_t}{r-g} \tag{10}$$

where g is the constant growth rate of dividends.[192]

This ratio is the basis of a common shorthand method of valuing firms, the Gordon Growth model in valuation literature.[193] That this method is based on assumptions that are almost always violated seems not to inhibit its use. One reason for this ubiquity is that the formula distills the most important elements of value into a handy relation which then becomes both a management guide and an appraisal rule: increase earnings growth, and the value of the firm goes up; increase the discount rate, and the value goes down.

Variations on the formula include a mid-year dividend model and the similarly ubiquitous price–earnings ratio as a guide to valuation.[194] For a continuous process (rather than discrete periods), calculus provides the formula for the present value of a constant, perpetual cash flow as:

$$PV(D_t) = \int_0^{\infty} (De^{-rt})\,dt = \frac{D}{r} \tag{11}$$

where D_t is cash flow at time t.[195]

---

192 The mathematics of this derivation follows that of S. LeRoy, Present Value, in *The New Palgrave: Finance* (New York: Stockton, 1987, 1989). This volume excerpts from *The New Palgrave: A Dictionary of Economics* (New York: Stockton, 1987, 1989). I have corrected formula (4) appearing in that text, which seems to have suffered a typographical error.

193 In the valuation literature, the formula is usually expressed as *d/(r–g)*, which differs from the derivation in the text above by a factor of 1+g. See, e.g., J. Abrams, *Quantitative Business Valuation* (New York: McGraw-Hill, 2001), Chapter 3.

The difference (which in practical terms is normally quite small, given that the perpetual growth rate parameter g should not exceed the growth rate of the economy as a whole) originates in the question whether to assume end-of-period dividends and whether the price includes the dividend for the current period or is ex-dividend. In this derivation, the price at time t does not include the dividend for time t. We also assume that the price and dividend are taken at the end of the period. Thus, the standard valuation literature formula will have, as the numerator, the next period's earnings.

The Gordon model is named after Myron J. Gordon, now Professor Emeritus of Finance at the University of Toronto. A recent paper of his cites as the reference his 1962 book *The Investment, Financing, and Valuation of the Corporation* (Homewood, IL: Irwin, 1962). See Joseph R. Gordon and Myron J. Gordon (1997), "The finite horizon expected return model," *Finan. Anal. J.*, 53, 3, 52–61, May/June 1997; also found at: http://www.chass.utoronto.ca/~gordon. Abrams, cited above, also mentions M.J. Gordon and E. Shapiro, "Capital equipment analysis," *Manage. Sci.*, 3, 102–110, 1956, and J.Q. Williams, *The Theory of Investment Value* (Cambridge, MA: Harvard University Press, 1938).

194 Abrams shows that the mid-year Gordon equation differs from the end-year equation by a factor of $\sqrt{1+r}$. If r is 5%, then this factor is about 2.5%.

Abrams also derives the relationship between the price/earnings ratio (using the *past* period's earnings) and the Gordon model multiple, which adjusts for the dividend retention ratio and the growth rate in earnings. See *Quantitative Business Valuation*, pages 78,79, 87-89.

195 A clear explanation is found in Alpha Chiang, *Fundamental Methods of Mathematical Economics*, 2nd. ed. (New York: McGraw-Hill, 1974), Section 13.5.

Relaxing assumptions: What happens if we relax the restrictive assumptions about transaction costs, default risk, pricing risk, interest rates, liquidity, and information? We then have a market which operates like most equity markets today and which contains a bid-ask spread, numerous investments of widely varying types, and a range of risk and reward opportunities.

In almost all cases, when using current earnings or dividends as the starting point, the Gordon Growth model is an invalid method of estimating the current price of a business enterprise.[195a] However, it neatly illustrates this principle: the value of a financial asset is based on the *present value of future dividends.*

Importance of Income Growth vs. Discount Rate

We can now return to the question posed earlier about the value of a firm: what is more important — the discount rate or the earnings growth?

Equation (10) provides the answer in a simplified setting. Using plausible values for discount and growth rates, we see that the earnings growth rate is at least as important as the discount rate.[196] Indeed, without the prospect of future earnings, the discount rate simply does not matter, and thereby a firm is rendered worthless.

For this reason, we take a balanced view of the economics of the firm. Projecting income growth should occupy at least as much importance as deriving a discount rate, and often much more.

Present Value of a Future Stream With Poisson Risk

Some business variables have a small risk of changing in each period, but if changed can cause a termination of business income, e.g., the default of a bond, the termination of a brand or franchise rights, or the destruction of crops or other products.

Introducing this type of risk into a stream of future cash flows obviously changes its net present value. Returning to the Gordon Growth model in Equation (10), we note that a perpetual stream generated by a constant growth rate and discounted by a constant discount factor can be neatly summarized in a present value formula. What if that cash flow stream is subject to termination at an unknown future time determined by a random process?

[195a] Among the assumptions often violated in its common use: perpetual constantly-growing earning stream, no discount or premium on NPV of cash flows, improper use of current earnings as basis, exaggerated growth rate, and wrong discount rate.

[196] Constraining these variables to a plausible range is vital because otherwise the denominator could approach zero. Indeed, one mathematical explanation for speculative bubbles is that, for a brief time period, investors perceive that the growth rate approaches the discount rate. Given the assumptions here of perpetual growth rates, this is absurd— which is why speculative bubbles eventually burst.

The first answer to this question is: we will deal with a mathematical expectation, not a simple sum. Fortunately, there is a formula incorporating risks modeled by a Poisson process, which can be used as a basis for some firm valuations. Although we will not derive it here, the *expected* net present value of a stream of income, with profit in year one π_1, discount rate r, and the periodic growth rate g, of the series of periods $t = 1, \ldots, \infty$, where π_1 is governed by a Poisson process, is:[197]

$$E(pv) = \frac{\pi_1}{r - g + \lambda} \tag{12}$$

where λ = mean arrival rate of Poisson Process.

The "mean arrival rate" of a Poisson process is the *lambda* parameter for the Process itself (which is the mean of the probability distribution), divided by the number of time intervals. For a Poisson process where *lambda* = 5, and 100 periods, the mean arrival rate would be 5%. A franchise termination may be modeled as a Poisson process with *lambda* = 1, meaning that only one event will occur, and we just do not know when it will occur. In such instances, the mean arrival rate can never be more than 1. Other business events can also be modeled as a Poisson process with different parameters.

Income Under Uncertainty

Uncertainty is a fact of life and certainly an essential factor in business planning and management. Because uncertainty is part of the context of the operation of a firm, investors take into account uncertainty when they consider investing in a firm. Thus, *both* the capitalization of a firm and its operation are affected by risk.[198]

---

197 This discussion is based on Patrick Anderson, "Valuation and Damages for Franchised Businesses," AEG Working Paper 2003-12, presented at the Allied Social Science Association convention, January 2004; and available at: http://www.andersoneconomicgroup.com.

The basic derivation of NPV formula for a series governed by Poisson risk is in A. Dixit and R. Pindyck, *Investment Under Uncertainty* (Princeton, NJ: Princeton University Press, 1994). A more rigorous mathematical discussion is in Darrell Duffie, *Dynamic Asset Pricing Theory, 3rd ed.* (Princeton, NJ: Princeton University Press, 2001).

Most of the literature on this topic involves default risk on bonds. A recent example is Joost Dreissen, "Is Default Event Risk Priced in Corporate Bonds?" University of Amsterdam, working paper March 2002; found at: http://www.defaultrisk.com, and http://www1.fee.uva.nl/fm/PAPERS/Driessen/corpbond_jdriessen.pdf.

198 Note that the role of risk in the value of a firm does not originate solely in the risk–reward preferences of investors in a marketplace. Even a sole proprietor running an all-cash business is concerned about risk.

This observation is hardly novel. All modern finance texts deal with at least simple models of risk when they describe the capitalization of the firm and investors' behavior. Unfortunately, the extensive discussion of risk-and-return in finance texts is usually not matched by a similar appreciation of risk in the operation of the firm.[199]

A Recommended Approach

Given that both the capitalization of a firm and its operations are affected by uncertainty, we recommend the following approach to modeling business variables over time:

1. The fundamental variables that underline an industry in which a firm operates (such as overall demand, prices, and perhaps regulatory restrictions) should be modeled with at least simple uncertainty. For example, the growth prospects of a business should be considered with at least a low, medium, and high growth trend for its industry.
2. In addition to the industry trends, the market share or other business-specific variables should be modeled with at least simple uncertainty.
3. The capitalization of a firm should be based on the risk to the business. Note that the "risks to the business" cannot be established without considering the variability in income and market share, which would be explicitly considered under the previous two recommendations.[200]

The importance of modeling the fundamental business variables with uncertainty motivates the introduction of a business simulation model in the section "A Business Income Model," which follows.

---

199 In particular, it is quite common in the valuation and finance literature to delve deeply into the role of risk in estimating the proper cost of capital for a firm, reflecting the great advances in modern portfolio theory. Many of these same texts give short treatment to earnings or revenue growth, often consigning it to a simple assumption.

200 Indeed the standard CAPM "beta" model so often used to estimate discount rates relies on a statistical relationship between stock prices (presumably based on earnings) for companies in one sector and companies in the market as a whole. Evaluating the business-specific risks will usually provide a much better estimate of the embedded risk to the equity investors than looking up comparable companies in a "beta book" or a similar method.

A Business Income Model

For many applications in business and finance, we will want to model the workings of a firm. In this section, we present a simplified business income model, which allows for powerful analysis of individual firms.

In particular, the model:

- Explicitly examines revenue and the variable and fixed costs to support business operations
- Separates operating expenses (and taxes) from financing expenses, which include the cost of debt (returns to lenders), dividends, and other distributions (returns to equity investors)
- Segregates various definitions of earnings into those consistent with most finance, valuation, and accounting references
- Allows for the fundamental drivers of the firm's business to be forecasted with uncertainty of various types

Earnings Drive a Firm

Such a model is, unfortunately, somewhat unusual. There are many models of individual firms in the accounting and finance literature. They are often much more detailed than this simplified model. However, they tend to focus on the accounting and financing detail, leaving the generation of income as a much less examined assumption.

As we presented in rigorous fashion above, there are two fundamental drivers of business value: earnings and the discount rate. Of these two, the more powerful is earnings. While this particular model includes parameters and equations for the capitalization of the firm, its focus is on modeling how the company makes money.

A Business Income Simulation Model

This model is presented in Simulink, the simulation modeling environment that is an extension of MATLAB. It is organized as follows:

- In the top-level view, we see a subsystem for earnings of the firm. This generates the variable for gross earnings. We use EBITDA as our measure of gross earnings. See Figure 10-2, "Income statement" and Figure 10-3, "Earnings subsystem."
- From this variable, the depreciation subsystem calculates depreciation and amortization. The difference between EBITDA and

depreciation and amortization is pretax operating earnings, before interest payments. See Figure 10-4, "Depreciation and amortization."

- From pretax operating earnings we calculate state and federal corporate income taxes on the firm. The result is net earnings, before interest. After subtracting the interest costs (calculated in the capital structure subsystem), we have net, after-tax earnings. See Figure 10-4, "Depreciation and amortization."
- In this simplified version, dividends are paid out of capital; a full simulation model would cause earnings in one period available for dividends in the next.

Running the Business Income Model

This model simulates how a company operates, using a greatly simplified approach. It is a template to use for many different analyses.

In order to run the model, you must:

- Customize, for the type of business under study. This will probably involve simplifying some of the blocks and subsystems while expanding others.
- Provide values for all the defined constant blocks in the model. These establish the base levels of key variables as well as growth rates. This could be done in an initialization file.
- Simulate the results over multiple years.
- Examine the results, correct any errors, and fine-tune the model to work properly and present the results properly.
- Run the final projections and save the results along with the important assumptions you made.

Application: Real Estate

An example of this approach, with an additional innovation, is shown in the following section, "Iterative Market Value and Tax Model."

Iterative Market Value and Tax Model

As will be discussed in Chapter 11, "Business Valuation and Damages Estimation," the market value of debt and equity are determined in the same

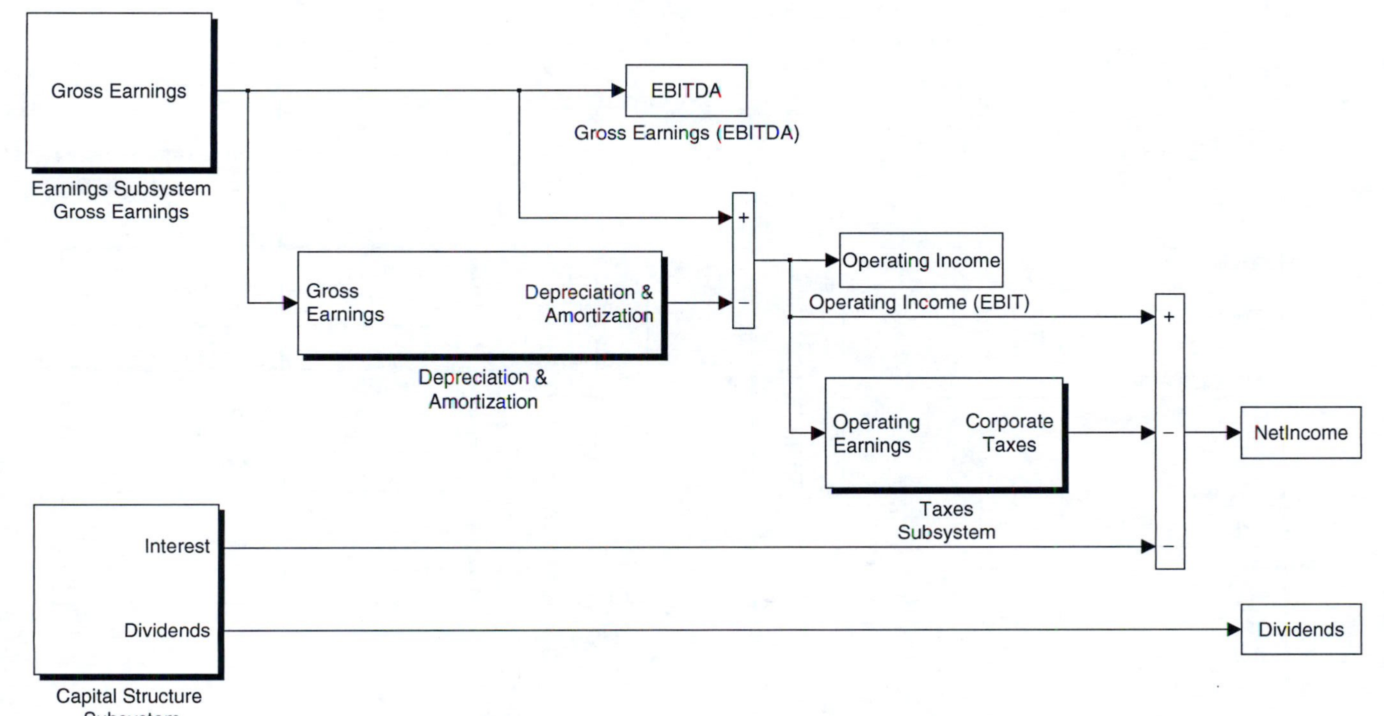

FIGURE 10-2
Income statement.

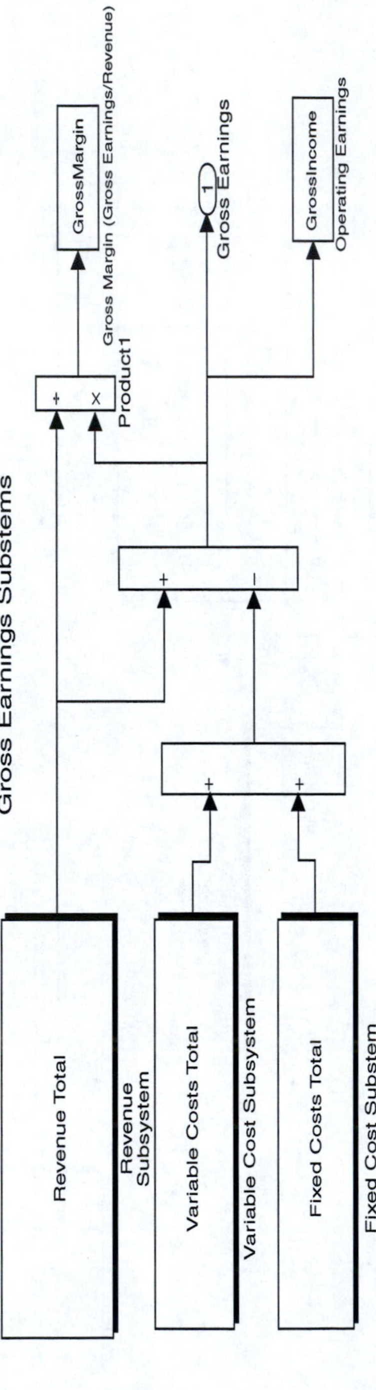

FIGURE 10-3
Earnings subsystem.

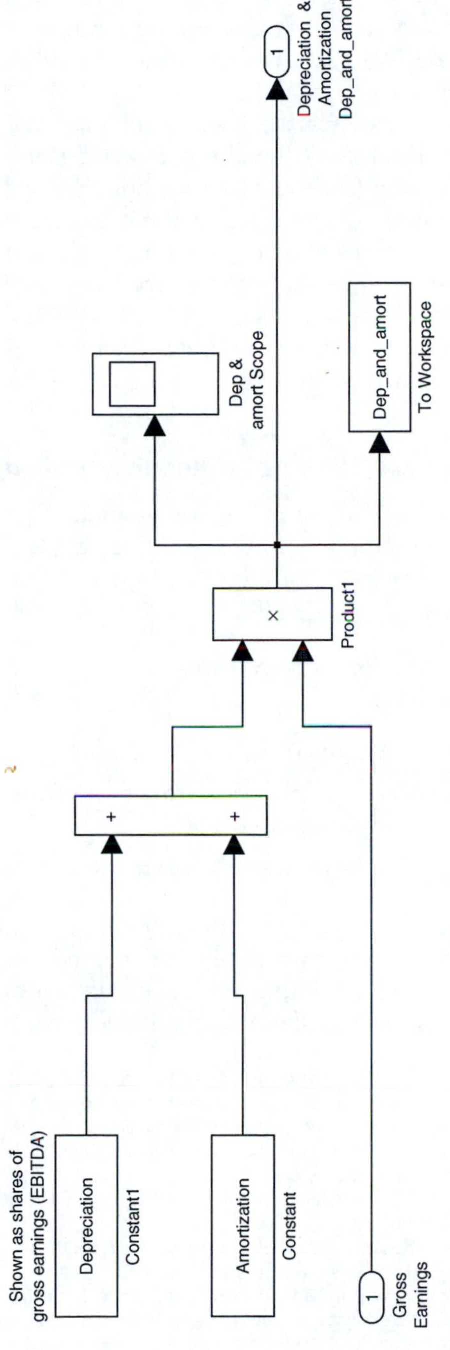

FIGURE 10-4
Depreciation and amortization.

market as discount rates.[201] A model determing all three would be a system of simultaneous equations.[202] Iterative models can be used to solve many problems of this type. We present such a model in this section, which can be adapted to many uses.

A similar situation exists for the taxation of most commercial properties. Normally, the best method to use in valuing an established, operating commercial or industrial property (such as an office building or factory) is an *income method*. As described in the chapter on this topic, income methods estimate the current market value of an asset by capitalizing a future stream of earnings.[203]

However, in order to forecast income, we must forecast expenses; one of the major categories of expenses for such entities is property taxes. As property taxes are normally *ad valorem* (levied on the value) taxes, we have a case of simultaneous equations.

Simplifying Assumptions Allowing an Iterative Method

One method for dealing with simultaneous equation problems of this type is to use an iterative method to find a solution, if a solution exists. In this case, we are convinced that a solution does exist because we can make the following assumptions:

- The market value must be positive.
- The tax burden on the property must be positive.
- The tax rate will be positive, but less than one.
- The tax burden is a simple linear function of the value, namely a rate multiplied by an assessed value.
- An initial guess for the tax burden and market value can be provided.

Such assumptions dramatically simplify the mathematics of the problem and allow us to use a straightforward iterative method. There are more sophisticated methods that can be employed, especially within MATLAB, but this approach makes use of the simulation capabilities of MATLAB and Simulink.[204]

---

201 To be precise, we would say the market value and market discount rates are set in the same market. The discount rates of an individual investor and the book value of debt and equity for a specific firm would not be determined in the market as a whole. However, the estimation of fair market value requires us to look at these variables as determined by the market as a whole and not individual investors.

202 See Chapter 11, under "Using Market Value to Weight Capital Structure" and "Iterative Methods to Weight Capital" on page 280.

203 See Chapter 11, under "Second Best Method: Income Method" on page 255; the preceding material in that chapter describes other methods as well.

204 The Optimization Toolbox within MATLAB contains a host of special tools. Simulink contains a number of algorithms to solve systems of simultaneous equations. Two references that describe methods of approaching equations like these, with applications in economics and finance are: M. Miranda and P. Fackler, *Applied Computational Economics and Finance* (Cambridge, MA: MIT Press, 2002) and P. Brandimarte, *Numerical Methods in Finance: A MATLAB-based Introduction* (New York: John Wiley & Sons, 2002).

An Iterative Model to Estimate Market Value and Taxation

We created a simplified business model that incorporates an iterative method of determining both market value of property and the *ad valorem* tax levied on the property.

The top-level view of the model is in Figure 10-5, "Iterative model: overview." The model is organized as follows:

- The operating expenses subsystem contains equations that forecast gross revenue and operating expenses, and output these variables. As one of the key expense categories is property taxes, an initial guess is used for this variable. See Figure 10-6, "Iterative model: operating expenses."
- The tax subsystem takes earnings (in this case, earnings before interest and income taxes) and estimates the market value of the property, given this amount of earnings. Using an *ad valorem* tax rate, it calculates the tax liability on a property of this value. See Figure 10-7, "Iterative model: tax subsystem."
- The difference between gross revenue, operating expenses, and property taxes is calculated by the sum block. As an allowance for future enhancements of the model, we created an input into the sum block for another variable. A Ground block is connected to this input, indicating that nothing is added or subtracted at this time.
- The resulting EBIT_iteration and tax_iterations variables are sent to the workspace at each iteration. However, the EBIT_iteration variable at time *t* becomes the estimate of earnings used in the tax subsystem at time *t*+1. A Unit Delay block causes the signal to wait a tick of the clock before being sent. This, of course, raises the question of the value this variable takes in the first iteration. Although not visible in the diagram, the block has an "initial condition" property which was set at our initial guess.
- An m-file initiates the model, simulates it, and requests a graphic. This script is reprinted in Code Fragment 10-5, "Simulation Callback for Iterative Tax Model" on page 240.

The graphical output, showing that value converges after just a few iterations, is shown in Figure 10-7, "Iterative model: tax subsystem." The line shows how the initial guess about the market value of the property is far too high, given the actual earnings from the property. After several iterations, the market value and *ad valorem* tax on that value converge to stable values. See Figure 10-8, "Iterative model: convergence of tax and value variables."

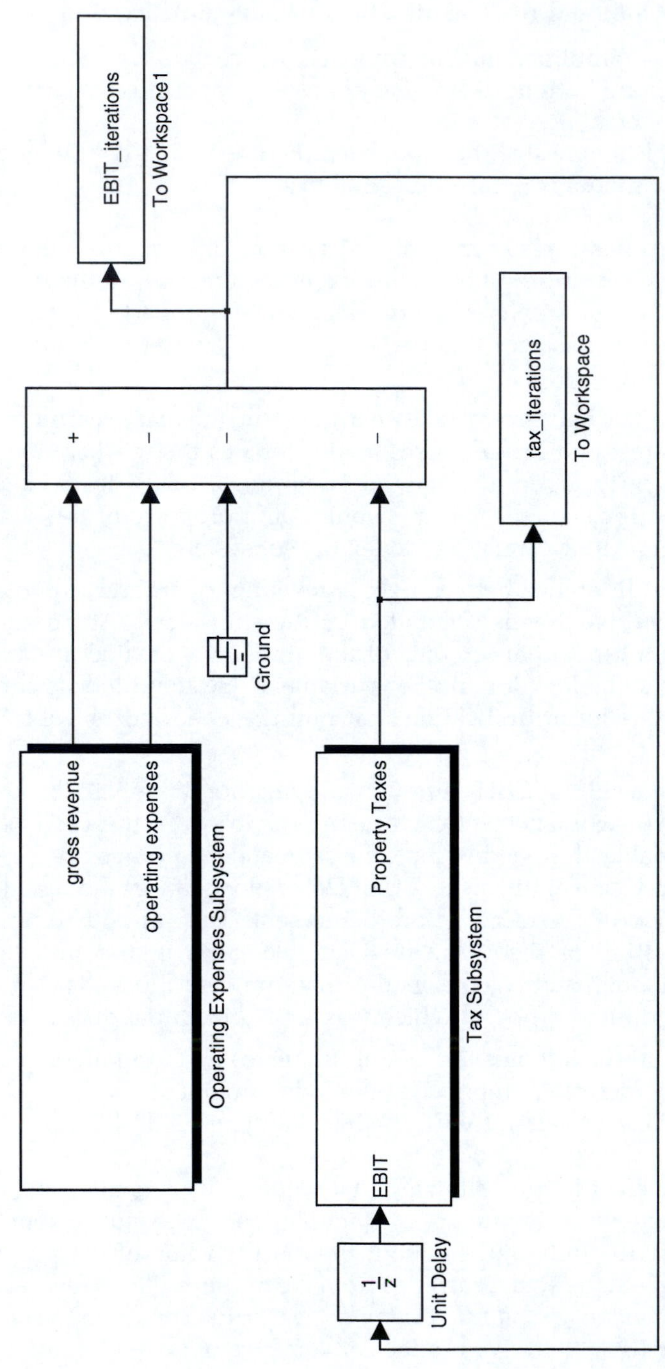

FIGURE 10-5
Iterative model: overview.

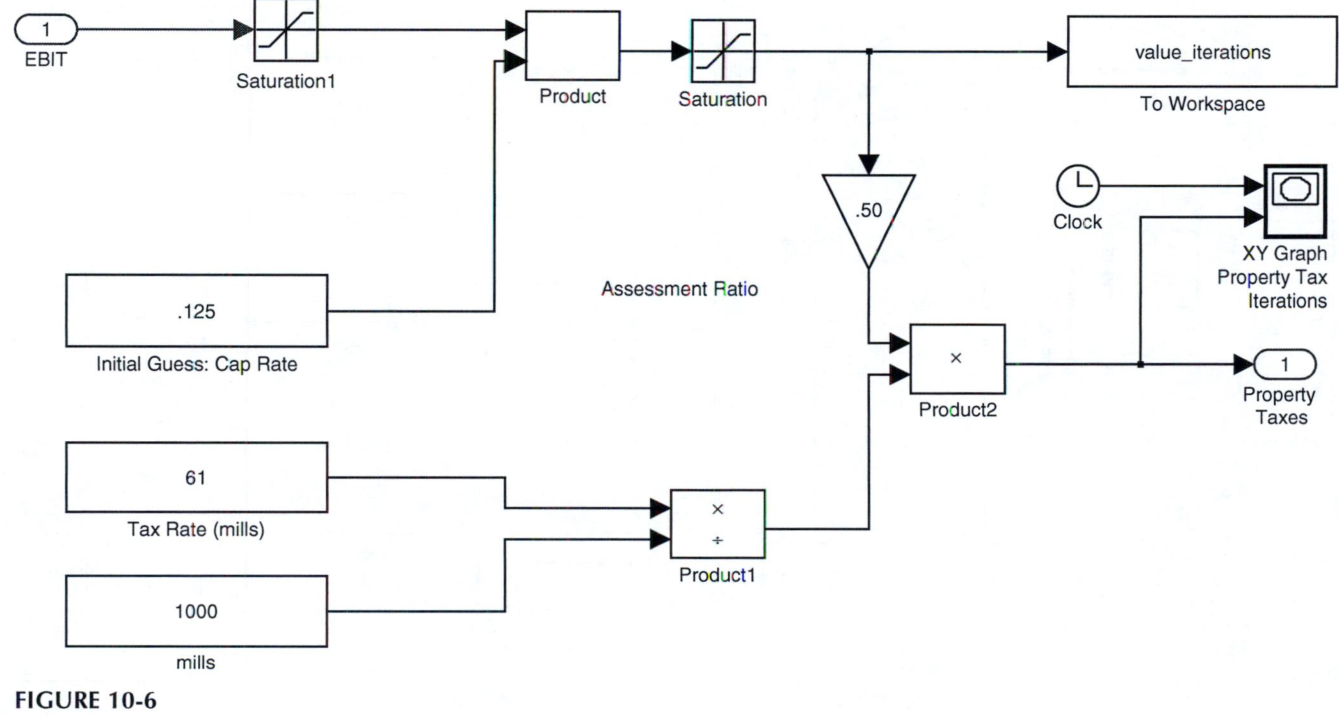

FIGURE 10-6
Iterative model: operating expenses.

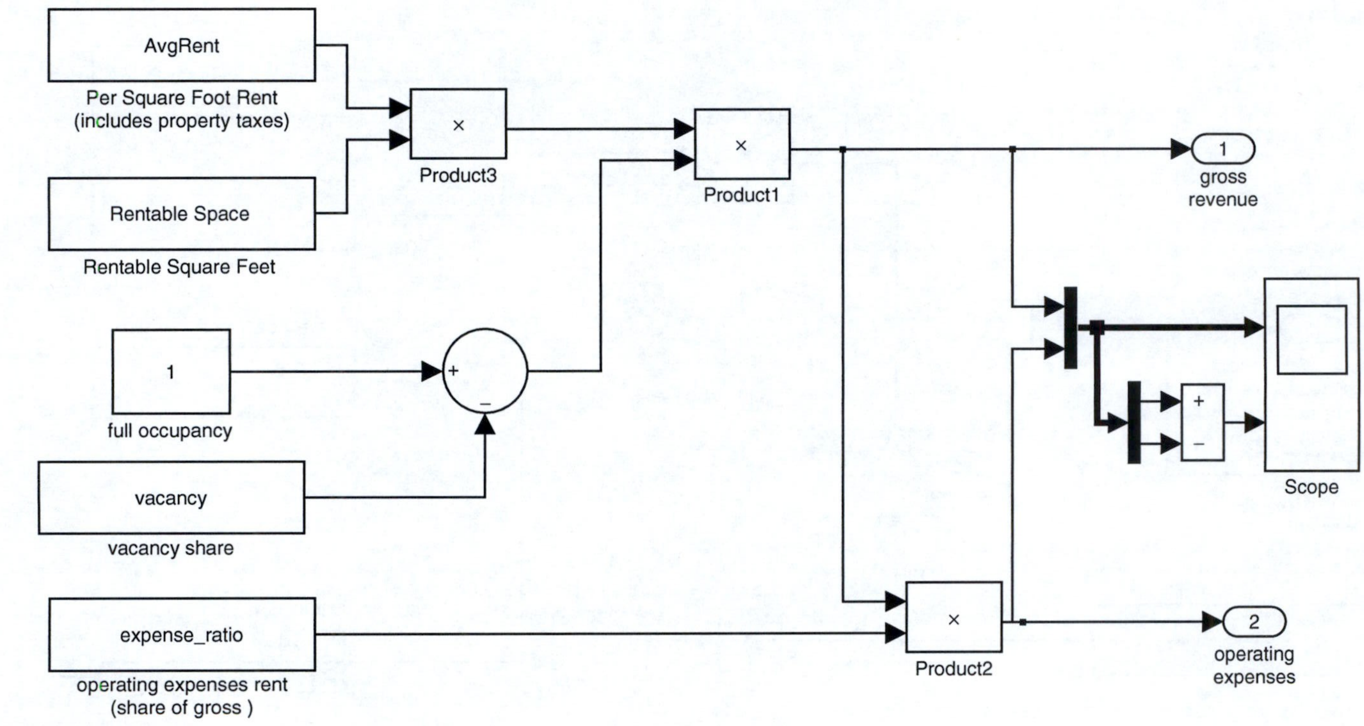

FIGURE 10-7
Iterative model: tax subsystem.

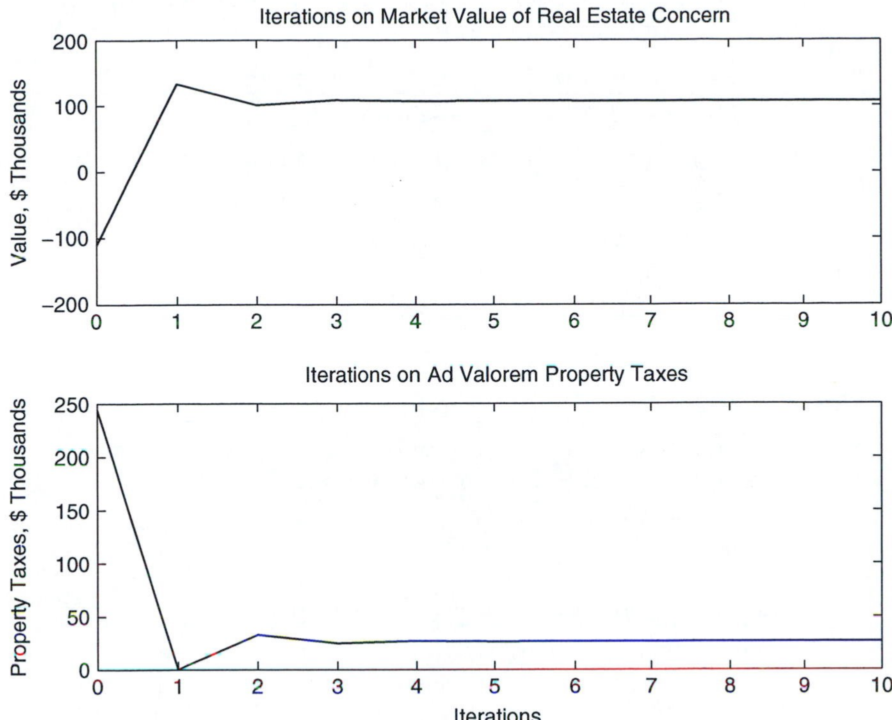

FIGURE 10-8
Iterative model: convergence of tax and value variables.

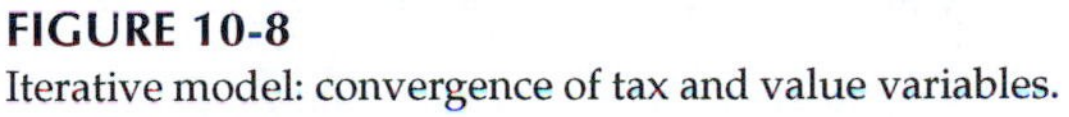

Appendix: Computer Code for Selected Utility Programs

Code Fragment 10-1. Trend Projection Function

```
function [projections, inputs] = TrendProject(base,
rate, deviation, numperiods)

%TREND PROJECTION Function, projects baseline, high,
%and low growth trends

%   Example:    a = TrendProject(100, .10, .05, 25)

%           [a, b] = trend_project
```

```
%   In the first example, 100 is the base revenue,
10% is the base growth rate,
%   and 5% is the deviation from that base
%  (calculated geometrically, as in
%   the high growth rate is 1 + deviation times the
base growth rate.)
%   Output matrix is a 3-columns x numperiods-rows
matrix; Second output is a
%   structure of inputs.
%  In the second example, a dialogue box opens up to
ask for inputs.
%   The function (under this option) causes one
%  figure to be created plotting three revenue
%  projections.
%
%   (c) 2003 Anderson Economic Group,
%   part of Business Economics Toolbox
%   Use by license only.
%-----------------------------------------------------
%   3. Run growth calculations three times (uses
subfunctions)
%-----------------------------------------------------

projections = calc_growth(inputs);

%   Display revenue matrix; columns are projections
and rows are periods
disp(projections);

...

%   ---------------------------------------------------
%   calc_growth
%   ---------------------------------------------------
function projections = calc_growth(inputs)
%   base case, high, low
```

```
for i=1:inputs.numperiods
    rev(i) = inputs.base * (1+inputs.rate)^(i-1);
    rev_high(i) = inputs.base *
(1+inputs.rate*(1+inputs.deviation))^(i-1);
    rev_low(i) = inputs.base * (1+inputs.rate*(1-
inputs.deviation))^(i-1);
end

projections = [rev_high' rev' rev_low'];

%    -------------------------------------------------
```

Code Fragment 10-2. Trend Projection Output

```
>> trend_project
Less than 4 arguments given; starting dialogue box.
--------------------------------------------------
500000.00      500000.00      500000.00
610000.00      600000.00      590000.00
744200.00      720000.00      696200.00
907924.00      864000.00      821516.00
1107667.28     1036800.00      969388.88
...
%output truncated
```

Code Fragment 10-3. Ito Process Projection

```
function data = ItoProject(method, drift, sigma, x0,
T, plotflag)
%
```

```
%   A function to simulate certain Ito processes,
%including Brownian motion,

%     Brownian motion with drift, and Geometric
%Brownian motion.

%   Arguments: method (see below), drift parameter
%(1.0 is no drift),

%     variance (standard deviation sigma), also
%known as "diffusion".

%   Optional: initial starting point x0; length of
%process T; and

%     plotflag, which if zero suppresses the figure
%containing the process

%     plot.

%   Methods: 'simple', 'drift', or 'geometric':

%     Simple Brownian motion (Weiner process) z, with
%increment dz

%         dz = eta_t*sqrt(dt); where eta is
%Normal(0,1).

%     Brownian motion with Drift rate alpha, and
%standard deviation sigma

%         dx = alpha*dt + sigma*dz

%     Geometric Brownian motion, produces lognormal
%returns:

%         dx = alpha*x*dt + sigma*x*dz

%         x has a constant drift rate alpha, and
%constant variance sigma^2.

%

%   Examples:

%         data1 = ItoProject('simple', 0, 1)

%         data2 = ItoProject('drift', 0.2/12, sqrt(1
%/12), 100, 12*50, 1)

%         data3 = ItoProject('geometric', 0.09/12,
sqrt(0.2/12), 100, 30*12, 1)

...
```

Code Fragment 10-4. Projecting Jump Processes

```
function inforisky = JumpProject(growth, discount,
lambda, x0, T, plotflag)

%   A function to simulate "jump" processes, subject
%to Poisson risk.

%

%

%   Examples:

%       data1 = JumpProject(.03, .05, 1/100, 100)

%       data2 = JumpProject(.03, .05, 5/100, 100,
%100, 1)

...


%   (c) 2003 Anderson Economic Group, http://
%www.andersoneconomicgroup.com

%   part of Business Economics Toolbox; for use by
license only.

%

%

%   net present values calculated from:

%   T total number of time intervals

%   t time index

%   g growth rate

%   d discount rate

%   x0 initial amount

%   lambda parameter for Poisson Distribution (mean
arrival rate)
```

Code Fragment 10-5. Simulation Callback for Iterative Tax Model

```
%   Initialize Iterative Ad Valorem tax and Value
Model
%
%   (c) 2003, Anderson Economic Group LLC
%   Part of Business Economics Toolbox; use by
%license only.

%--------------------------------------------------
%   1. Initiate
%--------------------------------------------------

Tstart = 0
Tstop = 10
modelname = 'iterative_taxvalue';
%   Variables
RentableSpace = 14032
AvgRent = 16.90
vacancy = .25
expense_ratio = .25
initial_value_guess = 1000000
upper_val = 10*initial_value_guess

%--------------------------------------------------
%   2. Simulate
%--------------------------------------------------
%
sset = simset('SrcWorkspace','current');
sim(modelname, [Tstart:Tstep:Tstop], sset, []);
display('Model Simulated.')
whos
```

```
line = '----------------------------';
display(line);
display('Results:')

tax_iterations
EBIT_iterations
value_iterations
display(line);

%--------------------------------------------------
% 3. Graph Results
%--------------------------------------------------

display('Graphing Results.');

figure,
subplot(2,1,1)
plot(tout, EBIT_iterations/1000),
title('Property Tax Iterations', 'fontweight',
'bold');
% Xlabel('Iterations');
Ylabel('Value, $Thousands');
title('Iterations on Market Value of Real Estate
Concern', 'fontweight', 'bold');

subplot(2,1,2)
plot(tout, tax_iterations/1000),
title('Property Tax Iterations', 'fontweight',
'bold');
Xlabel('Iterations'); Ylabel('Property Taxes,
$Thousands');
title('Iterations on Ad Valorem Property Taxes',
'fontweight', 'bold');
%   End of Callback
```

11

Business Valuation and Damages Estimation

Economics, Finance, and Valuation

The previous chapter introduced the *economics* of the firm. Using a firm's earnings and a discount rate, we introduced a model for the value of a financial asset such as stock in a company.

In this chapter we examine a subset of finance — the practice of valuing a particular asset, namely a business. There are numerous books on valuation, most of which are written from the point of view of an accountant or a professional appraiser.[205] Some of these are quite detailed in the topics they cover. However, their focus is often on standard valuation equations and discount factors rather than on the fundamental economics of the firm. Conversely, for a reader with a strong economics background, the emphasis here on understanding the underlying industry and business prospects of the firm will seem obvious.

Here, we approach valuation as an application of economics, and therefore treat carefully and systematically the actual *generation of value* rather than the evaluation of accounting statements. With such a grounding, we then present a quantitative methods to estimate the value of a firm.

Use of This Chapter by Readers of Different Backgrounds

For those using this book as an introduction to valuation, the material presented in this and the previous chapter will provide a firm grounding in the methodology of valuing a financial asset. For a practitioner, however, other references will be needed to properly complete a business appraisal or assess damages. For practitioners and other readers familiar with the finance and accounting references in valuation, this book will introduce certain rigorous quantitative methods to forecast earnings and estimate business value.

---

205 See the references listed in "Valuation Methods" beginning on page 251. Note that we consider these *good* references, although often focused more on discount rates than on earnings. See also the discussion on this topic in Chapter 10, "Applications for Business," under "The Value of a Financial Asset" on page 218.

What Is To Be Valued?

In business economics, the asset to be valued is often a business enterprise, a line of business within a larger firm, or a security such as a stock, bond, or option. However, valuation is hardly restricted to economists. Indeed, it is so common that virtually all adults practice it and most become quite good at it.

Example Valuation Tasks

Consider the following common tasks, all of which require the economic thinking that is the kernel of valuation:

1. A housewife goes to the grocery store to purchase goods for the home. She considers the available food items and supplies in multiple stores, deciding to buy some goods in advance, shift some purchases from store to store, and from one commodity to another substitute on the basis of her own assessment of relative values, the time value of money, and the cost of storing items for future use.
2. A father brings his daughter to a sporting event for which he expects to buy tickets. As they walk to the stadium, other fans (and professional scalpers) offer to sell tickets. Using his knowledge of past events of this type, he waits until the right moment before the start of the game and goes to the right spot outside the stadium to get the best seats for the money.
3. A judge determines that one business unlawfully deprived another of the ability to earn income in their new location. The judge must then establish a remedy, which by law and custom is normally the lost profits of the firm.
4. A business executive wishes to purchase another firm and consolidate the businesses into one large enterprise. What is the fair market value of these various firms?

Most people would identify the last example as a task requiring economic reasoning, and some would also identify the previous example. Yet *all* these examples involve the same essential economic reasoning that goes into the last example and is the basis for valuation. Professional economists or other experts may be called upon to handle the last two example tasks, but the accuracy of the valuation estimates in the first two examples are excellent goals for the most rigorous analysis required of professionals.[206]

---

[206] We will consider the accuracy and precision of estimates below. However, we note that experienced shoppers can readily detect sale prices that are real bargains as well as offer remarkably reliable estimates of the market prices for dozens and dozens of goods and services.

Difference between Market Value and Accounting Concepts

Now consider another example: an accountant prepares a balance sheet for a business, including schedules showing the assets (such as cash, accounts receivable, tangible and intangible assets, and any depreciation or amortization), liabilities (such as short- and long-term loans), and the equity investments and retained earnings. Is this a valuation exercise like the other examples above?

The answer is, generally, no. Accounting concepts are very important in business; indeed, they are vital for both trade and the organization of all businesses. However, accounting is based on the principle of *historical cost*, not current market value. Double-entry accounting was documented as far back as Renaissance Italy and helped Italian traders ply the seas, with investors, buyers, and sellers relying on a set of accounts.[207]

Accounting statements *account* for the expenditures and receipts of an enterprise. Historical cost is the elemental concept in accounting for past expenditures, since you cannot provide a set of accounts today without basing them on the expenditures and receipts that produced them. Historical cost is determined by the transactions of the firm itself. Summarizing those transactions over a set of periods produces the balance sheet, income statement, and other fundamental accounting records. To the extent a value of an asset is implied by these records, it is commonly called "book" value.

Market value is based on an entirely different principle, namely the capitalized sum of expected *future* benefits.[208] No buyer determines the price he

---

207 The first statement of accounting is generally credited to Luca Pacioli (1447–1517), a Franciscan monk and mathematician in what is now Italy. He published *Summa de Arithmetica, Geometrica, Proportioni et Proportionalite* in 1494, summarizing mathematical knowledge of the time. The book contained a section on "Details of Accounting and Recording" that described bookkeeping as used in Venice. Venice, the Adriatic city which was the center of the Republic of Venice in the 15th century, was a leading trading center along with Florence (also the center of a republic and a city on the Arno River), Genoa (whose republic stretched across the Tyrrhenian Sea on the West Coast of the Italian peninsula, to include the island of Corsica), and other Italian cities. See, e.g., Gary Giroux, *A Short History of Accounting and Business*, a "book in progress," available at: http://acct.tamu.edu/giroux/history.html; John Garrity and Peter Gay, Eds., *The Columbia History of the World* (New York: Harper & Row, 1972), Chapter 15. Giroux goes as far as crediting accountants with developing writing itself. While this is truly a heroic stretch — the index to *The Columbia History* fails to mention accounting but otherwise describes writing in ancient Sumer, Babylon, and Mesopotamia — it is clear that a good system of accounting was necessary for the industrial revolution, and that the ability to account for business transactions was a prerequisite for trade well before then.

208 The history of market value is as old as humans themselves, or at least dates from the time man was vanquished from the Garden of Eden. Man had to work to feed and take care of the family, and this required a careful evaluation of the amount of work needed to acquire food, shelter, and other needs. *The Columbia History* notes that the term *capital* from which *capitalized* and *capitalism* derive stems from the Latin *caput*, or head, coming in turn from a Babylonian term of the same meaning and sound. Debts and wages in ancient Mesopotamia were paid in silver, and the farmer was able to borrow during the seasons of sowing and cultivation and repay the debt (with a customary 33% interest rate) at harvest time. Depending on the abundance of the harvest and other economic conditions, the amount of silver necessary to purchase a measure of grain would fluctuate. See *The Columbia History of the World*, Chapter 1. Thus, the notions of investment capital, risk, reward, market value, and toil were well established in ancient times.

or she is willing to pay by adding up the historical costs of others. Buyers are interested in the benefits that will be returned to them should they make the purchase.

Therefore, the market value of an asset is determined by views about the future and, in particular, the views of other participants in the market. The demonstration of such views — actual market transactions — are the fundamental statements of value in economics.

Summary: Accounting vs. Economic Concepts

Table 11-1 summarizes the key comparisons between the accounting and economic concepts of value.

Book Value and Market Value

An accounting-based valuation of an asset — its "book" value — can be identical to its market value. However, such an event is more a coincidence than anything else. Theoretically, on the actual date of a transaction, the book value and market value are the same.[209] At all other times, however, the market value is likely to be different. Indeed, because market value is based on the expectations of future events, the market value of assets will typically fluctuate.[210] Indeed, one of the advantages of historical-cost accounting is that it creates some stability in the financial marketplace because companies can prepare fairly stable balance sheets periodically without "marking to market" all their assets.[211]

Thus, accounting statements can never be relied upon as the indicator of value in the marketplace.[212] Accounting statements have an essential, important role, but that role is not to predict market value.

---

[209] Even this observation requires some simplifying assumptions, such as assuming that the decision to buy and the purchase transaction are made at the same moment; that there are no distorting effects of sales or other taxes; that the purchase price was the same as the market price on that day; and there was no compulsion to buy nor were there undisclosed material facts about the asset.

[210] The most visible example of this is the stock market, where security prices will change minute to minute.

[211] For large companies with significant pension obligations, real estate, and foreign-exchange risks, the swing in apparent value from marking some of these assets to market every quarter — but not others — can mask more important business conditions. Certainly some parts of the balance sheet benefit from such treatment (such as financial derivatives), but a frequent mark-to-market of the *entire* balance sheet would be prohibitively expensive and confusing in many instances.

[212] That is not to say that book value has no usefulness. Without it, it would probably be impossible to account for a firm. Furthermore, among very similar assets, the ratio between book and market value can provide an insight into how the market is pricing certain assets.

TABLE 11-1

Concepts of Value in Accounting and Economics

Concept	Accounting	Economics
Statements of value	Book accounting records, including balance sheets and income statements	Market prices, including records of recent transactions of similar assets
Origin of value	Historical transaction records	Expectations of future benefits; supply and demand
Determinate actors	The firm itself	Others in the market
Determinate action	Transactions of the firm	Transactions in the market

The Role of Economist and Accountant in Business Valuation

The above discussion highlights the difference in approach between economics and accounting. Of course, a business valuation is not performed by an entire school of thought; it is performed by individuals. Most such individuals will be primarily trained in either economics or accounting. Which training is better suited for the task?

We concluded that accounting was never intended to provide the basis for market value and that standard accounting statements can never be relied upon as an indicator of market value. Furthermore, we argue that understanding the economic conditions in the industry is essential to understanding the growth prospects of a business and therefore its value. Economists are certainly trained for this task. However, the first place to start in valuing a firm is its accounting statements. The evaluation of costs, the calculation of profit margins, etc., will normally be done more knowledgeably and effectively by an accountant.

To properly value a firm, you must apply *both* intrinsic economic skills and accounting skills. Some individuals have both; many do not. In such cases, we recommend that a person with complementary skills be included on the valuation team. This suggestion is also made by Patrick Gaughan, who describes independently and in very similar terms the advantages and deficiencies of accountants and economists.[213]

In our experience, economists natively grasp the underlying growth prospects of the firm and more quickly see how changes in conditions (or interventions by other parties) affects the value of firms. However, the inclusion of an accountant skilled in evaluating accounting statements greatly reduces the chances of misreading the fundamental condition of the business and improves the accuracy of cash flow models.

[213] Patrick A. Gaughan, *Measuring Commercial Damages* (New York: John Wiley & Sons, 1999); Chapter 1.

Concepts of Value

Fair Market Value

When we say "value", we generally mean *fair market value,* defined as follows:

> The fair market value is the price at which the property would change hands between a willing buyer and a willing seller, neither being under any compulsion to buy or to sell and both having reasonable knowledge of relevant facts.[214]

It is interesting to note that, at least among taxing, finance, and valuation authorities, there is very little theoretical debate about the standard of value. The market of buyers and sellers — not the labor inputs in a commodity, its social worth, or its cost of production to the seller — defines value.[215]

The Gold Standard of Value: Actual Market Transactions

We present below some innovative ways to estimate value in the context of tried-and-true advice from other practitioners in the field. In doing so, we will always define value as fair market value, the price a willing buyer would pay to a willing seller with both having adequate information. An actual market transaction of this type is the gold standard of valuation.

Estimation Methods vs. Actual Value

The chapter could stop here if there was a ready market for all assets and they all traded frequently in standardized commodities. However, closely held businesses, various securities of small- and medium-sized companies, most investments in real estate, and many other assets do not. Therefore, techniques of estimating the value of these assets have been developed.

In this chapter, we describe three approaches to estimating value and provide quantitative models for the income approach. When actual market transactions are not available, projecting and then discounting the future income from an asset is often the best method of estimating its value in the open market.

---

[214] This particular statement occurs in the Code of Federal Regulations, Title 26, Chapter 1, Part 20, Section 2031-1. However, almost identical statements appear in almost all references to fair market value. See "Other Value Standards" on page 249 for other standards of value.

[215] On this point, the economics literature is the more obtuse. Ideas such as mercantilism and the labor theory of value floated about for centuries, which grounded value in nearly metaphysical qualities that may never meet the market. By comparison, the basis of accounting is historical cost — a specific fact with little metaphysical about it.

However, do not confuse an estimate using a specific method with an actual transaction price. All valuation methods are attempts to *estimate* the price to which a willing buyer and seller would agree. This distinction between an estimate of a parameter and the underlying parameter is normally quite explicit in disciplines such as statistics, econometrics, and economics.[216] Such a distinction is vital in financial valuation as well but is sometimes missing in valuation reports and the discussion of such reports. To be clear, when we use the term *value* as a verb in this book, we mean estimating the sale price upon which a willing buyer and seller would agree.

Other Value Standards

While we will discuss fair market value, defined by willing buyer–willing seller arm's length transactions, there are other variations of the term "value" which are occasionally used.[217] In particular:

1. *Intrinsic value*: This is the value of the securities of a company based on the underlying (or intrinsic) value of the company's assets.[218] Intrinsic value will often vary from the market price for a variety of reasons, including the fact that securities in a company are not the same as an actual fractional interest in all the company's assets.[219] Businesses are owned *through* securities, but those securities are *not* the same as a fraction of all the firm's assets.

---

[216] In statistics and econometrics, even the symbols for the variables have evolved to indicate whether a certain quantity is an (often unknown) parameter or an estimate for one. For example, the known sample mean of a random variable X is often expressed with a bar over the variable, and the Greek letter *mu* is often used to describe the unknown population mean. The relationship between the two is summarized as $E(\bar{x}) = \mu$ which means that the *mathematical expectation* of the sample mean is the population mean.

Similarly, a common practice in economics (especially since the advent of "rational expectations" models) is to incorporate expectations of consumers or workers in modeling their current behavior. The difference between those expectations (which are normally unknown) and the actual variables is quite explicit.

[217] An excellent reference on this topic is by Shannon P. Pratt, Robert F. Reilly, and Robert P. Schweihs, *Valuing a Business*, 3rd ed. (New York: McGraw-Hill, 1996), Chapter 2.

[218] For example, an investor may look at the underlying real estate investments owned by a REIT (a publicly traded Real Estate Investment Trust) and believe that the intrinsic value of those assets is higher or lower than the market capitalization of the firm. If the intrinsic value is higher, he or she may decide to buy stock in the hope that the rest of the market eventually feels the same way.

[219] I cannot improve on George Lasry's observation that "a thirsty stockholder in a brewery cannot walk into 'his' company and demand that a case of beer be charged to his equity account." George Lasry, *Valuing Common Stock* (New York: AMACOM, 1979); quoted in Pratt et al., *Valuing a Business*, Chapter 3. Pratt et al. also list court cases dating to the U.S. Supreme Court's 1925 decision in *Ray Consol. Copper v. United States* (45 S. Ct. 526) in which the Court declared that "the capital stock of a corporation, its net assets, and its shares of stock are entirely different things ...."

There are further, specialized meanings for the term intrinsic value in the investment community. The search for intrinsic value in stocks is a well-established technique that is sometimes called *value research*.[220] In addition, a portion of the market value of certain options is its intrinsic value.[221]

2. *Investment value*: This is the value a particular investor would place on an asset. Note that the investment value depends on the investor and, therefore, will not be the same for all investors.[222] The market value, however, will be the same.

3. *Fair value*: This term is inherently vague and, therefore, must be defined by a statement of "fairness." Indeed, the term is often used in laws and regulations in which the fairness criteria should be spelled out.

   There are some typical uses. One involves business combinations such as mergers or acquisitions, or transactions among equity holders in a business. The fair value here may not be the value set in an open market, where there are many buyers and sellers. Instead, the fair value is the price to which these parties would agree to buy and sell an asset without any compulsion.[223] Such prices may deviate from the fair market value because the asset owners or acquirers would be few in number, and their willingness to pay for a particular asset may not match the prices set for similar assets in the open market. However, fair value should be reasonably close to fair market value because arbitrage possibilities should keep the value to even an unmotivated buyer reasonably close to the market price. Furthermore, allowing fair value to be markedly different from market value would allow for abusive tax evasion.[224]

---

220 There are many other investment styles, including market-timing, asset-allocation, market indexing, technical or "chartist," and the tried-and-occasionally-true "dartboard" approach.

221 For a call option (the right to purchase an asset at a specified strike price K) on a stock with a spot (current) price of S, the *intrinsic value* is *max(S–K*, 0). The remaining portion of the option premium is its *time value*. See, e.g., N. Chriss, *Black Scholes and Beyond* (New York: McGraw-Hill, 1997); P. Brandimarte, *Numerical Methods in Finance: A MATLAB-Based Introduction*, (New York: John Wiley & Sons, 2002).

Options are discussed further in "The Real Options Method" on page 260.

222 This term is sometimes confused with intrinsic value. However, there is a difference; an investor may not wish to purchase stock in a company that he recognizes has substantial intrinsic value. The investment value to that investor of the asset is much lower than its intrinsic value.

223 See Financial Accounting Standards Board, Statement of Financial Standards No. 141, "Business Combinations," September 2000. The FASB Web site is: http://www.fasb.org.

224 For example, the break-up of a firm may require the various investors and lenders to place a fair value on the underlying assets. If the fair values are different from the market values, the taxes paid on gains or losses on these assets will be significantly different from what should be paid. In some cases, this difference is merely a deferral. In others, it would amount to an abusive tax shelter.

In addition, the disclosure of fair values of assets and liabilities, based on market values, helps prevent investors from being misled about the prospects or performance of a company.

Note that the Financial Accounting Standards Board, in line with the effort of regulatory bodies throughout the financial industry, is currently revising its statement of fair value to bring it closer in line with fair market value.[225]

4. *Liquidation value:* The value of a firm that is being liquidated is almost always lower than that of a going concern, and is clearly different from its fair market value as an ongoing business.[226] When assets are being liquidated, the conditions of the sale will often affect the market price significantly. For this reason, various types of liquidation value (such as "forced" and "orderly") are used to distinguish such a premise of value from the fair market value.[227]

Valuation Methods

The market value of any asset is defined as the price a willing buyer would pay to a willing seller in an arm's-length transaction, where both parties have adequate information.[228] In most cases, there is no purchase transaction or active market that establishes such a value, so methods of estimating the value of assets have been developed and refined.

Approaches to Valuation

There are three generally recognized methods of business valuation described in the fields of economics, finance, and accounting.

1. *Sales or market approach*: The most direct method, if data are available, is to review recent transactions for similar assets. The closer the asset being valued is to assets that have recently been sold on the open market, the better this method works.
2. *Income approach*: The economic income or discounted cash flow method is the basis for most financial analyses. It converts an expected future stream of benefits into a present-value amount.
3. *Asset-based approach*: This approach is based on evaluating the underlying assets being sold, looking for direct information on market values for these assets.

---

225 See FASB staff summary, "Disclosures about Fair Value," April 2003, citing FASB Statement 133; and "Financial Assets and Liabilities — Fair Value or Historical Cost" by Diana Wilis of the FASB staff, dated January 2002. Found at http://www.fasb.org.

226 The "willing buyer, willing seller" test is violated if one party is compelled to sell, which is normally the case in a liquidation.

227 See Pratt et al., *Valuing a Business*, 3rd. ed. (New York: McGraw-Hill, 1996), Chapter 3.

228 This definition is being repeated, but it is worth repeating.

> When better data on underlying assets are not available, accounting information based on the historical cost of assets and their depreciation over time may be considered. For valuation purposes, these accounting figures must normally be adjusted significantly.

All these approaches — when done properly — are consistent with the fundamental economics principle that the value of an asset is based on the capitalized stream of future benefits, adjusted for risk and other factors.[229] These practical approaches are recognized by the Internal Revenue Service,[230] appraisal authorities,[231] and business valuation authorities.[232]

Similarly, in the government activity in which values are most important — taxation — state laws and court decisions typically codify these

---

[229] This principle is recognized in economics as underlying present value theory, the term structure of interest rates, and the time preference underlying contracts (see, e.g., *The New Palgrave: A Dictionary of Economics*, London: Macmillan, 1987, 1991; various articles); in the newer field of finance (see, e.g., "Capital Asset Pricing Model" in *The New Palgrave*, cited above); and in specific studies of valuation that take as their basis finance and economics (see, e.g., Aswath Damodaran, *Investment Valuation* (New York: John Wiley & Sons, 1996). See also references cited in below.

[230] The seminal statement on this was Revenue Ruling 59-60, January 1, 1959. Later rulings 77-287 (restricted securities), 83-120 (closely held business), and 93-12 (minority shareholders) generally extend the principles stated in RR 59-60. Revenue Rulings can be found at: http://www.taxlinks.com.

For an excellent summary and reprinting of the revenue ruling texts, see Estate, Gift, and Income Tax Valuations, in J.R. Hitchner, Ed., *Financial Valuation: Application Models* (New York: John Wiley & Sons, 2003).

Newer and more general IRS references include IRS Publication 561, "Determining the Value of Donated Property," [Feb. 2000 edition cited], which describes three methods for valuing real estate: comparable sales, capitalization of income, and replacement cost new or reproduction cost less observed depreciation. The publication states that interests in a business should be valued by considering the fair market value of the assets plus an amount based on the "demonstrated earnings capacity of the business," along with other factors. See also IRS Publication 551, "Basis of Assets" (May 2002 edition cited).

The Code of Federal Regulations, which is now available online, summarizes many of the IRS rules. See Title 26, Vol. 14, Part 20; in particular 26 CFR 2031.0-9. Found at http://www.firstgov.gov (linking to the Government Printing Office at: http://www.access.gpo.gov/index.html).

[231] See American Society of Appraisers, Business Valuation Standards, which include standards (BVS III, IV, and V) for the Asset-Based Approach, Income Approach, and Market Approach to Business Valuation. These standards can be found at: http://www.bvappraisers.org/glossary/.

See also the *International Glossary of Business Valuation Terms*, which has been adopted by the American Institute of Certified Public Accountants, the American Society of Appraisers, and other Canadian and U.S.- based financial societies. The 2001 statement recognizes three approaches to valuation: the income approach, the market approach, and the asset approach.

[232] See, for example, Pratt, Reilly, and Schweihs, *Valuing a Business*, 3rd ed. (New York: McGraw-Hill, 1996), Chapter 3. The authors identify three methods of valuation as the income approach, the market approach, and the asset-based approach.

By comparison, A. Damodaran, *Investment Valuation* (New York: John Wiley & Sons, 1996), describes a "discounted cash flow" approach (essentially an income approach), a "relative valuation" approach (which appears to incorporate both market information and asset information), and a "contingent-claims" approach that relies on option theory. The contingent-claims approach is distinct from the others, and may over time be recognized as a separate approach. See "The Real Options Approach" on page 260.

approaches as acceptable methods of estimating the value of property so that it can be assessed for property tax purposes.[233] Taxation authorities, however, should not be considered unbiased interpreters of valuation theory; government entities have a vested interest in enhancing their own sources of revenue and therefore tend to adopt procedures, rules, and laws that increase the basis on which taxes are paid.[234]

Fundamental Equivalence of Methods

If done correctly and with good information, the values estimated using the market or economic income approach should be very close to each other and should represent a good estimate of the amount a willing buyer would pay a willing seller. In the absence of the necessary information regarding other approaches, asset-based approaches can be used, but only with substantial adjustments.

Deficiencies in the Asset Approach

The deficiencies of certain approaches to valuation, notably the use of historical-cost accounting records to suggest business values, has not gone unnoticed. The IRS, for example, abandoned valuation by the "excess earnings" method, which relies on historical-cost asset values, unless no other method is available.[235] Unfortunately, it is still widely used, resulting in a "plethora of misapplication."[235a]

---

[233] For example, Article IX of the Michigan Constitution requires the assessment and equalization of real and personal property in the state. Section 27 of the General Property Tax Act, PA 206 of 1893, codifies for Michigan the procedure for assessing the "true cash value" of property when put to "highest and best use." The state uses three methods of valuing property, all of which theoretically produce the same result. They are (1) the comparable sales, (2) reproduction costs less depreciation, and (3) capitalization of income.

Other Michigan authorities include Opinion of the Attorney General, No. 6092 of 1982; *Uniroyal v. Allen Park* (1984), 360 NW2d 156, 138 Mich App 156; *Somonek v. Norvell Twp* (1994) 527 NW2d 24, 208 Mich App 80 (on remand); and *Edward Rose Building Co. v. Independence Twp.* (1980) 462 NW2d 325, 436 Mich 620.

[234] The field of public choice in economics studies of the incentives that affect the behavior of government entities. However, the tendency of government agents to maximize tax revenue was observed at least as far back as the Bible. Charles Adams, in his detailed book *For Good and Evil: The Impact of Taxes on the Course of Civilization* (Lanham, MD: Madison, 1993, especially Chapter 3), recounts historical and religious episodes involving the practices of taxation originally told in the Old Testament books of Exodus, Hosea, Amos, Isaiah, Kings, and Chronicles; and histories such as Flavius Josephus' *Antiquities*.

[235] We do not describe this method here because it has no valid theoretical basis and no longer works in practical terms. However, most valuation references (especially older ones) still include it. For historical purposes, we note it was first developed in the 1920s by the IRS and then incorporated in Revenue Ruling 68-609. Although this Ruling has not been rescinded, the IRS has repeatedly criticized the method in practice, such as in Private Letter Ruling 79-05013.

For a history and explanation, see Pratt et al., *Valuing a Business*, 3rd ed., Chapter 13.

[235a] See Pratt et al., *Valuing a Business*, 3rd ed. Chapter 13.

New Approaches to Valuation

In recent decades, economists have developed much more sophisticated models for the value of assets. Much of this development has been in the sub-field of finance and investment, and has been focused on publicly traded assets such as stocks, bonds, and derivative securities. These models explicitly incorporate risk, and in some cases specifically include management policies.

The most developed of this new approach are "option value" methods, in which the risk of certain events is explicitly included in the value calculation. These methods have been widely used for financial assets, and more narrowly in business valuation. Another method, known as dynamic programming, has been largely confined to academic settings. We present both these methods in "Dynamic Programming and Real Options Methods" on page 259.

Recommended Order of Methods

We suggest below a recommended order of valuation methods. In suggesting this order, we attempt to maximize the use of information that represents actual transactions and the motivations of likely buyers and sellers, and minimize the amount of judgment, estimation, and use of unreliable historical-cost information. This order is defined for operating business valuation challenges and may not be applicable for other valuation tasks. By "operating" we mean generating income for investors by creating products or providing services. Estimating the value of an operating business often requires more analysis than estimating the value of a passive investment, or of a financial security like a stock or bond.

After a brief general discussion of each method, we provide economic analyses based on these approaches, including MATLAB routines, as well as examples.

Best Evidence: Market Price

In cases where information on actual market prices are available and the assets in question are genuinely comparable, actual transaction prices should be given the highest weight in an analysis.

When using market transactions as the basis for valuing an asset, be certain that the transactions contain all the elements needed for a true market value determination:

- Willing buyer
- Willing seller
- Arm's-length transaction

- Good information available to both buyer and seller
- Known sale price for the underlying asset

To truly understand the market values implied by asset sales, look carefully at the actual transactions.[236] Adjust for any creative financing, seller financing, partial sales, or other unusual terms that can obscure the true cost of the underlying asset to the buyer. Also be careful to disentangle the price assigned to the business enterprise (which is normally the wealth-producing asset, complete with important intangible assets) from hard assets like real estate, inventory, rolling stock, and other items that can be resold.

Second Best Method: Income Method

If a market transaction price is not available, an economic income or capitalized income method is usually the next best choice for estimating the value of an operating business.

Financial theory is based on the principle that value can be estimated by summing the present value of future cash flows to be derived from the asset and adjusting for risk and other factors.[237] A business enterprise, or a separable line of business within an enterprise, is an asset that can be valued with this approach. Cash flows in such a valuation model (often called a "discounted cash flow" or "capitalized income" model) are those derived from forecasted revenue and expenses.

Standard texts in business valuation rely on such an approach and describe various methods of identifying the proper cash flows, choosing the right discount rate, and adjusting for risk and other factors. For example:

> From a financial point of view, the value of a business or business interest is the sum of future economic benefits to its owner, each discounted back to a present value at an appropriate discount rate.[238]

The discount rate used to convert future cash flows into present value terms depends on current rates of return for similar assets and the risk associated with the specific asset.

Critical Assumptions of the Income Method

The income method relies on certain critical assumptions, including:

1. *Going-concern premise.* Generally, both accounting and valuation exercises are based on the assumption that the business is a "going

---

[236] If possible, review the purchase agreement and other relevant documents.

[237] We derive this under simplifying assumptions in Chapter 10, "Applications for Business," under "Deriving the Present Value of a Financial Asset" on page 219.

[238] Pratt et al., *Valuing a Business*, 3rd. ed. (New York: McGraw-Hill, 1996), Chapter 3, "Summary."

concern."[239] In some cases — most notably a bankruptcy — this principle is violated, and the assumption should not be used. Furthermore, there are other circumstances in which the going-concern assumption may not hold even for a firm that operates without bankruptcy.[240] Because of this, we believe the going-concern assumption should be questioned in every valuation assignment.[241]

If it is *not* a going concern, then a valuation method based on past performance will be unreliable, and probably would produce an estimate far in excess of the actual market value.[242]

2. *Historical results or other objective data predict future performance.* In most cases, the valuation of a business starts with the examination of the past year's financial performance. From that base year, growth assumptions and other assumptions about the future are used to project out the results.

   The examination of a base period is an excellent place to start. Clearly, past performance is the strongest indicator of future performance for a wide number of variables.[243] However, if there has been a significant change in the management of the firm, the marketplace for the firm's products or services, material change in the firm's accounting or other important policies, or a record of past failures to make required payments or report accurate financial and sales information, then reported past performance is especially suspect.

---

239 This point is glossed over in some valuation texts but made explicit in others. For example, Pratt, et al., cited previously, discuss the going-concern premise repeatedly. On the other hand, Damodaran, also cited earlier, does not even list it in his index, and discusses it only indirectly. It appears that valuation references with authors having accounting backgrounds are much more likely to state the point explicitly, while authors from a finance background assume that a knowledgeable practitioner would adjust the cash flow projections accordingly. As we consider the first principle of valuation to understand how a firm makes money, we favor the accountant's rule of making this check an explicit part of every valuation.
This topic is also addressed in "Note on Liquidation Value" on page 259.

240 Other cases where the going-concern assumption may be inappropriate are: when a franchised retailer or distributor has lost franchise rights to its major product; when an industry as a whole is suffering due to a dramatic change in conditions; when a major competitor, supplier, or customer is about to leave (or enter) the market; and when a key man or other vital asset to a firm has left.

241 The IRS, at least since Revenue Ruling 59-60, has required an evaluation of the prospective earning power of the company, and an assessment of the "good will of the business; the economic outlook in the particular industry; the company's position in the industry and its management." 26 CFR 20.2031-1.

242 To be precise, a well-considered income method will still work, but the analyst must be willing to put zeroes in the estimate of the future stream of income if the firm will no longer be producing and selling products.

243 Empirical research on both stock market values and prediction of real economic variables has shown that simply using the last period's value is usually a pretty good predictor of the next period's performance. Most predictions implicitly or explicitly are based on that model. Recall the discussion about Markov processes and projection under complex uncertainty; see "Projection: Complex Uncertainty" on page 212 in Chapter 10.

In some cases, the historical results of an entity are not the best basis for projecting future performance, but a separate set of variables are. For example, consider retailers active in multiple markets, many of which have recently had declining population and declining sales. If retailers can identify their target market near a new store location, and the demographic and geographic information suggest that the target market is growing rapidly there, then a faster-than-historical-average growth assumption would be reasonable.

In any case, the grounds for projecting future revenue growth in excess of recent periods must be substantial, and stated explicitly in the report.

3. *Market and product positioning are predictable.* Another critical assumption is that the product or service provided by the subject company remains consistent with its position in the past. If this is not the case, the effect of such a change must be explicitly considered, and the effect on revenue must be supported by analysis and other data.
4. *Management quality and focus will be maintained.* The income method typically assumes that management quality remains consistent and that the management focus will remain. If this is not the case, such a contrary assumption must be supported by other analysis.
5. *The regulatory environment will change relatively little.* If an industry is heavily regulated — and most industries are regulated in some fashion — then such regulations have a strong bearing on the market value of firms in that industry.
6. *The overall economic environment is predictable.* One does not need to predict small movements in interest rates or quarterly GDP to provide a solid economic base forecast for use in a valuation. However, be attuned to fundamental assumptions about the economy and the market that are embedded in any sales forecast.
7. *All other material, relevant conditions have been explored.* The items above apply to nearly every valuation project. However, most valuations require special attention to one or more significant factors that are not universally present in businesses. *Remember, think first about what creates value for the business.* Anything that affects the earnings of the business must be considered in your value estimate.

Third Best Method: Asset-Based Methods

An asset-based approach using historic accounting data normally cannot be expected to produce an adequate estimate of the value of a business.[244] For operating companies, the asset approach is less desirable than an income approach in terms of both time and accuracy considerations.[245]

However, asset-based approaches can be reasonable in the following situations:

1. The overall asset to be valued can be broken into discrete units, and the sum of their individual assets is reasonably close to the values of the entire asset.
2. Good market values, or values that can be estimated with an income approach, are available for each underlying asset.
3. There are no significant intangible, going-concern, or other value-affecting attributes to the assets that change because of a transaction.
4. The assets are commodities with a ready resale market.
5. The assets consist of substantial contingent claims or highly risky enterprises, in which an option-based approach best represents the actual market for the asset.[246]

Note on the Equivalence of Asset-Based Methods

The recommended order of preference in methods is based on an assumption that the entire business has a value that exceeds that of certain parts, and that the true market value of a firm depends on the combination of all the essential parts. In cases where the value of the business largely consists of the sum of separate, severable operating units, one can view each separable part as an individual valuation task and apply the best method for each.

Even in such cases, though, the value of the entity as a whole should be considered.[247]

---

[244] Among the various reasons why such an approach would fail is the fact that the asset being valued is the business itself. Although historic-cost methods are used to create a balance sheet for a business, such a balance sheet would not indicate the true value of a business. See "Difference between Market Value and Accounting Concepts" on page 245.

[245] It often takes far more time to individually value each asset than to value an entire entity. Furthermore, since the entity as a whole is often the reporting unit, the best data on past performance are often available for the entity. Finally, if an acquirer is purchasing an entity as a whole, that acquirer has demonstrated that the entity as a whole, not differentiated assets, is the target. While in some cases the acquirer will wish to keep certain operating units and sell others, the appropriate valuation method in those cases will still be an income or market approach to valuing the individual operating units.

[246] We discuss contingent-claims methods, or option theory methods, later. For the purposes of this discussion, we include such methods as part of the asset approach, although this is not clear.

[247] In some cases, the individual business units may be valued *higher* on their own than as part of a larger business entity. The history of mergers and acquisitions among publicly traded companies provides ample examples of mergers that resulted in a reduction in market capitalization, indicating that investors viewed the sum as being worth less than the parts.

See, e.g., Frank Easterbrook and Daniel Fischer, *Economic Structure of Corporate Law* (Cambridge: Harvard University Press, 1991), especially Chapter 7 and the references cited there; R.A. Brealey, S.C. Meyers, and A.J. Marcus, *Fundamentals of Corporate Finance* (New York: McGraw-Hill, 1995), Chapter 22.

Note on Liquidation Value

A fundamental accounting principle is that businesses are evaluated with the assumption that they are going concerns. In cases where such a going-concern assumption is violated, all or some of the firm's assets will often be liquidated. The liquidation valuation of a firm will normally depend on its "hard" assets, which include tangible assets that can be resold.[248] Intangible assets of a firm — such as the goodwill it has created over the years of business, the firm's reputation, the network of contacts, pipeline of future orders, and other vital aspects of an ongoing business — are often dissipated in a bankruptcy.[249]

Alternative: Real-Option or Dynamic Programming Methods

Although much less developed, in some cases newer methods may be superior to the traditional approaches. Two newer methods are discussed under the section "Dynamic Programming and Real Options Methods" that follows.

Dynamic Programming and Real Options Methods

In this section we describe a newer method for business valuation, the "real option" approach, which has become a practical tool in certain business valuations situations. We also introduce the dynamic programming method which, until now, has been largely confined to academic settings.

The Real Options Approach

Human beings have always valued risk, and fairly sophisticated mathematical and statistical treatises on the subject of games of chance appeared centuries ago. However, good tools for quantifying risk in financial and business terms have been developed over the past few decades. The seminal work by Black and Scholes that produced a workable option pricing model began a rush of new tools for pricing financial options. The insights gained can be applied to numerous other entities as well.

---

[248] For the purposes of this statement, we include assets that have value on their own (including valuable intellectual property such as software, or rights to income-producing assets based on intangible assets) as hard assets.

[249] While bankruptcy law is outside the scope of this book, we should note here that some bankruptcies allow a firm to reorganize and then emerge as a going concern again. These bankruptcies — normally referred to as "Chapter 11" after the chapter of the U.S. Bankruptcy Code entitled "Reorganization" — may allow some intangible assets to maintain their worth. On the other hand, "Chapter 7" bankruptcies typically result in the liquidation of the firm.

For some subset of business economics problems, the best approach to valuing an asset will involve explicit use of option pricing theory. This is sometimes called "contingent claims valuation,"[250] or "real options."[251] Here "real" is used to contrast the underlying *real* asset from the *financial* assets that are the basis for most derivative securities that are actively traded.

This newer class of methods does not fit neatly into the standard categorization of three approaches (market, income, and asset), but we tentatively place it in the asset category for the time being, recognizing that a new categorization scheme may be necessary.

Valuing Equity as an Option

One of the most intriguing insights emerging from the original 1972 article by Black and Scholes was sparsely discussed in it.[252] It was the observation that the equity in a corporation could be viewed as a call option on the value of the firm, with the strike price being the value of the debt. To motivate this, note that equity holders have the residual claim on the firm's assets. If the firm is liquidated, the proceeds will first pay off all bondholders. Anything left will go to the equity holders.

Of course, most shareholders do not want to liquidate the firm because they perceive its value as a going concern to be higher than its liquidation value. In such cases, their stocks are equivalent to "in the money" call options on the value of the firm. These securities can be sold and resold indefinitely as long as the market perceives the firm's value as higher than its debt.

Option Methods Produce Similar Value Estimates

Like all other valid methods, a properly completed option-method valuation, given good information, will provide a similar estimate of fair market value. Consider a simple example of this: the valuation of a well-established, profitable, publicly traded firm. Standard methods would estimate the fair market value of the firm as follows:

---

[250] A. Damodoran, *Investment Valuation* (New York: John Wiley & Sons, 1996) describes a "contingent claims" approach that relies on the option theory. See Chapter 17 and Chapter 18, particularly "Valuing Equity as an Option" in Chapter 18.

[251] A popularizer of this technique in the equity markets is M.J. Mauboussin, whose paper "Get Real: Using Real Options in Security Analysis," Credit Suisse Frontiers of Finance No. 10, June 1999, is one of the most influential articles on the subject.

[252] F. Black and M. Scholes, The valuation of option contracts and a test for market efficiency, *J. Finance*, 27, 399–417, 1972.

It is interesting to note that this article — one of the most influential in the fields of finance and economics in the past century — was rejected by the first two journals that received it. It is also interesting, though less entertaining, to review the intellectual history of the study of options. See Kevin Rubash, "A Study of Option Pricing Models," paper available on the Web site of Bradley University at: http://bradley.bradley.edu/~arr/bsm/model.html.

1. Using a market approach, the market capitalization of the stock and the market value of the debt would be summed up to produce an estimate of the enterprise value of the firm.[253]
2. Using an income approach, and then the earnings of the firm would be projected forward, and discounted by the weighted average cost of capital. The weights on the debt and equity factors would be set by the market value of debt and equity, which would be similar to the analysis performed for the market approach. The results should be close, though probably not identical, to that from the market approach.[254]
3. For the reasons described earlier in this chapter, an asset approach would be the least reliable, and it probably would be dispensed with.
4. An option method would involve the evaluation of the option-adjusted value of the equity and debt. The option-adjusted value of the equity, given that the firm is profitable and has strong prospects of remaining so, would be equal to its regular market value. The value of the debt would be similarly estimated. The sum of the two would be an estimate of the enterprise value of the firm.

Thus, for this company, the market approach and option approach would yield exactly the same result, and the income approach should yield a similar estimate.

Options Approach: Troubled Firms

The real options approach becomes especially useful when the going-concern assumption becomes untenable or uncertain. Consider the subgroup of firms that are near bankruptcy, or near the termination of important business arrangements that underlie their ability to earn profits in the future. For such firms, a standard discounted earnings model is a weak approach because the cash flows and resulting earnings in the future are so uncertain. An option-pricing model, on the other hand, will often provide a better value estimate.

---

[253] In many cases, a simplifying assumption that the market value of the debt is its face value is employed. However, for long-term fixed rate borrowing, preferred stock, and other securities, adjustments may be required. See also, for this approach and the income approach, the discussion "Using Market Value to Weight Capital Structure" on page 280, and "Iterative Methods" on page 280.

[254] If the results were not close, we should review all the assumptions and calculations, since the market value should not stray far from the intrinsic value (for a publicly traded firm). The prevailing discount rates, premiums, and discounts implied by market prices, however, do fluctuate. For this reason, we would not expect the two methods to produce exactly the same result.

Options Approach: Unusually Promising Firms

The flip side of using an options approach for troubled firms is their application to especially promising firms. In this case, the "real options" method calls for assessing the underlying assets for potential surprising increases in value, such as the introduction of new technology or a potential to dramatically improve productivity in new markets. In this case, the "volatility" in earnings will be greater than in most firms in the same industry; only much of the volatility will be generated by potential spikes upward in earnings.[255]

This approach does explain some of the deviations between underlying current earnings and stock market prices for firms considered to have unusually promising futures. However, it must be used with caution in valuing firms that will be sold, especially if the "unusually promising" management, technology, or other attribute will no longer be present after the sale.

Applying an Options Approach

While the Black–Scholes model was created for financial options, its insights — and the extensive development in the field since then — can be applied quantitatively to businesses today. MATLAB has extensive tools for use with financial options and derivatives of all types, including both extensions of the Black–Scholes model and binomial tree models that simulate risks over a discrete set of periods.[256]

In order to use an option pricing model, you must identify the various assets and any contingent claims on those assets. The contingent claims (options) must be identified by strike price and the events that could trigger the claim. At this point, the key variables are not detailed cash flows (as in an income model) but historical variation (volatility) of the value of the underlying assets, and time. For companies in troubled circumstances, the analyst can sometimes come up with more reliable estimates of these variables than of cash flows.

Keep in mind that much of option theory (as in portfolio theory) was developed with the assumption of symmetric variation, so be cautious in using standard formulas if you think the risks are one-sided.[257] The binomial tree method, discussed next, below may be a better method in such instances.

[255] See the paper by Michael J. Mauboussin, cited before.

[256] In the version available at the time of this writing, these capabilities are in the Financial Toolbox, with related functionality in the Statistics, Financial Derivatives, and Optimization toolboxes.

[257] See the discussion in Chapter 12, "Application for Finance," under "Mean-Variance Analysis" on page 304.

Binomial Trees

While the Black–Scholes model deserves its place in the pantheon of finance, it is not the only, or the best, model to use in many situations. For business valuations in cases where a firm is in or near bankruptcy, a simpler binomial tree model will often work quite well.

For starters, such a tree can be hand-drawn, with the branches signifying values, given the results of future decisions or other events. If probabilities can be assigned to each branch, then the combination of the probabilities associated with each event and the values that would result from each series of events form the basis for estimating the worth of securities in the firm. Even if a full option-pricing approach is not employed, such analysis can provide vital insight into the actual market values investors will place on a risky asset.

Valuing Financial Options

Given the importance of options in the financial markets, there are many references to their use and widely available software to calculate option values as well as related sensitivity and other parameters. MATLAB contains extensive option capabilities and is featured in excellent references on the subject.[258] As this topic is extensively covered in other texts, we will not address financial options further.

Dynamic Programming Methods

A method of solving complicated, multistage optimization problems called *dynamic programming* was originated by the American mathematician Richard Bellman in 1957.[259] Like most brilliant insights, the method promised a radical simplification of some intransigent problems. However, the method was difficult or even impossible to implement until quite recently. Two parallel developments now make it possible to use this method in business valuation:

1. The increasing use of dynamic programming in theoretical work, primarily in academic settings, and the associated advancements in technique

---

[258] See, for example, N. Chriss, *Black–Scholes and Beyond* (New York: McGraw-Hill, 1997) which presents the venerable Black–Scholes model, as well as Cox–Ross–Rubenstein binomial trees; and *Black–Scholes and Beyond Interactive Toolkit* (New York: McGraw-Hill, 1998), developed using MATLAB.

See also P. Brandimarte, *Numerical Methods in Finance: A MATLAB-Based Introduction* (New York: John Wiley & Sons, 2002).

[259] Richard Bellman, *Dynamic Programming* (Princeton, NJ: Princeton University Press, 1957) reissued by Dover, 2003. Bellman's introduction of nearly 50 years ago is still useful reading. It is interesting to note that Bellman anticipated a number of economic applications for his technique, and that the reasoning he expresses in his 1957 text parallels some of our arguments in this section.

2. The increasing speed in computing power, which makes possible the enormous number of calculations that are necessary to implement certain algorithms for solving dynamic programming problems

The technique is almost unknown in business valuation.[259a] We introduce it here as a method that has great potential to improve valuation, but at present is not well developed.

The Dynamic Programming Method

We present below a simplified derivation of the technique.[260] In this derivation, we model the management of a business as a multiperiod optimization problem.

1. A business is an organization which will live through multiple periods and for which a mixture of both reasonably predictable and unknowable events will occur. These events will present the management of the company with information, which can be summarized as data in a state vector. At each time period, holding the information available, the management takes certain actions, such as hiring, firing, purchasing, pricing, advertising, and selling. The variables can be represented as Markov processes, as the information available in one period summarizes all the useful information for the prediction of these variables in the next period.[261]
2. The challenge (the "optimization problem") presented to the managers of the company is to take actions in a manner that maximizes the value of the firm. If we take the value of the firm to be the

[259a] An extensive literature review found many instances of the theoretical use of dynamic programming in finance, especially since Robert C. Merton's classic, "Lifetime portfolio selection under uncertainty", *Rev. Econ. Stat.*, 51, 247-257, 1969; and Optimum Consumption and portfolio rules, *J. Econ. Theory*, 3, 373-413, 1971; Erratum, *J. Econ. Theory*, 6, 213-214, 1973. See D. Duffie, *Dynamic Asset Pricing Theory*, 3rd ed. (Priceton, NJ: Priceton University Press), Chapter 9.

There have been a handful of interesting applications for unusual aspects of a specific type of business, such as modeling the operation and termination of a nuclear reactor. In the references cited in following text, the method is used to value certain pure assets, such as a mine or natural resources, under greately simplifying assumptions.

[260] More complete presentations are in M. Miranda and P. Fackler, *Applied Computational Economics and Finance* (Cambridge, MA: MIT Press, 2002); A. Chiang, *Elements of Dynamic Optimization* (Long Grove, IL: Waveland Press, 1999); and L. Ljungqvist and T. Sargent, *Recursive Macroeconomic Theory* (Cambridge, MA: MIT Press, 2000). See the following notes for additional sources.

[261] See Chapter 10, "Application for Business," under "Markov Processes, States, and Uncertainty" on page 212. Because the Markov representation is fundamental to the development of this theory, and the action variables are decisions of the optimizer, certain dynamic programming equations are sometimes called "Markov Decision Processes."

expected future profits, discounted for time and risk, we can express this optimization problem in the following *functional equation:*

$$V(s, t) = \max_{x} \{ f(s, x) + \delta E(V(s_{t+1}, x_{t+1})) \} \tag{1}$$

Here, $V(s, t)$ is the value of the firm given the state s at the time period t. This value consists of two parts: the current profit of the firm $f(s, x)$ and the expected value of the firm in the next period after discounting by the factor δ.

The maximization problem involves the control variables or actions variable x, so the maximization operator references this variable or vector of variables. Because both the current profit and the future profits of the firm depend on the actions of the firm's management, the action variable is an argument to the profit and value functions.

Equation (1) is known as a functional equation because the expression $V(s, t)$ is not, strictly speaking, a function of just the variables s and t, but instead the maximization of a family of functions.[262] We will also refer to it as the *Bellman equation* for this optimization problem.

3. The existence of a solution to the optimization problem under the conditions we will normally encounter in business settings has been rigorously established.[263]
4. Actually solving the problem we have now stated has been the greatest difficulty in using this technique. Bellman himself called the problem the "curse of dimensionality" because the size of the problem is magnified exponentially by the number of variables and their possible paths over time. It is here that the advancements over the past decade in both analytical methods and numerical computing allow for the emergence of a practical dynamic programming method.
5. There are a number of methods for solving dynamic programming problems, each of which has certain advantages. These include:

---

262 This is best described in A. Chiang, *Elements of Dynamic Optimization* (Long Grove, IL: Waveland Press, 1999). Chiang notes that this is not a mapping from real numbers to real numbers, which would be a function. Instead, it is a mapping from paths to real numbers, the real numbers being the quantities being optimized. This is termed a functional, and sometimes distinctive symbols (such as $V[s, x]$) are used to distinguish it from a function.

263 Bellman himself provided many of these proofs. See also N. Stokey and R. Lucas, *Recursive Methods in Economic Dynamics* (Cambridge, MA: Harvard University Press, 1989). They note the important contribution made by L.M. Benveniste and J.A. Scheinkman, Duality theory for dynamic optimization models of economics: The continuous time case," *J. Econ. Theory,* 27, pp. 1–19.

a. Recursively solving the problem backwards from a known terminal value; this is known as "backward recursion."
b. Iterating on the values created at each step with variations in the *policy* created by the application of the control variables; this is known as "policy iteration."
c. The more traditional approach of specifying an initial set of values for all variables, calculating the value function at these points, and then iteratively searching for higher values until such searches yield no further improvement. This is known as "function iteration."

The advances in numeric computing techniques in economics has produced a small number of usable computer algorithms which perform the enormous number of calculations that are often required to solve dynamic programming problems. Many of these numerical algorithms are, at the current time, either experimental or newly developed and therefore cannot easily be applied to business economics problems. However, we expect progress in this field to continue at a good pace, and the recent introduction of usable dynamic programming utilities in MATLAB will enable determined economists to apply this technique to a variety of problems without further wait. In particular, the toolbox and related text authored by Miranda and Fackler provide an excellent introduction to the use of dynamic programming methods in economics.[264]

Distinguishing the Dynamic Programming Approach

Bellman's key insight began with segmenting an optimization problem into twoparts: the benefits this period, and the value of the expected discounted benefits in the future. For our purposes, we will separate the valuation problem into two parts: the return on an investment in the current period, and the eventual returns of both income and capital gains. While we are considering the market value of a firm, the technique was developed for use in optimizing a wide array of functions.

One of the advantages of the dynamic programming method in business valuation is its proper assertion of the primary importance of management policy. While traditional methods of valuation assume a very passive role for managers, dynamic programming puts management policy front-and-center.[264a] Although we are limited in our ability to quantitatively model

---

[264] M. Miranda and P. Fackler, *Applied Computational Economics and Finance* (Cambridge, MA: MIT Press, 2002); the related CompEcon toolbox is available from the MIT Press Web site. In addition, the Institut National de la Recherche Agronomique has developed a Markov Decision Toolbox for use with MATLAB. See the Web site of their Biometrie Intelligence Artificielle unit at: http://www.inra.fr/bia.

[264a] This is the case because we maximize over the control variables to solve the Bellman equation.

management policies, the dynamic programming method explicitly considered policy actions that are largely ignored in traditional cash flow models.[265]

A second advantage is that it better incorporates the complex uncertainty that exists in almost all business settings. A real, human manager will not set all expenses to grow smoothly, in the face of changing revenue. The dynamic programming method incorporates this explicitly compare this to the simple uncertainty that underlies most cash flow models.[266]

Mathematical Similarity of Methods

Both the real options and dynamic programming methods can be used to value business assets. If both produce a market value estimate that theoretically should be the same, then we would hope the two methods would be equivalent under a set of assumptions appropriate to the specific problem. The methods are similar mathematically, but are not identical. In particular, the source of discount rates and the availability of other traded assets are critical differences in assumptions.[267] These differences notwithstanding, given complete information, and some limiting assumptions, the two approaches produce the same answer.

Conclusion: Real Options and Dynamic Programming Methods

This section presents approaches that are either relatively new to the field of business valuation or almost unknown. Like most newer developments, these methods require more work and care to use in practical settings than other, more established methods. However, they both offer powerful analytical and practical advantages that, in some situations, should not be ignored.

Information Necessary To Value a Firm

We discussed above various methods of valuing a firm. In this section, we discuss a strongly related topic: what information is needed.

---

[265] To see this, look at a standard valuation text and note how the spreadsheets rarely even mention management policy. Of course, by forecasting expenses you are implicitly forecasting policy. However, such implicit policy forecasts are quite primitive, often not made explicit, and therefore likely to be both wrong and overlooked.

[266] If we return to the example of the manager's forecasting method in chapter 10, "Applications for Business," we see how the random-walk-with-drift approach is more sophisticated than assuming a trend growth with some deviation around it. (See Chapter 10, under "Simple Uncertainty: Deviation Around a Trend" on page 213.) However, many standard cash flow forecast models use exactly this assumption.

[267] See A.K. Dixit and R.S. Pindyck, *Investment under Uncertainty* (Princeton, NJ: Princeton University Press,1994), Section 4-3.

We organize the information necessary to value a firm into three groups: accounting information, economic assumptions, and management policies.

Accounting Information

1. *Historical sales and revenue:* This includes volume information as well as revenue. If there are different areas in which the products or services are delivered, or different lines of business, a segregation of those revenue streams and related costs would be helpful. If the valuation exercise is focused on one line of business, properly apportioned historical financial information is vital.
2. *Cost and profitability information:* The costs of the firm — both variable and fixed — as well as capital costs, interest and taxes, and noncash costs such as amortization and depreciation are required for an analysis using the income approach. If the firm is to be valued using a market or asset approach, then this information is still important but may not be decisive.

   Together with the other information collected, this must answer clearly the question "How does this firm make money?"
3. *Capital structure and discount rate information:* All companies must be capitalized in some manner, and the cost of that capital is the basis for the discount applied to future earnings. For large companies acquiring relatively small entities, this step may be straightforward. For closely held companies, it can be quite challenging. Note that the term "cost of capital" is often used loosely; so be certain you have independently estimated this critical variable and documented your assumptions.

   Of course, any recent changes in capital structure (including issuance of new stock, repurchases, options, substantial price changes, etc.) should be considered, as should unusual financing arrangements or the use of an ESOP or similar plan.
4. *History and form of business operation:* The type of business, its ownership, its management, and broad history are also essential information for valuing a firm. In firms with franchises or special licenses, a review of history will often reveal agreements carrying with them valuable rights.
5. *Any material, peculiar conditions:* Material events nondisclosed are a serious risk in the business of valuation. All the pencil lead in the world will not salvage an analysis of a firm that has its officers fleeing the ship; has had accounting policies abruptly changed; is losing money and does not know why; or is nearing bankruptcy or failing to pay its creditors.

Economic Assumptions

6. *Growth assumptions for the company:* As value is a forward-looking concept, the assumptions about the rate of growth in the future (even if it is negative) are critical. In our view, the forecasting of income growth is generally more important than the selection of a discount rate.[268]
7. *Economic conditions in the industry:* Always consider the industry in which the firm operates, its position in the market, and the size of its trade area. Good market research, even if it is rudimentary, will provide helpful limits on growth assumptions later on.
8. *Conditions in the geographic trade area:* For those companies operating in a specific trade area, the conditions in that trade area should be considered. For a retailer or wholesaler, is the population growing or declining? For firms operating on a nationwide basis, the trade area may indeed be the entire country, although even then certain areas of the country will often have significantly higher growth prospects than others.
9. *Value of franchises, licenses, and intangible assets:* The firm may have special franchise rights, licenses, trademarks, patents, or other intangibles that are essential to their future earnings capacity. These must be considered explicitly.[268a]
10. *Market transactions for similar businesses:* If market transactions involving similar enterprises can be identified, these should be given heavy weight.

Management Policies

11. *Management and management policies:* Who manages the firm is an essential aspect of the value of that firm. It may be that new management is expected to perform better than the previous management, or fix problems, or simply be a caretaker until a new owner is found. Management policies, such as

---

[268] See Chapter 10, "Applications for Business," especially under "Deriving the Present Value of a Financial Asset" on page 219, and under "Importance of Income Growth vs. Discount Rate" on page 222.

Note that, of our list of categories of information, at least 10 of the 12 categories apply primarily to forecasting income. While the risks of earning that income directly affect the discount rate (and therefore there can be no neat separation between the two), it is nonetheless instructive to simply count up the categories and note that more focus is on earnings growth than on the costs of capitalization. A review of the IRS regulations and Revenue Rulings provides similar guidance; see "Checklists for Information" in this chapter on page 270 and the notes to that section.

[268a] On the proper discounting for firms with cash flows subject to sudden termination, see P.L. Anderson, Valuation and damages for franchised business, AEG Working Paper 2003-12, available at: http://www.andersoneconomicgroup.com.

reinvestment practices, pricing strategy, marketing expenditures, and management appointments, are important. These vital assumptions should be considered explicitly.

12. *Important business plan assumptions:* If a company is planning a major expansion or has just completed one, will be losing one division or gaining another, is shifting production or changing processes, or is similarly changing the fundamental processes of its business, the valuation must recognize this.

Checklists for Information

The above list can form the basis for a checklist for requesting information. Checklists for documents are presented in a number of valuation references.[269] We emphasize in our list the fundamental drivers of value — including management, business plans, economic conditions, and franchises or other intangible assets — rather than a list of financial statements. This is, not coincidentally, also the approach followed by the IRS.[270]

Remember: Think First

Note that this list is one of categories and subcategories, not specific documents. In keeping with the first maxim of business economics — "think first, calculate later" — we suggest thinking carefully about the business before digging into a cashflow schedule. Recall that a number of now-defunct firms reported balance sheets healthy right up to the moment the market knew they were doomed. Alternatively, other firms limp along for years with significant

[269] See, e.g., "Research and Its Presentation," in J.R. Hitchner, Ed., *Financial Valuation: Applications and Models* (New York: John Wiley & Sons, 2003).

[270] Although Revenue Ruling 59-60 and the relevant sections of the Code of Federal Regulations are hardly models of efficient writing, they do require the IRS to consider the types of information listed above.

In terms of explicitly defining what the IRS must consider, Hitchner, cited in footnote 241, counts eight categories in RR 59-60, namely: (1) nature and history of business, (2) economic outlook for industry, (3) book value of the stock, (4) earning capacity, (5) dividend capacity, (6) whether the firm has goodwill or intangible value, (7) sales of stock, and (8) market price of stock in similar business. ("Selected Revenue Rulings," reprinting RR 59-60, Section 4.01 (a)–(h).)

This neat list does not appear in the IRS valuation sections of the Code of Federal Regulations but can be found in reprints of the original ruling. Revenue rulings can be found at: http://www.taxlinks.com.

Given the decades that have passed since 1959, and the advancements in finance and economics, RR 59-60's definition of the essential elements of business value were remarkably prescient. Compared with the eight categories from 1959, we would add — nearly a half-century later — *geographic trade area,* which could be part of the "economic conditions of the industry," although we consider it worth a separate category; *licenses and franchises,* which are probably subsumed in the goodwill category but have become much more important since 1959; an explicit *costs and profitability* category (reflective of the use of the income approach); and the uncounted *other material conditions* category. Reflective of the innovations in finance since 1959, we combine earnings capacity and dividend-paying capacity into one category.

problems. In either case, simply reviewing accounting statements would have given a valuation economist a false sense of the predicament of the firm.

Example Business Valuations

We now consider a small number of examples that illustrate the principles of valuation, and sketch analytical techniques that can be used widely. One application of these techniques is the estimation of lost profits for a business, which we also discuss later in this chapter.

Steps To Value a Business or Estimate Damages

Each valuation exercise should include the following steps:

1. Consider the economic role of the enterprise. This inquiry must answer the question "How does this firm make money?"[271]
2. Specify clearly the asset to be valued or the cause of the lost profits or other reductions in benefits for which damages are to be assessed. If only a portion of a business is to be sold, it is essential that the dividing line be clear in the valuation exercise. If you are estimating the value of the lost profits or other reduction in benefits to a business, you must similarly be quite explicit about what stream of benefits are being considered.

   Note that the valuation exercise is separate from ascertaining blame or deciding who are the "good" or "bad" parties in a dispute. That task may be more important than the valuation exercise, but it cannot unduly color the results of the valuation exercise.
3. Identify any trade restrictions, legal covenants, laws governing or restricting trade or competition, contracts or arrangements with suppliers and customers, and other factors that affect the market and potential profitability of the firm or line of business.
4. Select the best approach for valuing the business or estimating damages.
5. Value the business using the available data and the best approach.
6. Check the resulting value estimate against other sources, even if they are flawed.

[271] If the firm can be considered a bundle of assets, the intrinsic value of those assets should be considered. This is asking the question "How does this firm make money?" for each individual asset. The sum of the parts — and any advantages or disadvantages of their being combined — should also be considered.

7. Summarize the results, including the relevant assumptions and reasons for selecting certain methods.

Example: Beverage Distributors

Consider the value of a beer, wine, or liquor distributorship. Such a business is similar in type to other franchised businesses, as well as many retailers and wholesalers.

How Does It Make Money?

Such a distributor typically has an agreement with a supplier of alcoholic beverages, which allows the distributor to distribute specific brands of products in specific areas. From the difference between the wholesale price to the distributor and the price to the retailer, the distributor earns revenue. From that gross margin must be paid operating costs, capital costs, and taxes. The remainder is profits.

What Factors Govern the Future Stream of Income?

In the case of the distribution rights to a line of beer, the relevant stream of future benefits is defined principally by the following:

1. The market area in which the distributor has the rights to sell a specific brand of beer
2. The demand for that type of beer in that market area
3. The supply and price of substitute brands of beer and other beverages
4. The fixed and variable costs of distributing beer in that area, including the costs of promotion, marketing, and other tasks required of distributors by most suppliers
5. The selling price of that beer in that market area
6. The limitations in the state law and the franchise agreement on competition within the market area and on terminating the distribution agreement

What Is the Best Approach?

If market information (such as recent sales of similar distributors) is available, it should be given the strongest weight.

If not, the income method is the best and the only appropriate method. Asset-based methods would severely underestimate the business value of a profitable distributor, given that many of the key factors that ensure future income are intangible.[271a]

Shorthand Manner as a Comparison, Not a Method

As a check on our results, we could use imperfect but available "rule of thumb" information. In most industries, there is a shorthand manner of describing the value of assets. This is true in the alcoholic beverage industry as well; many industry participants discuss sales in terms of "dollars per case," meaning the sales price of the business divided by the volume of sales (in cases) for the trailing 12-month period.

A shorthand manner of *describing* a transaction, however, should not be confused with a method of *valuing* a business.[272]

Valuing the Distributor: Assumptions

In conducting this valuation exercise, we make a number of important assumptions which include:

1. The distributor is an ongoing enterprise.
2. The market structure in the area is not substantially disrupted by the change of the distribution rights for one line of beer, or from one distributor to another.
3. The consumer demand for specific lines of beer and the overall preference for beer relative to other competing beverages will not change substantially in the future.
4. No other legal, regulatory, geographic, income, or other disruptions of a substantial magnitude will occur in the near future.

Similar assumptions are often required in other industries as well.

Valuing the Distributor: Collecting the Data

We follow the strategy introduced in "Getting Data into a Model" on page 66, in which we collect data in a spreadsheet, import it into MATLAB for analysis and graphing, and then export a portion of it back for reporting. The following code fragment lists the variables that were collected for each business.[272a]

[271a] In this example, we are not considering the newer methods described in "Dynamic Programming and Real Options" on page 259.

[272] In this instance, the shorthand describes sales volume for the past, not earnings in the future. A buyer is only interested in *future* earnings and therefore will not base his or her decision on the shorthand measure.

[272a] As we are highlighting a technique for using data and algorithms systematically, rather than highlighting a specific algorithm, we do not explain each variable here.

Code Fragment 11-1. Input List for Valuation Function

```
{sales_rev_base, sales_case_base, ptr, sales_trend,
...
SalableAssets, discount_rate_liq_sales,
sales_decay_rate,
discount_rate_close, ...
earnings_pretax_share, earnings_pretax, ...
growth_low, growth_high, growth_mid, numperiods, ...
case_mult_low, case_mult_high, ...
income_tax_rate, eps_sub_ratio, sub_rate, ptr_sub, ...
wacc, wacc_low, wacc_high, lost_profit_alt};
```

Input sheets: Table 11-2 and Table 11-3 illustrate the worksheets that contain the input data for the business.

Table 11-2, "Cost-of-Capital Estimates for Small Beer Distributors, 2003," illustrates a cost-of-capital analysis for a specific industry — in this case, small, privately owned beer distributors, for the year 2003. Table 11-3, "Input Sheets for Individual Distributors," illustrates the input variables used for a sample company within this industry.

Creating the Structure

We now collect the information for a distributor, which includes a large number of variables, for use in our calculations. The method we recommend is to:

1. Collect the data on the input sheets with the relevant source and other notes.
2. Transfer that data into the MATLAB workspace using the Excel Link command or other methods.
3. Pack the information about each business into a data structure.

Using a structure collects all the data for this one distributor into a single object, which then can be saved, cleared from the workspace, reloaded in the future, and stored for future reference easily. We have presented a custom utility to do this (for information in this example) in Chapter 4, "Importing and Reporting Your Data," under "Example: Valuing Multiple Business" on page 79. The script we describe there collects all the necessary data from the

TABLE 11-2

Cost-of-Capital Estimates for Small Beer Distributors, 2003

Concept		Rates	
Equity			
	Equity cost for industry, simple CAPM[a]	11.1%	
	Less: discount for state laws or other factors[b]	–0.5%	
	Plus: micro-capitalization and franchise risk premium[c]	2.0%	
			12.6%
Debt			
	Prime rate[d]	4.7%	
	Premium for small businesses[e]	1.0%	
	Other adjustment	0.0%	
	Equals: pre-tax debt cost	5.7%	
	Times: 1-tax rate[f]	68.0%	
	Equals: after-tax debt cost		3.8%
Leverage			
	Share of equity[g]	70.0%	
	Share of debt	30.0%	
			100.0%
Adjusted Discount Rate			
	Weighted cost of capital[h] as adjusted		9.9%
Memo			
	Likely high discount rate[i]		11.9%
	Likely low discount rate		8.0%

[a]

Risk-free rate[1]	5.0%	
Plus: equity premium[2]	5.5%	
Times: industry beta[3]	1.10%	
Equals: industry cost-of-equity		11.1%

[1] Recent long-term Treasury note yields; historical averages are higher.

[2] Long-term risk premium over Treasury note yields; arithmetic average 1926–1990. Geometric average for same period is 7%. *Source:* A. Damodaran, *Investment Valuation* (New York: John Wiley & Sons, 1996.)

[3] Assumption based on wholesale trade of individual lines of nonessential commodities.

[b] Adjustment for less risk: minor brand.

[c] Differential for firms with capitalizations too small for public trading; increment for brand termination risk.

[d] Prime rate reported in Federal Reserve pub. E-2, note 6.

[e] Typical spread over prime, small creditworthy firms.

[f] Using an estimated average tax rate of 32%. *Source:* A. Damodaran. For food wholesalers statutory rate is 34%; lower rate allows for exclusions and noncorporate ownerships.

[g] Assumption for privately held firms; comparison figure for food wholesalers is 74%.

[h] Debt and equity costs, weighted by shares in leverage section.

[i] High and low are +/–20%.

TABLE 11-3

Input Sheets for Individual Distributors

Variable Name	Variables	Values
name_of_case	name fo firm	distA
Business Sales, Costs, and Profitability		
sales_case_base	Sales (cases) — base year	48,000
ptr	Price of case to retailer, average for year	$21.00
sales_rev_base	Calculated: sales revenue on case sales, base year	$1,008,000
earnings_pretax_share	Earnings pretax (after operating expenses, capital expenses)	4.95%
earnings_pretax	Calculated: pretax earnings	$49,896
Sales Growth and Projections		
sales_trend	Actual trend in annual sales (4-Year average rate of annual change)	–5.00%
growth_low	Expected annual revenue growth, low	–7.00%
growth_high	Expected annual growth, high	–3.75%
growth_mid	Expected annual growth, middle	–5.00%
numperiods	Number of periods to forecast sales at current profitability	7
Discount Rates		
wacc	Weighted average cost of capital, best estimate	10%
wacc_low	Weighted average cost of capital, low estimate	8%
wacc_high	Weighted average cost of capital, high estimate	12%
5-Year Lost Profits, Assuming Partial Substitution		
discount_rate_close	Discount rate (closing business) on profit	10%
sales_decay_rate	Sales decay rate	7%
eps_sub_ratio	Earnings pretax share; ratio of margin of substitute beer to current brands	0.675
sub_rate	Substitution ratio (portion of lost current brand substituted)	0.9
ptr_sub	Price to retailer, average, for substitute brands	17
Value-Per-Case Estimation Methods, Assumptions for Low and High Estimates		
case_mult_low	Case multiplier (low)	$5.50
case_mult_high	Case multiplier (high)	$11.00

workspace, checks for key variables, and then packs the information into a structure using a name provided by the analyst.[273]

Alternative Methods

As an alternative to a custom-programmed script such as this, you could also:

1. Keep the data in the workspace and perform any needed calculations on the variables. One drawback to this approach is the lack of saved information, unless a careful diary-style record is kept of both the data and the commands.
2. Save the data in standard numeric arrays (matrices), which might be saved all together in a `.mat` data file. The drawback to this approach is the confusion that results when you have more than one subject business.
3. Manually create a structure using the `struct` command or other methods. We also describe these methods in Chapter 4, "Importing and Reporting Your Data."

Estimating Business Values

Valuation Estimates by Multiple Methods

Once the data are assembled into a data structure, you can use multiple valuation formulas to provide plausible value estimates. We say "plausible" because no formula alone, even with good data, will always provide a good value estimate.

The multiple-method approach provides the analyst with a range of estimates, some quite plausible and some that should be rejected as improper.[274] The calculation of all these does not imply they are equally likely guesses; only that they are a set of calculations that could be performed on these data. A handful of these become, because the assumptions and methods are most proper for the case at hand, the estimates actually used. Each method is only briefly described below; the general approaches are described in "Approaches to Valuation" on page 251, as well as in "Steps to Value a Business or Estimate Damages" on page 271. In addition, a practitioner in this field should have one or more references besides this book.[275]

---

[273] This same method is included in Code Fragment 11-2, "Valuation Session for Beer Distributors," on page 290.

[274] In particular, if the business fails the going-concern test, the liquidation value estimate will become very important. For a true going concern, it is not important. However, for a firm that is struggling and may have to close its doors or restructure significantly, an income approach based on the past few years' results (especially if one of those years was a profitable one) will almost always lead to an exaggerated value estimate.

[275] See the references listed in the footnotes to "Valuation Methods" on pages 251 to 253, as well as the other works cited in this chapter.

Furthermore, the calculations done under each method are first-cut analyses, and would often require additional thought and adjustment.

Method 1: Liquidation

The liquidation method is appropriate when the firm is not a going concern or when the assets themselves will be sold quickly. Liquidation value should always be considered conceptually to be certain that the going-concern assumption is properly made or not made.

In addition, the liquidation value of the hard assets of a firm often become a portion of the price of a going concern. In such cases, the buyer pays some amount for the potential earning power of the business itself plus some amount for the inventory and other assets in active use.[276] The liquidation value of the firm's assets are often a first approximation for this latter category.

Method 2: 5-Year Lost Profits, with Adjustment

Using this method, the profits for a line of business are forecasted over a certain period, assuming that there is some substitution of sales for the discontinued (or sold) lines of business. The income from the substituted product line offsets some of the value placed on the line of business that is discontinued or sold.

This approach is appropriate to estimate damages for distributors who lose a certain product and therefore the profits that would result from it, but have the opportunity to earn profits on other products using the same facilities and business model. Note that the rate of substitution and the profits on the substituted lines are critical assumptions. Assuming zero substitution — an implausible assumption for most companies that remain in business — has the effect of valuing the entire lost profits stream without any offset.

Methods 3 and 14: Per-Case Multiples

The first step in our valuation methodology was to consider the economic value or intrinsic business value of the enterprise. This must answer the question "How does this firm make money?"clearly.[277]

Businesses do not make money on volume. They make money by receiving more revenue than they pay out in expenses. Therefore, while discussing the value of a firm in dollars-per-unit-volume terms may be convenient, you should not forget that *past volume is not the same as future profits.*

This axiom notwithstanding, it is common in numerous industries to discuss the market value of a firm by dividing it by some measure of volume. In the stock market, analysts may talk about "earnings per share." That is a

---

[276] Typically, the inventory of goods should be considered separately from the hard assets of the firm since the manufacturer or another buyer will often purchase or repurchase undamaged, fresh inventory at a modest discount. The other assets (such as equipment, vehicles, furniture and fixtures, warehouses, etc.) frequently have market values that are substantially discounted from their historic costs.

[277] See "Information Necessary to Value a Firm" on page 267.

convenient measure, but it does not mean that issuing more shares will result in more earnings. In many retail and wholesale trade industries, market value may be summarized on a per-square-foot or per-store ratio. Again, simply adding stores does not mean profits will increase proportionately. These metrics are, in reality, combinations of indicators for efficiency, current profitability, industry profit margins and competitive factors, all rolled up into one handy multiple.

In the beer distribution business, the common metric is "dollars per case," meaning the market value of a distribution business divided by the trailing 12-months sales on a case basis.[278] Using some judgment about past sales and a healthy skepticism about unverified claims of sale prices, this method relies upon directly estimating the market value of a distributor by using a multiple of the past year's case sales.

High and low comparison values: Because this metric is more reliable as a comparison than a direct estimator of value, we use a high- and low-multiple calculation to provide a plausible range.

Methods 5, 6, and 7: Capitalized Income

The standard methods within the income approach are based on discounting the future cash flows to a company. We present three variations here. The first is known as the free cash flow to the firm (FCFF) method and is often the best method when reliable financial statements and earnings projections are available, but no direct market value information on comparable firms can be obtained.[278a] The FCFF method is also commonly used to estimate damages that occur when a portion of a business is interrupted, preventing the firm from earning the profits on that portion of the business for a certain time.

Multiple assumptions: Because the assumptions used in an FCFF method are so critical, in this routine we automatically use low, middle, and high estimates for the key variables of earnings growth and weighted average cost of capital (WACC). These result in three different estimates, which are listed as different "methods" but are actually generated with the same algorithm, but with different assumptions. Note that the final estimate must have debt and equity weighted by market value, which is discussed below.

Related Free Cash Flow to Equity (FCFE) method: There is a related method known as free cash flow to equity, or FCFE. The key distinction is whether cash flows measured go to the firm as a whole or to the owners of the firm itself. These methods will produce equivalent estimates, provided consistent assumptions are used for both.

Using proper discount rate: Beware of using assumptions geared for one method in the other. In particular, note that the WACC is an estimate of the

---

[278] Most industries have a peculiar set of definitions to accompany these metrics. In this industry, a "case" is the liquid equivalent to 24 bottles of 12-oz beverages. Noncase sales (such as kegs or large bottles) are then converted to this standard measure.

[278a] Some references use "net cashflow."

implicit discount rate for the firm as a whole, not the equity holders. The use of a firm-wide cost of capital must be reserved for firm wide cash flow discounting. For FCFE calculations, the appropriate discount rate is that for equity holders only.

Using market value to weight capital structure: Another common error in valuation is to calculate a WACC, with the weights being the book value of debt and equity. The actual weights should be the market value of debt and equity. For firms with substantial intangible value, the market value of equity will typically be a much higher share of the total capitalization than the book value. Reading a balance sheet for a firm, based on historical cost accounting, and using those figures to weight the cost of capital will often result in a substantial overestimate of the market value of a firm.[279]

Recall that the value of equity will be based on investor's views about the capitalized value of future earnings of the firm. Since you are trying to estimate the results of those views (a market price), be aware that cost of capital and market value are determined simultaneously in the market as a whole. Many finance and valuation texts imply that one variable (usually WACC) can be obtained from a reference source and then simply plugged into a formula to produce — Voila! — the market value. If true market values are available, then from those market values we can calculate the discount rates implied or the earnings forecast implied, given certain other assumptions. However, if market value is not known, which is usually the case with private firms, then it cannot be used to simply calculate the weightings. Pratt provides an example in which the use of book values to weight the capital structure of a private firm — with all other variables being known — produces an overestimate of 30%.[280]

Iterative methods to weight capital: There is a practical solution to this quandary. First, use the book value to prepare a first guess of market value and use this to prepare a first guess of WACC. Use this, in turn, to prepare a second guess of market value. Use the second guess of market value to prepare the second guess of WACC and continue until the results are close enough to warrant no further iterations. These iterations are not difficult to perform manually on a standard spreadsheet,[280a] and the entire process can be programmed in MATLAB.[281] We present a similar application using a simulation model in Chapter 10, "Applications for Business," under "Iterative Market Value and Tax Model" on page 226. The Business

---

279 In the (normal) case where the cost of debt is lower than the cost of equity, understating the portion of the capitalization that is equity will result in an artificially low WACC estimate, resulting in an artificially high valuation estimate.

280 S. Pratt, *Cost of Capital*, 2nd ed. (New York: John Wiley & Sons, 2002), Chapter 7.

280a J. Abrams, in *Quantitative Business Valuation* (Irwin, 2001) presents a spreadsheet method in Chapter 6. Abrams earlier paper on this is "An iterative valuation approach," *Business Valuation Review*, March 1995.

281 Indeed, one of the powerful features of programming environments such as MATLAB, and a key distinction between them and spreadsheets, is that optimization (or simulation) can be performed natively. While a spreadsheet can be forced to iteratively calculate to solve a relatively simple problem like this, it is much better to use a tool designed for this use.

Economics Toolbox contains the function *Iterate Value* for this specific purpose.

Other Methods

Because every case is different, even a routine designed to estimate the value of a single type of firm in a single industry makes explicit allowance for some alternate measure of value or damages. You will note the alt (alternative) variable in the routine, which is there for this purpose. Such an alternative could be produced by another expert, by another measure, or through some yet undefined means. Its appearance in the calculation and comparison of plausible values allows it to be compared directly with the results of different methods.

Graphical Output from Valuation Calculations

The valuation function creates a set of figures. One of these is a bar graph showing all the plausible values calculated by the function, which is shown in Figure 11-1, "Plausible values from valuation command."

Summary of Valuation Session

Code Fragment 11-2, "Valuation Session for Beer Distributors" on page 290 in the Appendix to this chapter summarizes the complete session, starting with collecting and packing the input data, making the valuation calculations, and then saving the results.

Importance of Judgment

None of these methods, even with good information, is enough to accurately predict sales prices every time. Given *imperfect* information and the fact that the best economist is one that gets closest to what people think in their very human heads, it is especially important to realize the limitations of any one method.

On top of the limitations of a method, there are limitations on the ability of trained economists and other valuation analysts to select the proper discount rates, costs of capital, growth rates, expense ratios, and other factors. Therefore, it is vital that the analyst considers the weaknesses in data and methodology and makes appropriate allowances for his or her own limitations.

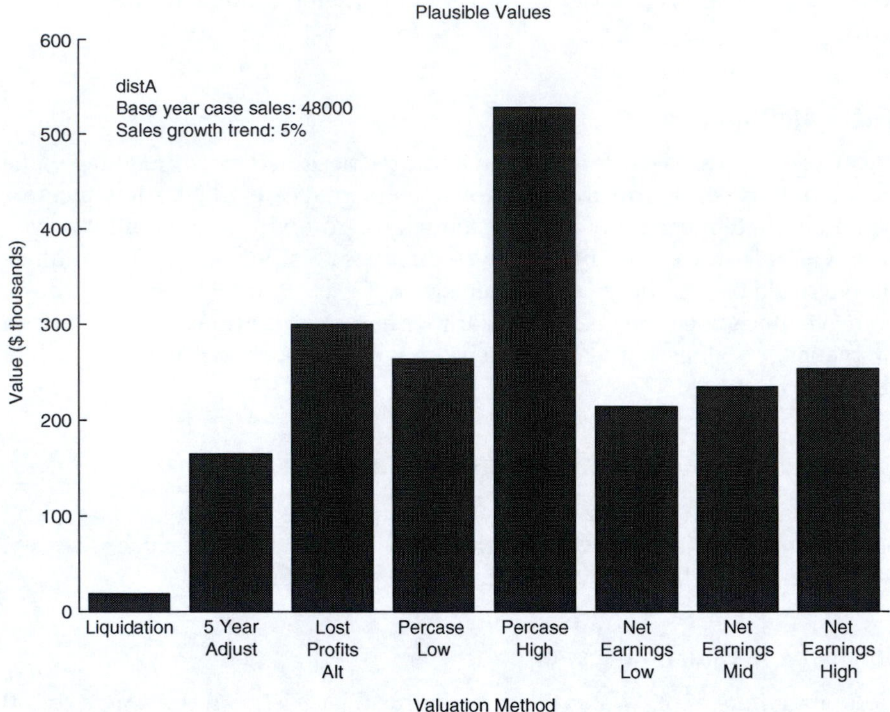

FIGURE 11-1
Plausible values from valuation command.

Rational Damages

The preceding discussion presumes that "lost profits" is the appropriate standard for damages in cases in which one business has, in violation of law or contract, deprived another of the ability to fulfill its business plans.

This use of the lost profits standard is well motivated and is normally the starting point for assessing the proper damages. However, we should consider the broader question of what should be the standard. In a number of areas, the law may establish a specific standard which sometimes will be higher than the actual lost profits.

In this section we discuss why such a level of damages could be consistent with the structure of the industry, using the results of recent inquiry into "optimal sanctions" by economists and others interested in law and economics.

Economic Structure of Corporate Law

The use of corporations and other forms of business persons has a long legal history. However, the underlying economic rationale behind the use of legally organized teams of people has only recently been the subject of intense economic analysis. This field of inquiry was largely begun by Ronald Coase, the 1991 Nobel Prize winner in Economics.[282] His famous, though opaque, "Coase Theorem" described the economic rationale for business organizations and the motivation for persons to voluntarily negotiate an efficient contractual agreement, provided property rights were secured.[283]

Subsequent to this event, a "law and economics" school of thought has arisen which attempts to apply economic reasoning to issues in contract, tort, and criminal law.[284]

Economics of Sanctions

One application of "law and economics" is the derivation of "optimal sanctions" or the best penalty to apply for a specific violation of law. One reference in the field explained:

---

[282] The prize was awarded "For his discovery and clarification of the significance of transaction costs and property rights for the traditional structure and functioning of the economy." See, e.g., University of Victoria (British Columbia, Canada), School of Economics, at: http://web.uvic.ca/econ/nobel.html.

The famous article that spawned the "theorem" was Ronald H. Coase, The problem of social cost, *J. Labor Econ.*, 3, 1, 1960. An equally influential article, which dealt with the firm, was R.H. Coase, The nature of the firm, *Economica*, 4, 386–405, 1937, reprinted in A.T. Kronman and R.A. Posner, Eds., *The Economics of Contract Law* (Boston: Little, Brown, 1979), pp. 31–32. See N. Foss, H. Lando, and S. Thomsen, The Theory of the Firm, in *Encyclopedia of Law and Economics*, 5610; University of Ghent, Belgium, 1996–2000; also available on the Web at: http://encyclo.findlaw.com/index.html.

[283] The "theorem" more explicitly holds that individuals will tend to negotiate an efficient outcome to most problems of resource allocation (including who pays what to whom for producing goods and services and how to deal with externalities such as effects of one person's actions on another), provided property rights are secured and transaction costs are minimized.

The existence of companies is supported by Coase's reasoning by noting how the costs of transactions and information became prohibitive for individuals who attempted to arrange individually all their economic transactions. Thus, the most efficient way (the way that wastes the least amount of resources) to arrange things is often to combine different persons into a company and appoint a manager to run the company.

See, e.g., D. Friedman, The World According to Coase, *The Law School Record* (n.d.) University of Chicago Law School; found at: http://www.daviddfriedman.com/Academic/Academic.html.

[284] See, e.g., R. Cooter and T. Ulen, authors of *Law and Economics*, found at: http://www.cooter-ulen.com; and the *Encyclopedia of Law and Economics*, cited previously.

> We use here the economics of sanction. The objective of a legal rule is to deter certain undesirable behavior without deterring (too much) beneficial behavior. Rules should minimize the sum of losses from (a) undesirable behavior that rules permit, (b) desirable behavior that the laws deter, and (c) the costs of enforcement.[285]

The desirable and undesirable behavior and the costs of enforcement will vary with industry. We therefore look at some example industries to see if optimal sanctions in those industries are any different from straightforward lost profits.

Optimal Damages and Franchise Disputes

Many industries, including beer and wine distribution, automobile retailing, chain restaurants, video rentals, and other retailers are set up as franchises. In many such instances, the manufacturer (or equivalent franchisor) grants exclusive trade areas to distributors, who are effectively prevented from representing the brands of competing franchisors. Such a structure creates a vulnerability for the franchisee, as most franchise agreements require the approval of the franchisor before selling the business to another person. Without the consent of the manufacturer, the business becomes almost worthless.[286]

The legislature in states such as Texas, Wisconsin, Michigan, and Illinios have adopted the "fair dealing" laws to prevent abuses by manufacturers in such cases, particularly in the automobile retailing and beer and wine distribution industries. In doing so, the legislature sometimes establishes a damage remedy that is the full market value of the firm. Such a sanction would generally be larger than the damages calculated on a lost profits basis and indeed may be a few times larger. Does such a remedy have a strong rationale?

Under the "law and economics" scrutiny outlined above, in which optimal sanctions should be determined as those that minimize the total costs to society, a damage remedy should be higher than proven lost profits in order to deter manufacturers from abusing their power. It is not clear that the full market value of the firm is the proper sanction; the ratio between actual lost profits and market value of the firm could vary substantially. However, the use of a standard above that of simple lost profits is clearly justified.

Antitrust

Federal law sets a treble-damages sanction for certain antitrust violations. In this case, the difficulty in proving violations of antitrust laws and the

---

[285] Frank Easterbrook and Daniel Fischel, Optimal damages, in *The Economic Structure of Corporate Law* (Cambridge, MA: Harvard University Press), 1991.

[286] By "business" here we mean the value as a profitable, going concern, selling the Franchise-brand products.

severe damage such violations can have on both employers and consumers has resulted in a sanction deliberately set high as a deterrent.

However, sanctions that are too high can be counterproductive. They encourage the use of the law as a tool for harassment and even extortion. The threat of losing a court case with high damages, even if small, can be used to pressure companies to make payoffs to litigants who, if the legal process was completely fair and cost-free, would walk away with nothing. In the antitrust arena, the federal government's long-running cases against Microsoft and, earlier, IBM, illustrate the difficulty of proving an antitrust case against a determined and well-financed company. In the IBM case it also shows how the costs of the litigation itself can drain resources away from employers who need them to pursue investments in new markets and technologies.

Punitive Damages

In some cases, juries are allowed to recommend punitive (or "exemplary") damages as well as compensatory damages. In recent years, higher courts have restricted some of the more outrageous examples of runaway punitive damages.

The risks of punitive damages probably have a salutory effect on certain companies that are tempted to behave poorly. However, they also have the effect of hurting employers who are trying to produce good products pursue innovation, or simply provide products that are misused. We note the controversy primarily to advise economists and other readers of this book to avoid using economic methods to suggest punitive damages, unless there is an explicit reason for doing so. If punitive damages are indicated as a matter of law, then the judge or jury can often establish these. Avoid mixing in your own unlabeled estimate of punitive damages with the compensatory damages you actually calculated.

The Discount Rate

Many aspects of valuation and damages assessment involve the use of a discount rate. Such a rate converts a multi-period stream of money into a lump sum, or vice versa. Obviously, the selection of this rate has a significant effect on the overall valuation or damage award, particularly if the stream being discounted (or projected into the future) is long.

General Principle: Match Rate with Purpose, Duration, and Risk

The general principle for selecting a discount rate is to match the rate with the purpose of the discounting, as well as the time frame and the appropriate risk of the cash flow stream.

If the purpose is to discount future earnings from equity investments in large firms, then the return on similar securities is a good place to start. Conversely, if someone is carefully saving money in a liquid investment, then discounting the future earnings from that pool of money with a stock market yield would be inappropriate.

Risk and Return

The fundamental financial principle that risk must be compensated should be kept in mind when selecting a discount rate. Equities earn, on average, higher returns than high-grade bonds, which in turn earn higher returns than U.S. Treasuries. The risk profile of the asset in question should match the underlying securities for which a return has been calculated for use as a discount rate.

Data Sources

There are a number of sources for discount-rate information. The U.S. Federal Reserve Board and its member banks (particularly the St. Louis Fed) provide extensive information on interest rates and some information on the equity markets.[287] The best-known provider of historic equity market information is Ibbotson Associates, whose annual statistical guides are widely used in valuations.[288]

Data Are Not Methods

While historic information is important in estimating a discount rate, keep in mind that data — even good data — are no substitute for sound thinking and the proper method. There are many valuation and damage reports circulating with extensive discount-rate analysis but with extremely poor analysis of the underlying business. Understand the business first, then discount its earnings.

---

[287] The Federal Reserve System Web site is at: http://www.federalreserve.gov. The Fed H.15 release includes interest rate data for Treasury securities, commercial paper, high-grade corporate bonds, and state and local bonds.

The St. Louis Fed Web site is at: http://www.stlouisfed.org. The St. Louis Fed has a history of independent analysis of economic data, particularly on monetary policy. It provides the online FRED II database and the well-written and graphically excellent *National Economic Trends and Monetary Trends* publications.

[288] The Ibbotson Web site is at: http://www.ibbotson.com. The *SBBI Yearbooks* contain historic return information on stocks and bonds, including risk premia by size and industry. Ibbotson also provides estimates of stock-price betas in the *Beta Books*. Before naively using these data, note the guidance below and above in this Chapter.

Small- and Medium-Size Businesses

A common mistake is to assume that investments in privately held businesses are like investments in the stock market, and therefore an index like the S&P 500 or the Dow Jones Industrials is an appropriate indicator of a small firm's cost of equity capital. In fact, unless you are discounting another portfolio of investments in large-cap, New York Stock Exchange-listed companies, it is probably the wrong discount rate.

Most companies are small businesses that do not issue publicly traded stocks. They are often closely held and the owners may have some degree of direct or indirect personal responsibility for the business. These companies are often quite profitable and employ the majority of private sector workers in America. However, equity investors in these firms must face significant additional risk and restrictions when compared with owners of publicly traded securities. These include:

- Limitations on marketability.
- Higher risk in general.
- Other covenants, such as partnership agreements or loan covenants, that restrict the ability to sell interests.

Why then does anyone invest in small businesses? There are obviously benefits, which include:

- The ability to work in a protected or niche market.
- The ability to exploit particular skills.
- The ability to earn higher returns which are sufficient to compensate for the risk and restrictions.
- The hard-to-quantify desire to be your own boss.

Furthermore, we have noted that some owners of "small" business keep them very strongly capitalized, borrowing far less than the leverage bank lenders would allow. Does this mean these individuals are naive investors? Be careful before using assumptions that may seem warranted for publicly traded firms when analyzing small firms.

For all these reasons, investments in most companies should not be discounted using the returns from an index of large-cap, publicly traded companies.

Capital Asset Pricing Model (CAPM) and Small Businesses

CAPM is a powerful organizing principle in the field of finance. There are many data sources that explicitly or implicitly assume the CAPM model. While CAPM and its modifications are excellent starting points for understanding the market value of a large, publicly traded company, it often fails as an approach for smaller businesses.

This failure is largely due to both a lack of data on these companies and a theoretical framework that assumes a large number of potential investments with a range of risk and reward characteristics. CAPM is a good starting point for a discussion of institutional investor behavior and the behavior of investors with significant funds and a long time horizon. However, CAPM assumptions generally fail in the case of small business investors who also work at the firm in which they have invested.

Consider a prototypical investor in this sector. He or she has two clear options:

1. Decide to continue as an employee and invest additional funds in the stock market.
2. Invest those funds to become a part-owner of a small business and accept the restrictions that are normally attached to owner-operators of small firms.

Those are choices that evidence a range of risk and reward; however, they are not choices that define a continuum of investments. This does not mean that the CAPM model does not apply to small businesses; it does mean that substantial adjustment and a search for relevant data will often be required.

A Note on Discount and Capitalization Rates

The economics literature uses the term discount rate in a general sense as the key parameter converting a stream of future benefits to a lump sum today.[289] One of the uses of a discount rate is to capitalize a stream of earnings into a present-value equivalent. In such cases, the discount rate used is also called a capitalization rate or "cap rate."

When using a discount rate to capitalize a future stream of benefits, the economics literature makes little distinction between the terms "capitalization rate" and "discount rate."[290]

However, the terms are not always used interchangeably in the finance world, particularly in those areas where the literature is dominated by terms originating in accounting rather than economics. For example, some articles

---

[289] By general, we mean of broad application. The derivation of discount rates in the economics literature is quite rigorous, starting with the utility theory of the consumer and including both time and risk.

[290] For example, see Stephen F. LeRoy, Present Value, in *The New Palgrave Dictionary of Economics*, Vol. 3 (New York: Stockton, 1991), in which the "discount rate" is used explicitly to "capitalize" future earnings. The *Palgrave* does not even have an entry for capitalization rate.

See also Hal Varian, *Microeconomic Theory*, 3rd ed. (New York: Norton, 1992), Chapter 20. Varian's seminal microeconomic text presents a rigorous discussion of intertemporal utility to derive the role of the interest rate and explicitly describes asset markets, the Capital Asset Pricing Model, and even the Arbitrage Pricing Model. Yet there is no entry in the index for "capitalization rate," and the discount factor and various rates of return are all described consistently with the notion of a disount rate being used to capitalize future benefits.

state that capitalization is a specialized form of discounting, which implies a single discount rate and growth assumption in perpetuity.[291] A variation on this — embedded in a definition by the American Society of Appraisers — is a definition of capitalization that confines its usage to one shorthand mathematical formula.[292] This is much more restrictive than separate definitions in the same document, which indicate that discounting and capitalization are both methods that convert future streams of income to a present value.[293]

Some authors confuse the use of a discount rate with the use of a particular theory or approach for estimating the discount rate, such as the CAPM.[294] It is not uncommon to see the capitalization of income method described as a totally different method from discounted cash flow, even when the cash flows in question and the discount rates are quite similar.

One influential authority in this field uses the convention that a capitalization of income model means that a single rate was used to discount future earnings and to predict future changes in income, while a discounted income approach means that variations in earnings or discount rates over time could be involved.[295] Under this convention there is no mathematical difference, but the nomenclature of "capitalization" implies a restrictive set of assumptions.

Finally, the seminal work in public sector cost–benefit analysis, Edward Gramlich's *Benefit–Cost Analysis of Government Programs* (Englewood Cliffs, NJ: Prentice Hall, 1981) devotes the entire Chapter 6 to the discussion of discount rates and even derives the capitalization rate shorthand formula described in a following note. However, the term capitalization rate does not appear in the index, and no distinction is made between using a discount rate for multiple periods and a capitalization rate when the mathematically equivalent shorthand formula is available.

291 For example, see Randy Swad, Discount and capitalization rates in business valuations, *New York CPA Journal*, October 1994; found at: http://www.nysscpa.org/cpajournal/old/16373958.htm.

292 American Society of Appraisers, Business Valuation Standards; "Income Approach," and "Definitions;" reprinted in S.P. Pratt, *Valuing a Business*, 3rd. ed. (New York: McGraw-Hill, 1996).

Using a capitalization rate of 10%, a perpetual stream of income is worth 10 times the initial year's income. Of course, using a discount rate of 10% produces the same result. Indeed, as shown in many economics (and some valuation) texts, the long form equation for discounting repeated cash flows over multiple periods using the same discount rate reduces to this shorthand formula.

Furthermore, a similar shorthand mathematical formula (PV = current earnings/factor) can be calculated for many combinations of future growth and discount rate assumptions, not just under perpetual discounting and constant growth assumptions. Indeed, the ubiquitous "price/earnings ratio" used in discussions of stocks is nothing more than a quick calculation of this type of equation, using restrictive (and often unreasonable) assumptions.

293 See their definitions of capitalization and discounting.

294 For example, the Swad article cited above *defines* the discount rate as a derivation from a CAPM equation.

295 Pratt et al., *Valuing a Business*, 3rd. ed. (New York: McGraw-Hill, 1996), Chapter 9, "Relationship between Discount Rates and Capitalization Rates." Pratt cites the American Society of Appraisers BVS standards, although his book is careful to describe mathematically how discount rates and capitalization rates (using his definition) are related.

See also Pratt, *Cost of Capital*, 2nd. ed. (New York: John Wiley & Sons, 2002), Chapter 4.

See also, e.g., Darrell Koehlinger and Robert Strachota, Rates of return — direct capitalization versus discounting, in the online journal *Valuation Viewpoint;* found at: http://www.shenehon.com/Library/valuation_viewpoint/ratesof.htm:

"Discount Rate = Capitalization Rate + Change Over the Holding Period."

A Recommendation

All authorities in this field note that the selection of a discount rate must match its intended purpose. However, the confusion in nomenclature can lead to errors in application and a failure to understand the methods being used.

Economists will likely cling to the notion that discount rates can be used to discount and capitalize streams of income, and then proceed to debate heartily the discount rate to be used. However, be aware that many readers of the accounting literature perceive these concepts differently, and be careful to note the definition of these terms in reports you write.

Appendix

Code Fragment 11-2. Valuation Session for Beer Distributors

```
%   Beer Distributors Example Session,

%     I. Get Data from Excel Sheets, via Excel Link

%        Pack into data structures

whos;      %Check to see what is in
           %workspace.

data_bd;   %Script to check data and
           %create structure; script will
           %ask user for name of structure;

%Note: this must be done separately for each data
%structure.

%In book example, this was done twice, for "distA"
%and "distB"

%clear     %clear workspace and put in second set of
%data

%data_bd   %pack second data structure
```

```
save distributors.mat distA
save distributors.mat distB -append

clear

%   II. Load data file
load distributors.mat
whos

%   III. Valuation Commands
%   Results, and figures, will then be in workspace

[valuesA, valdataA] = val_bd(distA)

[valuesB, valdataB] = val_bd(distB)

%   IV. Save Valuation Results to .mat File
%   Note: functional form of SAVE

save('distributors.mat',  'valuesA',
'valdataA','valuesB', ...
    'valdataB', '-append')

clear
```

Code Fragment 11-3. Valuation of Beer Distributors Function

```
function [values, valdata] = val_bd(valinputs);
%VALUATION of BEER DISTRIBUTORS by multiple methods
function,
%   For use with valuation data input structure
%   created by DATA_BD or similar script, or
%   ... manually.
%   part of Business Economics Toolbox
%   Use by license only.
```

```
% Utilize various valuation methodologies--see
% subfunctions below

% Method 1: Liquidation

Val_Liq = val_liq(SalableAssets,
discount_rate_liq_sales);

%-----------------------------------------------

% Method 2: 5-year lost profits, with adjustment

% Projects sales (cases) over five years, assuming
% wind-down of business and substitution;

% full-profit sales at rapidly declining rate

% substitution sales at lower profit margin (than
% imported beer)

sub_time = [1:5];

[sales_sub sales_adj_current sales_adj_sub ] =
SalesSub(sub_time, sub_rate, sales_case_base,
sales_decay_rate, ...

    sales_trend, name_of_case);

[val_5_adj, cf_adj, earnings_sub] =
EarningsSub(eps_sub_ratio, earnings_pretax_share,
ptr, ptr_sub, ...

    sales_adj_current, sales_adj_sub,
discount_rate_close);

%   Note: directly entered vw calculations use field
%   in structure: lost_profit_alt

%-----------------------------------------------

% Method 3: Per Case Model (low)

Val_PerCase_Low = sales_case_base*case_mult_low;


%-----------------------------------------------
```

```
% Method 4: Per Case Model (high)

Val_PerCase_High = sales_case_base*case_mult_high;

%--------------------------------------------------

% Method 5: FCFF - low---uses conservative growth
% assumptions, and high discount rate

%note: trend_project produces three cash flows in
%the first output matrix;

%we use the middle one (without the deviation
%indicated by the third argument)


[CF_low, CF_low_info] =
TrendProject(earnings_pretax, growth_low, .05,
numperiods);

CF_low = CF_low(:,2);

Val_FCFF_low = pvvar(CF_low, wacc_high);

%--------------------------------------------------

% Method 6: FCFF_mid--used middle growth projection,
% and directly-estimated WACC discount

[CF_mid, CF_mid_info] = trend_project
(earnings_pretax,sales_trend, .05, numperiods);

CF_mid = CF_mid(:,2);


%   calculate wacc as mean of low and high estimates

% wacc = mean([wacc_low, wacc_high]);

Val_FCFF_mid = pvvar(CF_mid, wacc);

%--------------------------------------------------

% Method 7: FCFF - high--uses optimistic growth
% projections, and low discount rate

[CF_high, CF_high_info] = trend_project
```

```
(earnings_pretax, growth_high, .05, numperiods);

CF_high = CF_high(:,2);

Val_FCFF_high = pvvar(CF_high, wacc_low);

%-------------------------------------------------

%----------------Functions------------------------

%-------------------------------------------------

function Val_Liq = val_liq(SalableAssets,
discount_rate_liq_sales);

Val_Liq = SalableAssets*(1-discount_rate_liq_sales);

% Note: Saleable Assets (non-inventory)


%-------------------------------------------------

%-

function [sales_sub, sales_adj_current,
sales_adj_sub ] = SalesSub(sub_time, sub_rate, ...

    sales_case_base, sales_decay_rate, sales_trend,
name_of_case)

tindex=size(sub_time, 2);   %create index for number
of periods

for i = 1:tindex

    sales_trend_current(i) =
sales_case_base*((1+sales_trend)^i);

    sales_adj_current(i) =
sales_trend_current(i)*((1-sales_decay_rate)^i);

    sales_adj_sub(i) = (sales_trend_current(i)-
sales_adj_current(i))*sub_rate;

end

sales_sub = sales_adj_current + sales_adj_sub;

%   graph sales in sub scenario
```

```
%   Note: cell array, for use in plotting

sub_info{1,1} = name_of_case;

sub_info{2,1} = ['Sales Decay Rate, Substitution
Scenario: ', num2str(sales_decay_rate*100), '%'];

sub_info{3,1} = ['Sales growth trend: ',
num2str(sales_trend*100), '%'];


figure

subplot(2,1,1)

plot(sub_time, sales_adj_current, 'r', sub_time,
sales_adj_sub, 'b');

legend('current', 'substitute');

title('Case Sales of Current and Substitute
brands');

subplot(2,1,2)

plot(sub_time, sales_sub, 'g', sub_time,
sales_trend_current,'c.-', 'linewidth', 2);

legend('Substitution Scenario', 'Current Sales
Trend');

title('total case sales, substitution method');

set(gcf, 'Name', 'Substitution Sales', 'Tag',
'Substitution');

text(0.05,0.30, sub_info,'units', 'normalized',
'fontangle','italic','fontname','Times New
Roman','fontsize',12);


%-----------------------------------------------


function [val_5_adj, cf_adj, earnings_sub] =
EarningsSub(eps_sub_ratio, earnings_pretax_share,
ptr, ptr_sub, ...

    sales_adj_current, sales_adj_sub,
discount_rate_close);

%   calc net earnings in sub scenario; valuation


%   earnings rate on substitute brands, calculated
%   as a ratio of the
```

```
%   earnings rate on current brand; note ptr is
%   different for sub brand

%   also

earnings_sub = eps_sub_ratio *
earnings_pretax_share;

%   earnings cash flows, current and sub brands

% cf_adj_current  =
% earnings_pretax_share.*sales_adj_current*ptr;

%cf_adj_sub      =
earnings_sub.*sales_adj_sub*ptr_sub;

%   earnings stream totaled from both brands

cf_adj = cf_adj_current + cf_adj_sub;

%   valuation as discounted sum

val_5_adj = pvvar(cf_adj, discount_rate_close);

%   Graph cash flow

CFS = [cf_adj_current' cf_adj_sub'];

figure;

bar(CFS, 'stacked');

legend('Cash flow from current brand', 'Cash flow
from substitute');

title('Earnings from Current and Substitute
Brands');

set(gcf, 'Name', 'Substitution Scenario Earnings',
'Tag', 'Sub Earnings');
```

12

Applications for Finance

Finance and Economics

Finance is a field within economics, which is devoted to the study of the capitalization of enterprises and households. By capitalization we mean the various forms of financing and operating, including borrowing (debt) and direct investment (equity). How an enterprise is financed affects its growth prospects, its current profits, and its fair market value.[296]

Interest in finance has grown rapidly in the past few decades, and today there are many times more workers employed in occupations more related to finance than economics.[297] Unfortunately, that growth in interest has not coincided with a growth in the understanding of the economic underpinnings of financial transactions.

In this chapter, we will focus on finance applications within business economics and, in particular, look at the economic underpinnings of business transactions.

The applications we will consider include:

1. Consumer lending, using home mortgages as an example
2. Depreciation of assets, using automobiles as an example
3. Investment portfolios and, in particular, the application of the mean-variance portfolio analysis techniques
4. Other applications such as growth models and credit risk

Chapter 10 and 11 deal directly with finance applications for business.

[296] We discussed earlier two concepts that, along with risk, determine the fair market value of a financial asset: the expected future earnings and the discount rate. (See Chapter 10, under "Earnings or Discount Rate?"). If we broaden the concept of financial assets to include other assets such as depreciating assets (automobiles and homes financed by mortgage loans, both of which provide a stream of noncash benefits), we again see the strength of this insight. Most of this chapter will focus on the discount rate portion of the equation.

[297] This is especially true when you count accounting as a specialty within finance. Many accountants, especially those with significant experience, have some form of economics training and a very good grasp of the dynamics of business. However, accounting in general does not require significant economics education. Furthermore, as we discuss in the chapter on valuation, the historic cost principle of accounting is inconsistent with the market value principle of economics. This and other differences in approach lead to conflicts in the way accountants and economists approach common business questions.

Consumer Lending

Mortgage Analysis

Mortgages are a common consumer lending product. The dollar amounts involved are substantial enough to justify detailed analysis for most consumers, and for all lenders. As an example consumer finance application, we discuss in this section an analytical routine that examines the financing of a typical house purchase.

The mort_calc function described below accomplishes a number of tasks that provide consumers with useful information about potential mortgages. Some of these (such as the calculation of the monthly payment) are fixtures of many applications. Others, such as graphing of amortization schedules and the estimate of total monthly PITI (principal, interest, taxes, and insurance), are not.[298]

This function does the following tasks:

1. Checks to ensure that all variables are present and that the interest rate is entered correctly. This is particularly important with interest rates. Interest rates are often quoted in percentage terms (a mortgage rate of 9.5 typically means a 9.5% rate on the loan), while most computer programs (MATLAB included) look for interest rates in decimal form. Thus, it is a good idea to check any interest rate entered to ensure that it is in the decimal form. The "if ... error ... end" block in the function accomplishes this.
2. Calculates the principal from the home value and the down payment. A more sophisticated application might include other fees, points, etc. Note that we approach the loan amount as a secondary consideration; the primary consideration is the price of the property.
3. Calculate the periodic payment on the mortage loan. We have used the MATLAB payper function here.
4. Calculates the amortization, using a different MATLAB function other than that used for calculating the payments. We have used the MATLAB amort function here.
5. Performs an audit check to ensure that the two different calculations are identical. Such checks are essential in any model in which amounts that should be the same are calculated more than once. Always check, even if you are sure that you designed the model properly.

[298] The PITI estimation is critical for determining the eligibility of borrowers for loans. Most lenders consider the ratio of PITI to gross income when underwriting loans, generally requiring gross income to be more than four times PITI.

6. Creates labeling information to use in the graphic. You will note that, throughout this book, we recommend creating the labeling information separately from the plot command when possible. This allows the information to be identified easily and used consistently. Note that the information created here is used twice, once in the plot and once in the structure of key information (see the next step).
7. Creates a structure of values and information to provide to the user and then assign the structure into the base workspace. MATLAB structures are a powerful way to organize disparate types of data and include descriptive labels along with the data.
8. Creates the plot. Finally, using the variables created previously and the labeling information, it creates the graphic.

The MATLAB code is given in Code Fragment 12-1, in the Appendix to this chapter.

Graphics for Mortgage Analysis

The function produces a graphical description of the mortgage balance and the amortization (principal and interest portion) of the loan payments over time. To ensure that the viewers of the graphs understand the terms of the loan being analyzed, it also includes text describing those terms, as well as notes on the date printed. See Figure 12-1, "Mortgage analysis."

Porting this Tool to the Web

In Chapter 16, "Bringing Analytic Power to the Internet," we describe the functionality typically offered by consumer lending institutions on their Web sites. We note that, in the first wave of offerings, these Web sites served as electronic brochures. In the second wave, they provided additional, simple calculators and qualification rules. Third-wave applications which appeared quite recently provided users with a richer experience.

Using the analytical tools described in this book, we might wish to provide a consumer with the following enhancements to the features offered on typical mortgage lending sites:

1. The ability to have more complex analysis performed on various mortgage options
2. The ability to store key data in a document that can be shared with other lending institutions
3. The ability to see rich graphics describing the various options

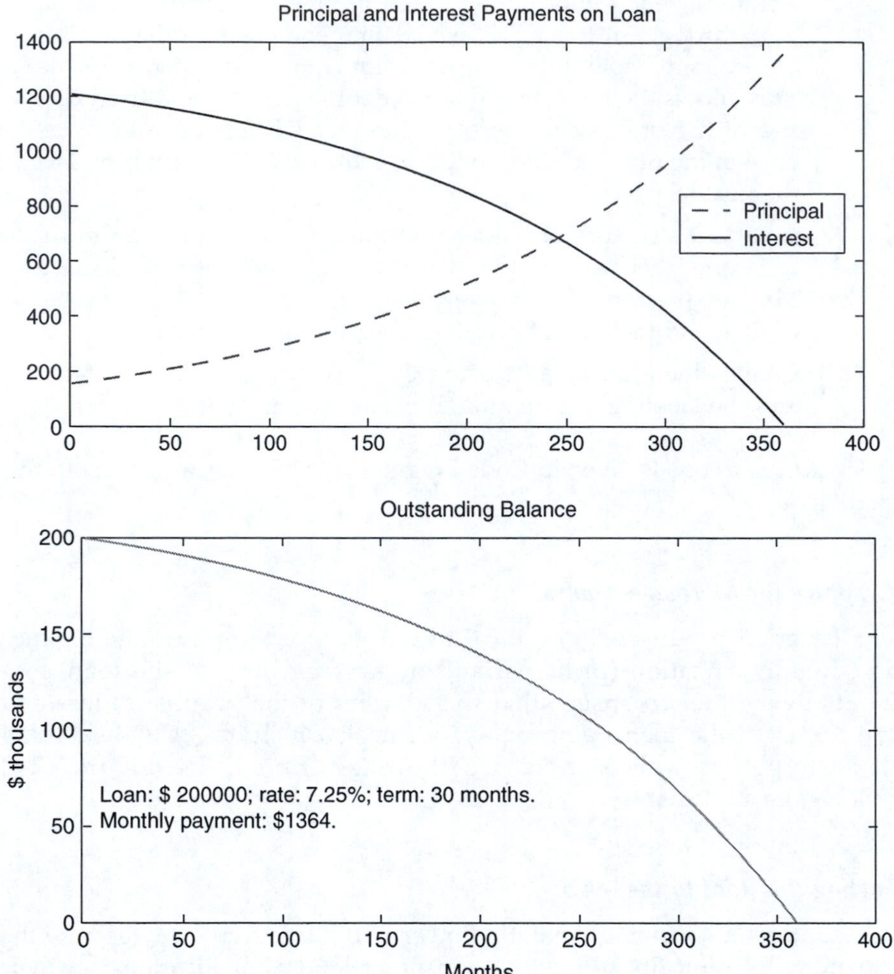

FIGURE 12-1
Mortgage analysis.

MATLAB Web Server

We describe in Chapter 16 how MATLAB's exceptional analytical capabilities can be used on a Web server. Thus, if the data were available for the application above, we could run it on a Web server and provide the graphics and output to a user accessing the application through an ordinary Web browser.

For simple analytics, including the calculation of loan payments, there will often be more robust alternatives to using a MATLAB Web server. However, for advanced analytics and in environments where the users are more limited, a MATLAB Web server application could extend the reach of customized applications to a selected audience around the globe.

Document Exchange

An additional enhancement would be to capture information of a potential loan in an XML-formatted document, which could then be used to exchange and compare information among various sites. We describe the use of XML (rather than the HTML that is commonly used to describe Web pages) in Chapter 16, under "MATLAB and XML" on page 416.

Depreciation

One of the more common analyses performed for companies with large tangible assets is a depreciation analysis. This often takes one of two forms: depreciation for tax purposes and depreciation for book purposes. The fact that most companies perform at least two depreciation calculations is an indication of its arbitrary imposition in the tax code.

Of course, the market value of those assets is not likely to coincide with a depreciation schedule imposed by a tax authority, or by an accounting convention. As market value drives the economics of an enterprise (and the wealth of a household), their managers often want to know the actual (market) value of assets, and compare them with the tax or accounting value.

Auto Depreciation

The U.S. Internal Revenue Code sets a harsh depreciation schedule for most vehicles, a straight-line depreciation over 5 years, and then normally imposes a half-year convention that effectively adds another year.[299] For many users, the actual depreciation allowed is even less.[300]

The auto depreciation function we created for the Business Economics Toolbox calculates depreciation in multiple ways:[301]

---

299 This applies under the "modified accelerated cost recovery system" which has been in effect for business property placed into service from 1987 through the date of the book. For the most accessible but technically accurate description, see the *Master Tax Guide*, Chicago, CCH, various years, Section 1236; or see Internal Revenue Code, Section 168; or IRS publication 946, available on the digital daily Web site at: http://www.irs.gov.

300 There are "luxury car" limitations on the allowable amount of an automobile's cost that can serve as the basis for depreciation; there are limitations on the use of itemized deductions; and there are other limitations based on the business usage share. In addition, the tax laws change from time to time.

301 Of course, we are not providing tax advice to readers of this book. However, we are suggesting tools that will estimate the market value, tax depreciation, and other variables, given a set of assumptions that the user will choose.

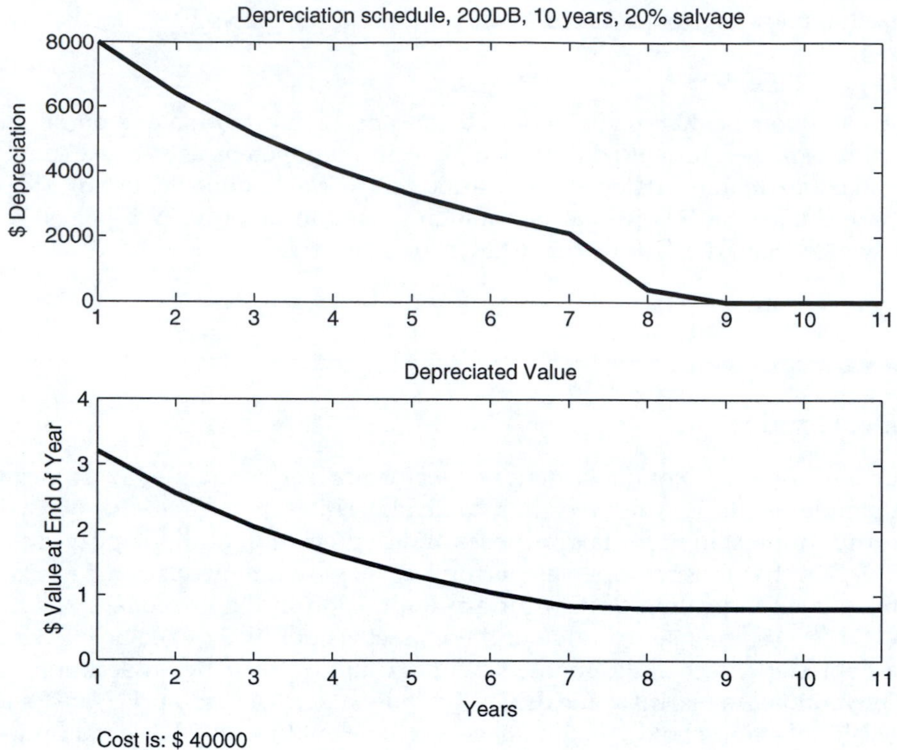

FIGURE 12-2
Auto depreciation: double declining balance.

- The double-declining balance (DDB) method, using both a full-year and half-year convention
- The straight-line method using a half-year convention
- A slow depreciation schedule for long-lived vehicles

The function also produces three graphs and summarizes the data in a structure. See Figure 12-2, "Auto depreciation: double declining balance," Figure 12-3, "Auto depreciation: half-year convention, Modified Accelerated Cost Recovery System (MACRS)," and Figure 12-4, "Auto depreciation: long-lived vehicles." See also Code Fragment 12-2 in the appendix to this chapter.

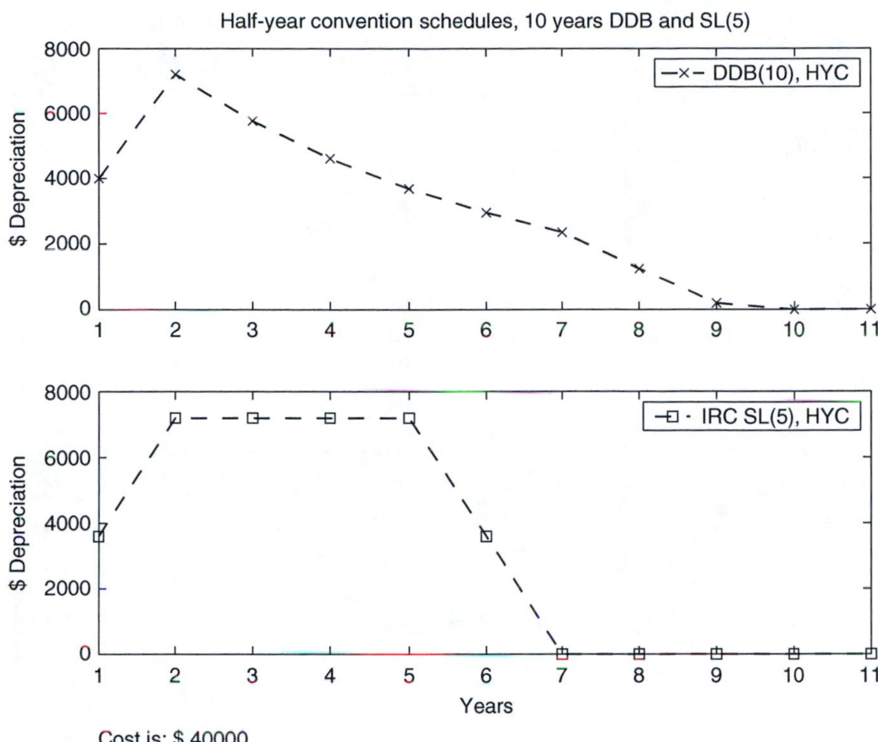

FIGURE 12-3
Auto depreciation: half-year convention, Modified Accelerated Cost Recovery System (MACRS).

Investment Portfolio Analysis

The analysis of investments, particularly the analysis of portfolios of different equity and bond investments, is a field that is particularly well-suited for the tools described in this book.

The use of MATLAB to analyze options and perform other computationally intense tasks has been the subject of a number of books.[302] In this chapter, we will not deal with the analysis of financial options,[303] but will describe

[302] See, for example: Neil A. Chriss, *Black Scholes and Beyond: Option Pricing Models* (New York: McGraw-Hill, 1997).
Paolo Brandimarte, *Numerical Methods in Finance: A Matlab-Based Introduction* (New York: John Wiley & Sons, 2002).
We also discuss option theory in Chapter 11, under "Option Methods" on page 260.

[303] Chapter 11, "Business Valuation," describes "real options" analysis.

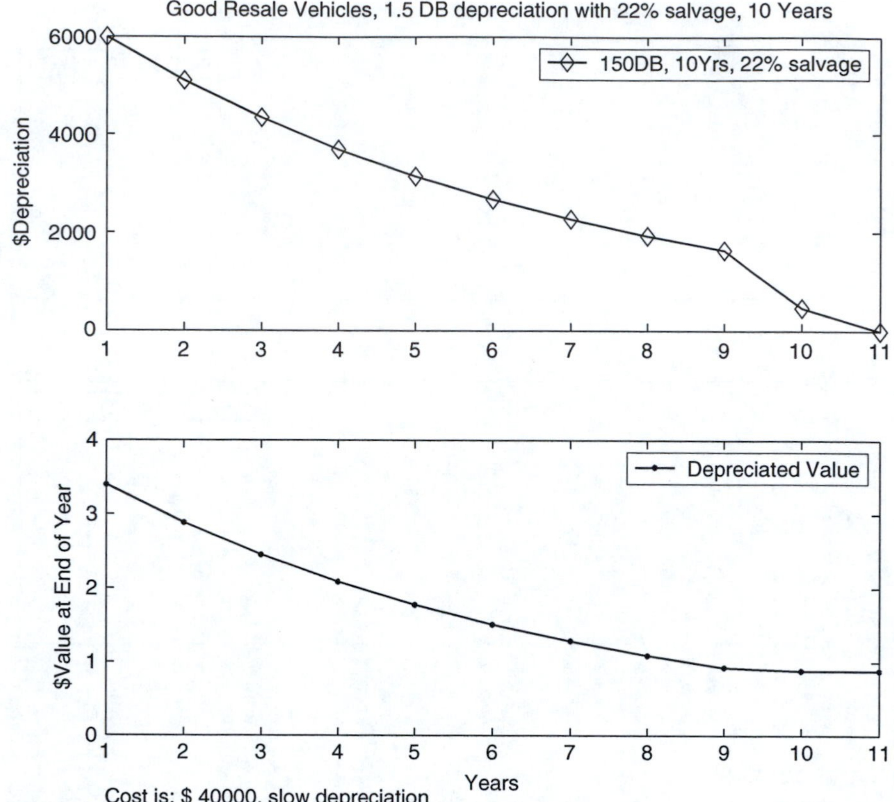

FIGURE 12-4
Auto depreciation: long-lived vehicles.

the use of MATLAB to perform the classic mean-variance portfolio analysis, expanded and implemented in a fashion that allows the application to automatically acquire data, analyze it, and plot it.

Mean-Variance Analysis

There are numerous references for portfolio theory, and we will not attempt to duplicate their contents. However, it is important to understand the core approach that underlies most applications: the *mean-variance* approach to analyzing investments. This approach, pioneered by Harry Markowitz and extended by William Sharpe, is based on the axiom that individuals attempt to increase their rewards and minimize their risks.[304] In a classic extension

[304] H. Markowitz, Portfolio selection, *J. Finance*, 7(1), 1952; *Portfolio Selection: Efficient Diversification of Investments* (New Haven: Yale University Press, 1959); reprinted, New York: John Wiley & Sons, 1970.

of microeconomic theory to the financial markets, they describe how individuals can optimize their investment portfolios, trading off risk for reward.

The mean-variance approach is characterized by the use of the mean return as the indicator of reward and the variance in returns as the indicator of risk. This approach has great intuitive appeal and very effective mathematical qualities. It earned Harry Markowitz, William Sharpe, and Merton Miller the 1990 Nobel Prize in Economics.

Of course, the classical derivation of the theory ignores transaction costs and makes strong assumptions about the availability of information as well as other limiting assumptions.[305]

Portfolio Analysis in MATLAB

MATLAB offers a financial toolbox with numerous routines designed to implement techniques in modern portfolio theory. The functions in the toolbox provide assistance in pricing securities, calculating interests and yields, charting financial data, computing cash flows, analyzing derivatives, and optimizing portfolios.

The GetStocks Function

We provide the GetStocks function as part of the companion Business Economics Toolbox. It combines a number of separate tasks:

- Acquiring data on past prices of one or more securities from a data provider, using the Internet
- Calculating return data from the price data
- Calculating portfolio statistics, including the covariance matrix for the data series
- Graphing the price series

The code for the GetStocks function can be modified to get other stocks, depending on data providers, as well as to perform other financial analysis

304 W. F. Sharpe, Capital asset prices: a theory of market equilibrium under conditions of risk, *J. Finance,* 19(3), 1964.
For a historical perspective and personal remembrance by Sharpe, see Revisiting the capital asset pricing model, *Dow Jones Asset Manager,* May/June 1998; also published at: http://www.stanford.edu/~wfsharpe/art/djam/djam.htm.
An excellent summary is also contained in H. Markowitz, Mean Variance Analysis, in *The New Palgrave: A Dictionary of Economics* (New York: Stockton, 1991).
William Sharpe uses MATLAB for economic and financial analysis and has made available an electronic work in progress on the topic entitled Macro-Investment Analysis. It can be found at: http://www.stanford.edu/~wfsharpe/mia/mia.htm.

305 Among the strong assumptions are that investors' concept of risk is captured in the variance, which implies a quadratic, symmetric utility function for investors, as well as ignoring the limited liability of stockholders.

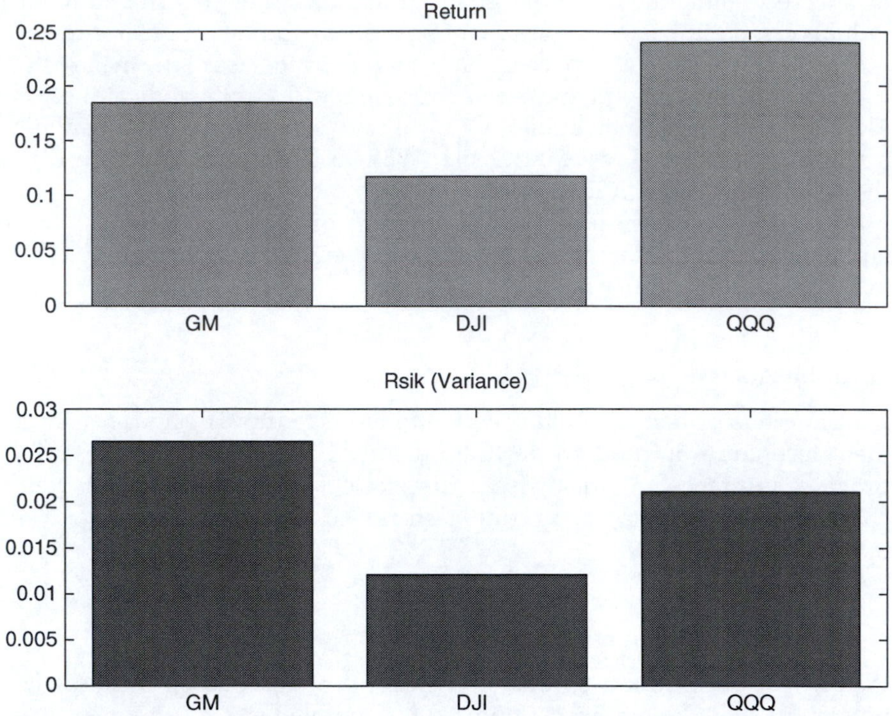

FIGURE 12-5
Return and volatility percent analysis.

and graphing tasks. The code is given in Code Fragment 12-3 in the Appendix. Graphs produced by the function are shown in Figure 12-5, "Return and volatility per cent analysis," and Figure 12-6, "Comparison of price changes."

Other Topics in Finance

We cover topics in finance in other chapters of this book and highlight a few of these below:

Portfolio Theory and Economic Diversification

The insights from portfolio theory are applied to the study of economic diversification in Chapter 9, "Regional Economics," under "Diversification and Recession Risk" on page 189.

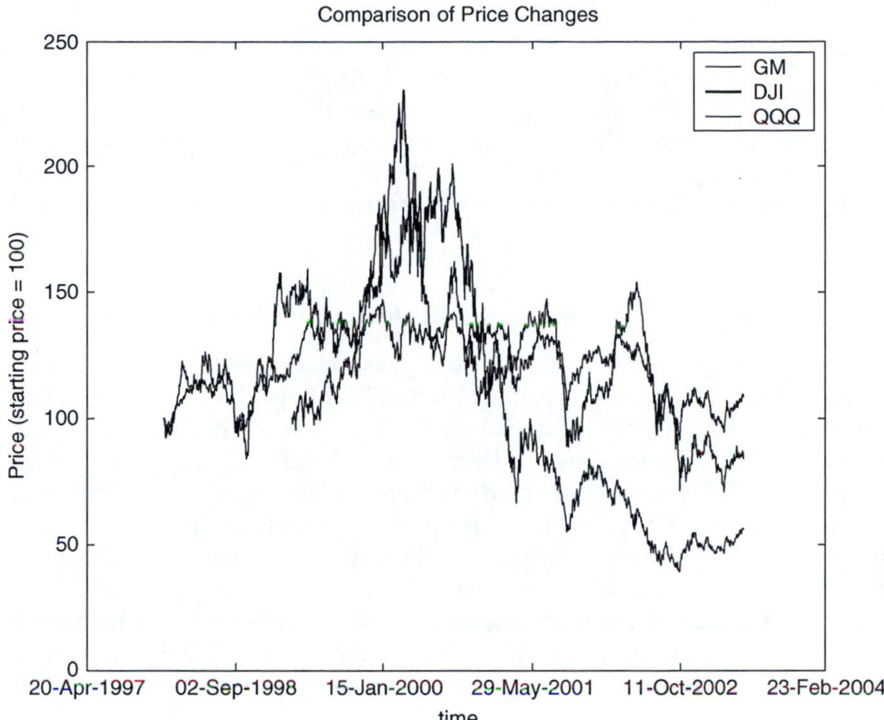

FIGURE 12-6
Comparison of price changes.

Growth Models and Simulink

Simulink can be used to model complex financial and economic systems, including the capitalization of an enterprise as it grows or shrinks in size. We describe some of these tools in Chapter 10, "Application for Business," on page 207 and in Chapter 11, "Business Valuation and Damages Estimation," on page 243.

Specific Functions for Projecting Growth

Because the forecasting of future economic and business conditions is vital in business economics and finance, we present specialized tools for this purpose. In particular:

- We describe growth blocks to be used in Simulink models in Chapter 10, "Applications for Business," on page 207.
- We provide MATLAB functions to forecast growth in Chapter 10, "Application for Business," under "Projecting Trend Growth" on page 208.

Credit Risk

We provide an example Simulink model that uses fuzzy logic to assess credit risk, in Chapter 15, "Fuzzy Logic Business Applications" under "A Credit Risk Fuzzy Logic Simulation Model" on page 307.

Credit risk directly affects the discount rate the market places on loans and indirectly on the business entities that make those loans. Therefore, the methods demonstrated in this example application can be adapted to a wide variety of problems in finance.

Capitalization Rates, Earnings, and Ad Valorem Taxation in Real Estate

The proper capitalization of long-lived assets is particularly important because the longer the amortization period, the larger impact interest costs have on current earnings. In addition, the longer the duration of a loan, the closer the lender comes to shouldering equity-level risk in a firm.[306]

Real-estate investments are particularly long-lived, as well as relatively expensive and illiquid. For these reasons, the capitalization of real-estate assets (including home mortgages) receives such attention from consumers, investors, and the federal government.[307]

For a discussion of methods to properly value real estate that is taxed on an *ad valorem* basis, see Chapter 10, under "An Iterative Model to Estimate Market Value and Taxation" on page 231.

---

306 See the discussion on valuing equity and debt as an option, in Chapter 11, "Business Valuation and Damages Estimation," under "Option Methods" on page 260.

307 It is not an accident that the federal government chartered Fannie Mae in 1938, Freddie Mac in 1970, and other quasi-governmental institutions for the purpose of helping homeowners finance their homes. These entities rely on explicit or implicit government sponsorship to create a more liquid, lower-cost mortgage market. The Internal Revenue Code's continued allowance of the mortgage interest deduction, as well as a property tax deduction, is another indication of the broad support homeownership enjoys.

Appendix

Code Fragment 12-1. Mortgage Calculator

```
function [piti, schedule] = mort_calc(value, down,
rate, term);

% mortgage, taxes, and insurance calculator

% Produces monthly PITI (principal, interest, taxes,
% and insurance) payment,

% and schedule in structure form.  Also plots
% amortization schedules.

% Enter value of home, down payment, annual interest
% rate (decimal format),

% and term in years.  Example: mort_calc(250000,
% 50000, .07, 30)

% PLA 1 Dec 00; 20 Dec 00 ver 3; 23 Sept 2001; 27
% Sept 2001; 10 Nov 01 ver 5.3

% copyright 2001 Patrick L. Anderson; Limited rights
% granted under license

  echo off;

format bank;

if nargin == 0

  action = 'default_values'

elseif nargin == 4

  action = 'custom values'

else error('Must enter correct values, or use
defaults.');

end

% ------------------------------------------------

switch lower(action)
```

```
case 'default_values';

  value = 250000;

  down = 50000;

  rate = 0.0725;

  term = 30;

end;

% ----------------------------------------------------

if rate>1

  error('Please enter interest rate as a decimal;
for example, .08 for 8% per annum.');

end

p = value - down;

if p<0

  error('Value of house must be greater than down
payment.')

end

% Calculations of Payments

% ----------------------------------------------------

  disp(action);

     pmt = payper(rate/12, term*12, p);

     taxes = .037 * value * 1/12 * 1/2 * .90;

     ins = (1470/600000) * value/12;

     piti = pmt + taxes + ins;

% ----------------------------------------------------

% whos;

% ----------------------------------------------------

% amortization calculations
```

```
[PRINP,INTP,BAL,P] = amortize(rate/12, term*12, p, 0);

payment = PRINP(1) + INTP(1);

month_index = [1:1:length(PRINP)];

% Report on audit check; Do payment and
% amortization calculations agree?

%   ---------------------------------------

if pmt == payment

   disp('Amortization and loan payment calculations
agree.')

else disp('Check to ensure payment and amortization
agree.')

end

%   Labeling Information

printdate2 = ['printed on: ',  datestr(now, 26),
'.'];

format bank;

loandata1 = ['Loan: $', num2str(round(p)), '; rate:
', num2str(rate*100), '%; term: ', num2str(term), '
months.'];

loandata2 = ['Monthly payment: $',
num2str(round(pmt)), '.'];

loandatacell = {loandata1, loandata2};

format short g;


%   Create structure of values and vector of
%   amortization information; put in workspace

%   ----------------------------------------------------

schedule = struct('loan', p, 'payment', payment,
'rate', rate, 'piti', piti, 'term', term,
'principal', PRINP', 'interest', INTP', 'balance',
BAL', 'month_index', month_index');
```

```
assignin('base', 'schedule', schedule);

disp(loandata1);

disp(loandata2);

disp('The following values are in the structure
"schedule" in the workspace:');

disp(schedule);

disp('Monthly principal, interest, and balance are
also available in workspace as "X"');

X = [schedule.principal schedule.interest
schedule.balance] ;

assignin('base', 'X', X);

%   Plotting---------------------------------------

figure(10),

%   set figure window title (visible) and figure
%   property tag (invisible)

figuretag10 = ['Projection for Mortgage'];
%Insert name on this line

set(gcf, 'Name', figuretag10, 'Tag', figuretag10);

%   first subplot

subplot(2,1,1),

plot(schedule.principal, 'r--'), hold on;

plot(schedule.interest, 'b'), hold off;

legend('Principal', 'Interest');

title('Principal and Interest Payments on Loan',
'fontweight', 'bold');

%   second subplot

subplot(2,1,2),

plot(schedule.balance/1000, 'g');

Title('Outstanding Balance', 'fontweight', 'bold');

xlabel('Months');

ylabel('$thousands');
```

```
%       insert key text into current axis
text(0, -.15, printdate2, 'units', 'normalized',
'fontangle', 'italic');
text(0.05, .30, loandatacell, 'units', 'normalized',
'fontangle', 'italic');
```

Code Fragment 12-2. Auto Depreciation

```
function [dep_sched] = auto_dep(cost);
%   Auto depreciation as function of original cost

%   of vehicle;
%   assuming DDB(10 years), with 20% salvage;
%   also produces SL(5) with half-year convention
%   and 10% salvage;
%   [Based on standard Internal Revenue Code MACRS
%   depreciation for cars; does
%   not incorporate "luxury" vehicle limitations or
%   expensing.]
%   Also produces 150DB 10 year "slow" depreciation
%   to 22% salvage.
%   Example: [a, b] = auto_dep(40000)
%   ...produces three graphs, depreciation series,
%   and a structure called
%   "dep_sched" with additional data.
%
%   PLA, 21 May 2002
%   (c) 2003 Anderson Economic Group. Part of
%   Business Economics Toolbox.
%   Use by license only.
%   key assumptions
if nargin<1
```

```
cost = 40000

disp('---------Missing original value.  Using
$40,000 as a default.---');

end

salvage = .2 * cost;

salvage_irc = .1 * cost;

salvage_b = .22 * cost;

life = 10;

life_irc = 5;

years = [1:1:(1+life)];

factor = 2;     %use to create declining balance
factor; 2 means Double Declining Balance

factor_b = 1.5; %no factor for IRC; use straight-
line; 1.5 for "slow" depreciation

%   ---------calculate depreciation 4 ways:---------

%   first way

%   note: add zeros at end of series to make sure
%   same length when plotting

dep_series_a = depgendb(cost, salvage, life,
factor);

dep_series = [dep_series_a, 0];

cumdep = cumsum(dep_series)

value_series = repmat(cost, 1, length(years)) -
cumdep;

%   second way

dep_series_hyc = [(.5 * dep_series_a ), 0] + [0,
(.5* dep_series_a)];

%   third way

dep_irc = depstln(cost, salvage_irc, life_irc);
```

```
dep_series_irc_a = [ repmat(dep_irc, [1, life_irc]),
zeros(1, (life-life_irc)) ];

dep_series_irc_hyc = [(.5 * dep_series_irc_a ), 0] +
[0, (.5* dep_series_irc_a)];

%   fourth way--good resale vehicles ("slow")

dep_series_b1 = depgendb(cost, salvage_b, life,

factor_b); %trouble here?

dep_series_b = [dep_series_b1, 0];

cumdep_b = cumsum(dep_series_b)

value_series_b = repmat(cost, 1, length(years)) -
cumdep_b;

whos;

%   Note: MATLAB description:  Depreciation =
%  depgendb(Cost, Salvage, Life, Factor)

%

% Cost: Cost of the asset.

% Salvage: Estimated salvage value of the asset.

% Life: Number of periods over which the asset is
% depreciated.

% Factor: Depreciation factor. Factor = 2 uses the
% double-declining-balance

% method.

%

%    SL = DEPSTLN(COST,SALVAGE,LIFE) calculates
%   straight-line depreciation

%-----------------------------------------------------

%   Plots

figure(1),

subplot(2,1,1),

plot(years, dep_series);
```

```
title('Depreciation schedule, 200DB, 10 years, 20%
salvage', 'fontweight', 'bold');

% xlabel('Years');

ylabel('$Depreciation');

subplot(2,1,2),

plot(years, value_series);

title('Depreciated Value', 'fontweight', 'bold');

xlabel('Years');

ylabel('$Value at End of Year');

text(0.0, -0.2, ['Cost is: $' num2str(cost)],
'units', 'normalized', 'fontangle', 'italic');

figure(2),

subplot(2,1, 1), plot(years, dep_series_hyc, 'mx');

title('Half-year convention schedules, 10 years DDB
and SL(5)', 'fontweight', 'bold');

legend('DDB(10), HYC');

% xlabel('Years');

ylabel('$Depreciation');

subplot(2,1, 2), plot(dep_series_irc_hyc, 'ms');

legend('IRC SL(5), HYC');

xlabel('Years');

ylabel('$Depreciation');

text(0.0, -0.2, ['Cost is: $' num2str(cost)],
'units', 'normalized', 'fontangle', 'italic');

figure(3),

subplot(2,1, 1), plot(years, dep_series_b, 'g');

title('Good Resale Vehicles, 1.5 DB depreciation
with 22% salvage, 10 Years', 'fontweight', 'bold');

legend('150DB, 10Yrs, 22% salvage');

% xlabel('Years');

ylabel('$Depreciation');
```

```
subplot(2,1, 2), plot(value_series_b, 'g');

legend('Depreciated Value');

xlabel('Years');

ylabel('$Value at End of Year');

text(0.0, -0.2, ['Cost is: $' num2str(cost) ', slow
depreciation'], 'units', 'normalized', 'fontangle',
'italic');

%--------------------------------------------------

%   Create structure to report (transpose into
%   column vectors)

dep_sched = struct('cost', cost', 'years', years',
'full_year', dep_series', 'half_year',
dep_series_hyc', 'irc', dep_series_irc_hyc', 'slow',
dep_series_b' );

dep_sched;

%   record fieldnames within schedules

names = fieldnames(dep_sched);

%   create cell array of same data

cell_sched = struct2cell(dep_sched);

% %   assign data to base workspace, so it can be
% accessed

% assignin('base', 'schedules', schedules);

assignin('base', 'names', names);

assignin('base', 'cell_sched', cell_sched);


% %   need to have all rows of same length for
% following array:

% printing_sched = [schedules.years
% schedules.full_year schedules.half_year
% schedules.slow];

% disp('---------Years,   full-year,      half-year
```

```
% convention depreciation.------------');
% disp(printing_sched);
```

Code Fragment 12-3. Get Stocks Function

```
function [stocks] = GETSTOCKS(varargin)
%   Retrieve stock price data over time from
%   Internet information
%   provider; calculate return and volatility
%   information; plot.
%   Returns a structure a with data.
%

% Example:  a = getstocks('Jan 1 1998', 'may 15
% 2003', 'd')
%   "fetches" GM,  Dow Jones Industrials, and Nasdaq
%   100 information.
%
%   PLA 16 May 2003; 6 June 03
%   last revision: added datetick command
%   (c) 2003 Anderson Economic Group; use by license
%   only.
%   Part of Business Economics Toolbox, provided to
%   purchasers of Business Economics Using MATLAB, CRC
%   Press, 2003.
%
%   Uses: MATLAB datafeed toolbox, financial toolbox,
%   fts toolbox
%   Requires:   (1) Active internet connection.
%               (2) Data provider service, such as
%                   Yahoo, Bloomberg, IDC; These
%                   data providers may change
```

```
                    service, parameters, etc.
%                   without notice. This service is
                    not part of the
%                   toolbox.
%
%   Theory reference: mean-variance analysis of
%   Markovitz and Sharpe
%   See W.F. Sharpe, Macro-Investment Analsyis,
%   http://www.stanford.edu/ ~wfsharpe/mia/mia.htm

echo on;

format bank;

%   Check for input data----------------------------

if nargin <3;
    %   Insert date range
    startdate = datenum('Jan 1 1999');
    enddate = datenum(today);
    periodicity =  'd'; %'d' for daily values.
    disp('Using default dates; from January 1, 1999
to most recent day.');
else
    startdate = datenum( varargin(1) )
    enddate = datenum( varargin(2) )
    periodicity = varargin{3}           %note: direct
% cell reference, ensures character array output
end

%   names of stocks--edit here to retrieve different
%   stocks----------

stocks.aname = 'GM';

stocks.bname = '^DJI';

stocks.cname = 'QQQ';
```

```
%   periodicity information-------------------------
stocks.startdate = startdate;
stocks.enddate = enddate;
stocks.periodicity = periodicity;

%---------------------------------------------------
%               Analysis
%---------------------------------------------------
stocks = FetchData(stocks);

stocks = CalcReturn(stocks);

stocks = PortStats(stocks);

[h] = ChartData(stocks);

%---------------------------------------------------
%                Functions
%---------------------------------------------------
function stocks = FetchData(stocks)
%   Connect to data source-------------------------
connect = yahoo;

%   Once connection established, get data using
%   "fetch" command------
%   "fetch" here gets a two-column vector;
%   first column is dates, second is prices------


%   Get General Motors data;

stocks.a1 = fetch(connect, stocks.aname, 'Close',
stocks.startdate, stocks.enddate,
stocks.periodicity);

stocks.b1 = fetch(connect, stocks.bname, 'Close',
stocks.startdate, stocks.enddate,
stocks.periodicity);
```

```
stocks.c1 = fetch(connect, stocks.cname, 'Close',
stocks.startdate, stocks.enddate,
stocks.periodicity);

%   Get Order Correct-----------------------------

%   Confirm right order--2nd column is prices, 1st
%   is dates.

%   Most recent price information is at top of vector
%   from "fetch" command;

%   But... TICK2RET assumes the first row is the
%   oldest observation, and the last row is the most

%   recent.

%   EWSTATS also assumes most recent price is last
%   observation.

%   Therefore, must "flip" vector so last item is
%   last observation.

stocks.a = flipud(stocks.a1);

stocks.b = flipud(stocks.b1);

stocks.c = flipud(stocks.c1);

%-----------------------------------------------

%-----------------------------------------------

function stocks = CalcReturn(stocks)

%   using financial and statistical theory,
%   calculate distribution stats

%   ticktoret: converts price data ("tick") to
%   returns (decimal form)

%    [RetSeries, RetIntervals] =


%    tick2ret(TickSeries, TickTimes)

%   EWSTATS: Expected return, covariance, and number
%   of observations; from return time series.

%   Note: "expected" is calculated using
%   probabilities; equally-weighted

%   observations implies expected return =
```

```
%   mean(return). Our implementation
%   of EWSTATS provides no weights, so the
%   "expected" return is the arithmetic mean.

[stocks.aret, stocks.aInt] = tick2ret(stocks.a(:,2),
stocks.a(:,1));
[stocks.bret, stocks.bInt] = tick2ret(stocks.b(:,2),
stocks.b(:,1));

[stocks.cret, stocks.cInt] = tick2ret(stocks.c(:,2),
stocks.c(:,1));

%--------------------------------------------------

%---------EWSTATS portfolio analysis----------------

function stocks = PortStats(stocks)

% Find shortest series---------------------------

lengthstocks = [length(stocks.aret),
length(stocks.bret), length(stocks.cret)];
minlength = min(lengthstocks);
trimmedstart = lengthstocks - minlength + 1;
%must start on first observation

if sum(trimmedstart)>0
    warning(['Must trim series to identical lengths;
trimmed observations total: ']);
    disp(trimmedstart - 1)
end

%   Trim series to minimum similar length for return
%   series

RetSeries = [stocks.aret(trimmedstart(1):end),
stocks.bret(trimmedstart(2):end), ...
        stocks.cret(trimmedstart(3):end)];
DecayFactor = 1          % 1 is default, and is
                           linear average
```

```
WindowLength = 30        %   use 30 for daily data


[stocks.ExpReturn, stocks.ExpCovariance,
stocks.NumEffObs] = ewstats(RetSeries, ...
                      DecayFactor, WindowLength);

%   Summarize data on return, volatility, end prices

stocks.n_stocks = 1:3;


disp('---Series, End Prices, Return (percent), and
Volatility (std. dev. of return, percent)---');

stocks.labels = {stocks.aname stocks.bname
stocks.cname};

disp(stocks.labels);

end_prices = [stocks.a(end) stocks.b(end)
stocks.c(end)];

disp('--expected return, per cent--');

disp(100*stocks.ExpReturn)

disp('---periodicity of data, and number of
effective observations----');

disp(stocks.periodicity);

disp(stocks.NumEffObs)

disp('---Return volatility (Covariance) Matrix------
----------------');

disp(stocks.ExpCovariance)



echo off;

                                        %Display Key

%---------------------------------------------------
%---------------------------------------------------
function [h] = ChartData(stocks)
```

```
%   creates figures; returns vector of figure
%   handles

%   Create bar chart--------------------------------

%   Return bar chart

h(1) = figure,

subplot(2,1,1);

bar(100* stocks.ExpReturn, 'g');

title('Return, per cent', 'fontweight', 'bold');

set(gca, 'XTickLabel', stocks.labels);

%   volatility bar chart

subplot(2,1,2);

bar(100 * diag(stocks.ExpCovariance), 'r');

title('Volatility (Variance), per cent',
'fontweight', 'bold');

set(gca, 'XTickLabel', stocks.labels);

%   key info

text(-0.2, 0.5, ['periodicity: ',
stocks.periodicity]);

%   Create indexed price chart (note: not trimmed) -----

stocks.aopen = stocks.a(1,2); %first data pair are
                              (date, price)

stocks.bopen = stocks.b(1,2); %first data pair are
                              (date, price)

stocks.copen = stocks.c(1,2); %first data pair are
                              (date, price)


h(2) = figure,

plot( stocks.a(:,1), 100*(stocks.a(:,2))/

stocks.aopen, ...

    stocks.b(:,1), 100*(stocks.b(:,2))/stocks.bopen, ...
```

```
    stocks.c(:,1), 100*(stocks.c(:, 2))/stocks.copen );
title('Comparison of Price Changes', 'fontweight',
'bold');
legend(stocks.labels);
ylabel('Price (starting price = 100)');
xlabel('time');
datetick('x', 'QQ-YYYY');

% ---------------FTS Interactive Chart--------------
%   can be omitted if FTS toolbox not installed
%   create financial toolbox "object"
stockt = fints(stocks.a);

% chart using fts chart command
% figure,
chartfts(stockt);
title([ char(stocks.aname), ' Interactive Chart']);
Xlabel('Closing Price');

h(3) = gcf;

%-----------------------end-----------------------
```

13

Modeling Location and Retail Sales

Introduction: The Spatial Dimension in Economics

This chapter explicitly and rigorously develops the spatial dimension in economics. We start by reviewing the weak development of the spatial dimension in contemporary academic economics training and compare that with both the rich historical tradition and the primary importance of location in applied economics.

From this point, we follow the development of gravity models of spatial interaction from the time of Isaac Newton to today and then apply them to an important challenge in business economics — the analysis of retail sales. We discuss the fundamental tools of the analysis of retail sales, including the definition of trade areas, penetration rates, and other measures, and the use of Geographic Information Systems (GIS). Finally, using examples drawn from actual work in competitive retail markets, we present specifications and estimation methods for models of retail sales.

The Location Poverty of Modern Economics

Most academic economics texts talk very little about location. Indeed, to the extent location is mentioned at all, it is in the context of being within a national or state border, or between urban and rural areas.[308] Rarely is the vital importance of proximity to customers or competitors discussed.

An academic discussion of the supply curve within a market or company often assumes a constant cost of acquiring or delivering goods.[309] However, for much of the U.S. economy, including most retail, housing, construction,

[308] For example, most macroeconomic texts start with a closed model of the U.S. economy and later move to an open economy. In such studies, the only location factor is whether you are on one side of the border or the other. Furthermore, prices, tax rates, and other variables are often assumed to be constant through out a country.

[309] A perusal of an intermediate microeconomics textbook illustrates this point quite well. Advanced texts are usually more abstract, including those cited in Chapter 5, "Library Functions for Business Economics," under "Comparative Statics" on page 98.

lending, transportation, agriculture, tourism, and mining, location is one of the most important factors in generating and sustaining earnings. Indeed, it is not hard to find subsectors that are almost entirely dependent on differences in location or delivery costs.[310]

Similarly, the typical academic discussion of demand may make mention of how customer demand is affected by the total costs involved in the purchase of a product or service. However, the effect of location on the overall cost involved in a purchase is often ignored.

One recent effort to reconsider the role of location in economic thought put it succinctly: "How do economists routinely deal with the question of how the economy organizes its use of space? The short answer is that mostly they do not deal with the question at all."[311]

Location and the Birth of Economics

On this topic, modern business economics could learn from the past. Adam Smith began his *Wealth of Nations* by a discussion of labor, and the division of labor. He devoted the third chapter of his book to how "the division of

---

The expections prove the rule. Harold Hotelling's "Stability in competition," published in 1929 in the *Econ. J.*, 39, and Steven Salop's "Monopolistic competition with outside goods," *Bell J. Econ.*, Spring 1979, are benchmark studies of competition when consumers have different transportation costs to individual retailers. Spatial distribution of consumers and retailers is, however, usually ignored even in hedonic models of demand.

310 There are many obvious "location, location, location" subsectors such as real estate, convenience stores, and resorts. However, the more instructive examples involve day-to-day tasks in which location is not the most obvious factor. For instance, consider the following two examples:

(1) The benefits of faster delivery have created the entire express mail industry, in which customers pay five or ten times the standard postal rate to receive a small package within one day. The success of the industry can be attributed primarily to the costs of doing business with distant parties and, therefore, to the value of a new technology that makes these costs smaller. (Of course, there are other supporting causes, including the negative effects of the original government monopoly on the postal system's efficiency and incentive to innovate and the technological and managerial efficiencies of the leading express mail companies.)

(2) Bookstores continue to flourish, even as the Internet age provides a low-cost alternative for those merely interested in purchasing books. Indeed, if cost-per-book were the only factor in consumer decisions, traditional bookstores should (in line with the exuberant expectations of the denizens of the Internet investment bubble) have disappeared by now. They must be providing some experience for which customers gladly pay a premium, in terms of time for travel, search costs, and prices of books. As these bookstores congregate in urban and suburban strongholds, their proximity to their customers' working and living locations is clearly an essential factor in the demand for their product.

311 M. Fujita, P. Krugman, and A. Venables, *The Spatial Economy* (Cambridge, MA: MIT Press, 1999), Chap. 2.

See also Richard Arnott, Spatial Economics, in *The New Palgrave: A Dictionary of Economics*, "Spatial economics has remained outside the mainstream of economics because space cannot be fitted neatly into competitive equilibrium models of the Arrow–Debreau type."

labor is limited by the extent of the market."[312] Smith's discussion of 200 years ago contains the essential insight we will use in the discussion of retail sales — how the size of the market and the costs of servicing that market limit commerce and wealth.[313] Smith's discussion of those ancient countries that had developed significant wealth — India, China, and some in the Mediterranean region — notes how the ability to expand their markets through transportation was critical.

Developments after Smith similarly focused on the production side of the economy. In a classic work, the German economist J.H. von Thunen proposed in 1826 a model that explained how agricultural production was efficiently organized when farmers cultivated and harvested crops in areas surrounding a city. The city provided the mechanical and specialized functions that were too expensive for individual farms to purchase. The costs of transit became the critical variable, and the cost differentials caused by the differences in travel led to a pattern of different crops being cultivated at different distances from the city.[314]

Business Economics and Location

The explicit incorporation of location as a primary variable is a key distinction between theoretical economics and applied economics. Because location is extremely important in the business world, the theory and practice behind modeling location and retail sales have developed largely in the world of business economics.[315] This chapter describes the theory behind such models

---

312 A. Smith, *An Inquiry into the Nature and Causes of the Wealth of Nations*, 1776, E. Cannan, Ed. (Chicago, IL: University of Chicago Press, 1976), Book 1, Chap. 3.

313 Smith finds that specializing in one trade provides greater wealth, but that without commerce, there is no ability to specialize. His discussion is based on the production (supply) side of the economy. Most of this chapter will assume that we live in a society where manufacturing and agriculture take a much smaller portion of the economy than in Smith's day, and that transit across regional boundaries for goods and services is relatively easy (that is, relative to 1776). These differences mean that the regional constraints on commerce and wealth on the supply side of the economy are much smaller than in 1776 and indeed much smaller than in 1876 or 1956. In particular, a worker who lives in a small town today may have an opportunity to work for high wages in a specialized trade in a nearby mid-sized or large town.

314 J.H. von Thunen, *Der Isolierte Staat in Beziehung auf Landschaft und Nationalokomie* [translated: *Isolated State*], 1826; 3rd ed. (Stuttgart: Gustav Fischer, 1966); cited in R. Arnott, Spatial Economics, in *The New Palgrave: A Dictionary of Economics*.

The best explanation of the von Thunen model is in Fujita et al., *The Spatial Economy* (Cambridge, MA: MIT Press, 1999). These authors also cite C.M. Wartenberg, trans., *von Thunen's Isolated State* (Oxford: Pergamon Press, 1966).

315 Even A. Sen and T.E. Smith, whose review of the limited academic literature on spatial economics is extensive, note that the first systematic efforts to develop estimation methods forgravity models were those of U.S. highway planners, starting in the 1960s and 1970s, and lament that these pioneers have "perhaps not been adequately recognized for [their] innovativeness and for the fact that [they] clearly laid the groundwork for the contemporary gravity model." A. Sen and T.E. Smith, Introduction: Classical Gravity Models, *Gravity Models of Spatial Interaction Behavior* (New York: Springer, 1995).

presenting two derivations, and also how such models are extremely useful in the analysis of retail sales.

What is "Retail"?

By "retail sales" we mean transactions with a consumer which result in a service or goods being provided at a specific site. Thus, sales of household goods, automobiles, and food would be considered retail sales. The same approach may also be useful for consumer banking, services such as dry cleaning or medical doctors, and entertainment venues such as casinos, racetracks, theatres, and stadia. On the other hand, the manufacture of such goods, and the wholesale trade that got them to the point of sale, would be outside the definition of retail sales.

A Note on GIS

In part of this chapter, we make use of Geographical Information Systems (GIS) a major development in the ability of economists to explicitly incorporate geography and distance in their analyses. We do not explain in this book how to use GIS software, including the Mapping Toolbox that is available with MATLAB. However, the information is presented here in a manner that should allow the reader to understand what was done and to successfully commission someone trained in GIS to perform a similar analysis.

Applying Economic Theory to Sales Location

The Economics of Time

The foundations of microeconomics provide ample support for the notion that retail activity will decrease, given increased distance from a retail outlet. This arises primarily from the fact that covering distance takes time, and time is valued by consumers and producers alike. The economics of time is based in utility theory, which is at the basis of microeconomics. Once you recognize time as an input into the utility function of a consumer, a model of microeconomic behavior will suggest that consumers shop closer to home or work in order to minimize the welfare loss due to travel time.[316] As discussed in "Location and the Birth of

[316] We assume that covering distance generates costs in both time and money. As indicated above, it is surprising how often this observation is *not* made in contemporary academic microeconomics.

Economics" on page 328, pioneers such as Adam Smith and von Thunen were able to explain location patterns and their relationship to productivity and wealth in the 18th and 19th centuries. While most derivations of consumer demand from utility theory omit location as a factor, location can and should be directly included in applied work.[317]

Distinction with Time Preference and Other Concepts

The economics of time has been developed in other areas of applied economics, such as finance and econometrics. However, the economics of time we discuss in this chapter focuses on the value *consumers* place on their time. We can distinguish this from the common uses in other fields as follows:

- The time value of money in finance arises from consumers' *time preference* for consumption. Fundamental microeconomics indicates that consumers must be compensated for delaying consumption, and from this originates a time value of money.[318]
- In econometrics and other applications of statistics, the seasonal adjustment of many data series reflects the intrayear cycles of human behavior in the market, as well as natural phenomena.
- Changes in behavior that occur because of unexpected, one-time changes in income, tax policy, or demand are normally distinct from the behaviour of consumers who value time.[318a]

In this chapter, we will concentrate on the economics of time as it affects the geographic choices of consumers and producers.

Increased Distance Means Increased Costs

The careful economist will avoid assuming that the cost of increasing the distance from a consumer to a producer is merely a wage rate multiplied by the time it takes to make a one-way, or even two-way, trip. When consumers choose a producer, they are often estimating more than the costs for a single transaction. For example, consumers choosing a dry cleaner may wish to learn about a provider's prices and services, find a convenient location along their driving routes, and maximize nearby shopping

[317] Some of the approaches to incorporating location in utility functions include using distance as a cost (the approach generally followed here), using distance as a type of subjective "spatial discounting," and considering distance a form of disutility. A brief survey of these is in A. Sen and T.E. Smith, Introduction: Theoretical Approaches to Gravity Models, *Gravity Models of Spatial Interaction Behavior* (New York: Springer, 1995).

[318] To be precise, we are speaking here only of the pure interest portion of the discount rate used in financial markets, and excluding such factors as risk, liquidity, etc.

[318a] Of course, we have the luxury of this presumption only as long as governments have yet to figure out how to tax time!

opportunities. Their selection will be based on all these variables, and the time variable considered will often be the cost of multiple trips over a long period of time.

Consider also a consumer choice of an automobile dealer. We might conjecture that the amount of money involved in a purchase of a new automobile is substantial enough that even a small discount would induce a consumer to travel a good distance. In fact, analyses of sales in quite competitive markets — even when dealers compete for sales by lowering prices — show the majority of sales go to the closest dealer, not the "lowest price" dealer. How do we reconcile the conjecture with the observation?

Once you consider the number of times a consumer will visit a dealer, the apparent cost penalty per mile becomes much larger. If a consumer kicks tires on one trip, makes a purchase agreement on a second, picks up the vehicle on a third, and returns for warranty service for four additional times during the next year, he will have made a total of seven round trips to the dealership. Therefore, driving ten extra miles to save a few dollars on the purchase price quickly turns into 140 miles.[318b]

Newton's Law of Gravity

A universally known metaphor for social behavior is the physical attraction between two masses, which we call gravity. The closer two objects are and the larger their masses, the stronger their attraction is to each other.

Isaac Newton stated the law of gravity after a fabled apple hit him on the head. The law can be cogently stated as:[319]

> ... any particle of matter in the universe attracts any other with a force varying directly as the product of the masses and inversely as the square of the distance between them.

The gravity equation defines the attractive force (*F*, expressed today in units called Newtons) in terms of the sizes of two objects (their masses) and the distance between them. The gravitational pull is equal to the product of the gravitational constant (*G*) and the product of the masses (m_1 and m_2) divided by the square of the distance between the masses (*r*):

[318b] That's 10 miles each way, 7 trips. If the cost-per-mile in the first year is $1.00, and the consumer values his or her time at $25/h, and the average driving speed is 45 mi/h, the cost penalty in the first year is over $200.

[319] The excerpted explanation is from the entry Newton's Law of Gravity found at: http://britannica.com. See the following footnote for links to Newton's own writings.

$$F = \frac{G \cdot m_1 \cdot m_2}{r^2} \tag{1}$$

Isaac Newton put forward the law in 1687 and used it to explain the observed motions of the planets and their moons, which had been reduced to mathematical form by Johannes Kepler early in the 17th century.[320]

The Social Gravity Model: Antecedents

The same principles behind the physical definition of gravity can be applied to social gravitational models. Social scientists began thinking about human behavior in terms of the physical laws of gravity in the 19th century.

H.C. Carey's summary of social sciences in 1858 described the tendency of a human to "gravitate to his fellow man" and used the gravitational pull to explain railway traffic and migration.[321] E.C. Young in 1924 stated that human behavior "does not lend itself to exact mathematical formulation" but stated a formula for migration that was based on the physical law of gravity:

$$M = k \frac{F}{D^2} \tag{2}$$

where migration M is directly proportional (using the constant k) to the intensity of attraction F and inversely proportional to the square of the distance D between two communities.[322] Over the next few decades, this was extended by other social scientists to a more general social gravity model of the form

$$T_{ij} = GP_iP_j(d_{ij})^{-2} \tag{3}$$

[320] Newton published his *Philosophiae naturalis principia mathematica* or *Principia* in 1687. Its treatment of gravity owed much to Johannes Kepler, who had previously described how planets move through space. In particular, Newton modified Kepler's Third Law of planetary motion, which decribes how planets move in ellipses. See, e.g., "History" pages published by the University of St. Andrews, found at: http://www-gap.dcs.st- and.ac.uk/~history/Indexes/HistoryTopics.html, especially entries for Newton, Leibniz, and Theories of Gravitation. For a description of Newton's modification of Kepler's equation, see the entries on Kepler and Newton in *Online Journey Through Astronomy*, found at: http://csep10.phys.utk.edu/astr161/lect/history/newtongrav.html.

[321] For a brief historical review, see A. Sen and T.E. Smith, Introduction: Classical Gravity Models, *Gravity Models of Spatial Interaction Behavior* (New York: Springer, 1995). The quoted phrase is from Vol. 1 of H.C. Carey's 3-volume essay on the social sciences, *Principles of Social Science* (Philadelphia: Lippincott, 1858).

[322] E.C. Young, *The Movement of Farm Population*, Bulletin 426, Cornell Agricultural Experiment Station, Ithaca, New York, 1924, p. 88; quoted in A. Sen and T.E. Smith, Introduction: Classical Gravity Models, *Gravity Models of Spatial Interaction Behavior* (New York: Springer, 1995).

in which the interaction T_{ij} among two communities (denoted by the subscripts i and j) is proportional (using the constant G) to the product of their populations and the inverse of the intervening distance squared.[323]

The Quadratic Human?

The centuries have been kind to Isaac Newton's law of gravity, which states that the gravitational force between physical masses varies inversely with the square of the distance between them. Should this law of physics have anything to do with the social behavior of humans?

The advances of the social sciences over the past century indicate that it does. Indeed, the recognition that much observable human behavior varies not linearly, but with the *square* of a relevant factor has been both powerful and durable. Many of the most robust, well-developed, and commonly used statistical techniques in the social sciences are based on the assumption that humans value deviations from the expected result in proportion to the square of that deviation.[324] This "quadratic" loss function is complemented by the common use of a quadratic utility function in microeconomics. In finance, the mean-variance framework that underlies modern portfolio theory is similarly based on a quadratic risk preference. Methods to solve quadratic equations predate even the development of modern number systems, with algebraic and geometric methods documented for the ancient Babylonians and Euclid of Alexandria in ancient times, and Indian and Arab scholars in the first millennium A.D., before the publication of the *Summa* in Italy in the Renaissance era.[325]

---

[323] J.Q. Stewart, An inverse distance variation for certain social influences, *Science,* 93, 1941; quoted in A. Sen and T.E. Smith, Introduction: Classical Gravity Models, *Gravity Models of Spatial Interaction Behavior* (New York: Springer, 1995).

For a clearly written, nontechnical description of gravity models, see geographer M. Rosenberg's entry on gravity models found at: http://www.about.com.

[324] These methods include the the use of the variance and standard deviation to describe a distribution (both based on the square of the deviation from the mean) and all extensions of these, including analysis-of-variance (ANOVA), ordinary least squares regression models, and variations of these methods such as nonlinear least squares. Other methods, including those based on the chi-square distribution, are formed on a similar assumption.

Part of the motivation for this widespread usage is the tractable mathematics of squared deviations (e.g., they are always positive); part is the strong empirical background, starting at least as early as the work on regression of Sir Francis Galton (1822-1911) along with that of his contemporary Karl Pearson who introduced the correlation statistic. Galton anticipated this long ago. "There seems to be a wide field for the application of these methods to social problems," he wrote in "Kinship and Correlation," *North American Review* 150 (1890). Found at "Sir Francis Galton as Statistician," at: http://www.mugu.com/galton/statistician.html.

[325] See the "History" pages published by the University of St. Andrews, found at: http://www-gap.dcs.st-and.ac.uk/~history/Indexes/HistoryTopics.html; especially the history of algebra. The *Summa de Arithmetica, Geometria, Proportioni et Proportionalita* was written by Lucia Pacioli and published in Venice in 1494.

Observations about quadratic human behavior have resulted in theories about the length of time in a meeting,[326] the relationship between environmental resources and population,[327] and other phenomena, including the location of population and industry.[328] These theories are not always correct,[329] or even always precisely quadratic, but their prevalence implies something significant. Sometimes the use of the exponent 2 is a handy approximation to human behavior that is clearly nonlinear, sometimes it is a mathematical convenience, and sometimes it matches actual data; quite often, it appears to be more than one of these.

Is there a law of human nature that states that we have a quadratic loss function? We agree with Young's observation that "human behavior doesn't lend itself to exact mathematical formulations" and state that the answer is,

---

[326] Shanahan's law is, "The length of a meeting rises with the square of the number of people present."

[327] Thomas R. Malthus wrote, famously, "Population, when unchecked, increased in a geometrical ratio, and subsistence for man in an arithmetical ratio." T.R. Malthus, *An Essay on the Principle of Population: A view of its Past and Present Effects on Human Happiness* published in 1798. Malthus's dire prediction that overpopulation would outstrip the world's ability to feed itself has been proven, both theoretically and empirically, to be false. Human productivity, imagination, and work have consistently expanded the supply of goods necessary to feed, clothe, and house the population of the world.

The Malthusian notion that population growth inevitably portends starvation — or even deprivation — has a modern analogy in the "limits to growth" theories. The seminal author refuting such notions was the late Julian Simon (1932–1998) whose many books include *The Ultimate Resource* (1980) and *The Ultimate Resource 2* (1996) which is now available at: http://www.juliansimon.com. See, in this latter book, Chapter 5, for citations of public statements made in modern times anticipating mass starvation as food supplies became inadequate to feed a growing population, and the epilogue where he charitably credits economists starting with Adam Smith and David Hume and leading to Freidrich Hayek with understanding the fundamental role that human productivity plays.

Along with his scholarly books, Simon is remembered for his famous 1980 wager with Paul Ehrlich on the question of whether population growth and resource scarcity would drive prices of natural resources up (the view of the "limits to growth" theory then popular) or down. After 10 years, with the earth's population growing by 800 million, the prices of the commodities chosen by Ehrlich (tin, chromium, copper, nickel, and tungsten) had all fallen, and Simon had won the bet. See, among other appreciations, Rev. R. Sirico, The ultimate economic resource, *Religion and Liberty,* 8 (3), June 1998, published by the Grand Rapids, MI, Action Institute and also available at: http://www.action.org.

[328] See the discussion on von Thunen's model, which implies concentric circles of agricultural crops around a city (a circle having an area proportional to the square of its radius), in this chapter under "The Location Poverty of Modern Economics" on page 327.

[329] Malthus was simply incorrect. However, there have been darker uses put to tools of social science, and this review of the intellectual transition from the natural to the social sciences would be incomplete without at least noting this. In particular, the statistical techniques developed by Galton would, in conjunction with the theories of evolution propounded by Galton's cousin Charles Darwin, be used to support the eugenics movement and its efforts to improve the human race through better breeding. Following this, many horrors of the 20th century were claimed to be based on "scientific" evidence about classes and races.

The modern confrontation between the seeming promise of the natural sciences and the morality and world-view based on religion and theology was ignited by the intellectual achievements of Newton and others cited here. It still simmers. See "Science and Religion" in J.A. Garraty and P. Gay, Eds., *The Columbia History of the World* (New York: Harper & Row, 1972); and P. Johnson, *Modern Times: The World from the Twenties to the Eighties* (New York: Harper & Row, 1983), especially the introductory chapter, "A Relativistic World," and the final chapter [in the original edition], "Pamplisets of Freedom."

fortunately, no. Indeed, we will relax the assumption that the exponent on the distance–cost factor for retail consumers is 2 and allow human-generated data to provide an answer to this question.[330] However, before going further down the road of examining consumer behavior, we should at least pause for a moment and reflect on how much of human behavior appears to roughly follow a quadratic loss function.

The Social Gravity Model: Applications

We described previously how the basic social gravity model relates strength of a bond between two places to the product of their populations divided by the distance between the two cities squared. The original application of this model to examining retail behavior came in the seminal book *Law of Retail Gravitation* published in 1931.[331] In his book, William Reilly applied the physical gravity model directly to social phenomenon and created the retail gravity model based on the equations of Newton and Kepler.

Retail Sales Adaptation of the Gravity Model

Using techniques and data that would have been difficult to imagine in the time of Newton, and quite a leap even for Reilly, we can find out how well their insights predict 21st century behavior.

Adapting Newton's physical gravitational model to retail sales assumes that attraction to a center increases with the size of the center and proximity to the center. This means that, everything else being equal, customers are more attracted to a large retail center than to a small retail center; and they are more attracted to a nearby retail center than to a faraway retail center.

Applying the model to consumer behavior, we assume that one of the masses in the physical model is the retail location, and one of the masses is the customer. We further adapt the model for retail sales (and take a more cautious approach to describing human behavior) by assuming the distance (D) taken to an exponential power (k) that could be different from 2. This results in the following gravity model equation:

330 A. Sen and T.E. Smith, who have extensively documented the origins of gravity models, indicate that the use of the exponent 2 in social gravity models has always been a matter of contention, and note that starting in the 1950s, researchers began treating the exponent as a parameter to be estimated statistically, using available data. We follow this path, though note the prevalence of the quadratic behavior assumption in many statistical, microeconomic, and finance methods. A. Sen and T.E. Smith, Introduction: Classical Gravity Models, *Gravity Models of Spatial Interaction Behavior* (New York: Springer, 1995).

331 W.J. Reilly, *The Law of Retail Gravitation* (New York: Knickerbocker Press, 1931).

$$S = \beta \cdot \frac{1}{D^k} \tag{4}$$

where sales (S) equals a constant (β) multiplied by the inverse of distance (D) to a constant power (k). The k parameter determines how steeply the attraction diminishes as distance increases.

Although the physical law of gravity specifies this parameter as 2, and much of the early work on social gravity models accepted this as the likely amount, we will commonly estimate this parameter from the available data.

Uses for Distance–Sales Analysis

Analyzing the distance–sales relationship is an essential tool to understanding retail markets, where "retail" is broadly defined to include goods and service providers that operate from one or more defined geographic points.

The uses for this analysis include:

- Identifying trade areas for retailers, service providers, and other firms
- Forecasting the ability of a downtown area or other retail center to attract customers, residents, or workforce from surrounding areas
- Projecting the potential sales from the opening of a new retail location
- Assessing the effect of a new retailer on the sales of existing, similar retailers in the same market area
- Estimating the potential market for specific segments of a market, such as estimating the demand for housing within certain age or income segments.

All these uses, and related ones, can be accomplished using the techniques presented in this chapter.

Customer Location Data

We cannot jump headlong into an explanation of the techniques of sales–distance analysis without first discussing the importance of data.[332] In

[332] This is an application of the maxim "Know your data" described in "Maxims of Business Economics" on page 15.

this section, we discuss the sources of such data and, in the next, the first step in using them.

An analysis of the sales–distance relationship relies on extensive, high-quality data. Such data can be tedious to acquire; however, once collected, they offer valuable insight for strategy decisions and performance analysis. In the following text, we weigh the benefits and challenges of both individual customer location data and regional sales data, and discuss the options for measuring the distances between customers and retail centers.

What Is Location?

We normally assume that the location of the consumers is their resident address and the location of the retailer is the storefront. In most retail sales analyses, this assumption is a safe one. However, we should consider briefly some questions to address when specifying the location of each element in the pair:

1. If the consumers normally come from home, then the residence is their location. However, if the consumer is typically traveling (as is the case in airports, resort areas, or in vacation communities), the consumer location must be specified differently.
2. If consumers visit a retailer from their work location or visit on a commute to or from work, the location specification is more complicated. Sometimes, a simplifying assumption is made that the number of consumers commuting from the area of the retailer is about the same as those commuting to the area. This is not likely to be true in most cases.

An approach we have taken is to explicitly note that some consumers will reside elsewhere, commute to a work location, and make purchases when visiting from work. This sales–distance relationship is established from the home–retailer and home–work–retailer trips. However, unless additional information can be brought to bear, the analyst will normally not be able to disentangle the two behaviors.[333] In such cases the analyst may sometimes use aggregate data as the basis but note the missing information and any material distortions that may be caused by it.

Sources of Customer Location Data

Individual customer location data should include individual customer purchases and the customer's distance from the retail center.

[333] One of the ways additional information can be brought to bear is to identify, using other information, two classes of customers and estimate the relationship separately for each.

Many companies regularly collect this data and may, or may not, realize it. For example, auto dealers have access to the addresses of past customers through registration information; banks know the addresses of account holders; sports teams know the addresses of season ticket holders; and travel agents know the addresses to which to send plane tickets.

The more extensive the data source, the better. The examples in this chapter use data in the form of a list of sales and the distance of each sale from the retail center. "Finding Distances" on page 345 explains various methods that can be used to determine the distance between the retail location and the customer's address.

If you are using customer location information for multiple retail locations within a market area, you should record the distance between the retailer and the customer, even if that retailer is not the closest to the customer's location.

Although these data are the most accurate to use in the analysis, their use is not always feasible. If the data is not already collected, it will likely be very difficult to compile for a short-term project. Furthermore, many establishments are unable to gather this information on their customers. For example, many coffee-shop patrons would be taken aback if they ordered a *latte* and the coffee shop employee asked for their address. The same difficulty holds true for most retail stores.

Initial Analysis of Regional Sales Data

The location data are frequently compiled as the number or percentage of customers by geographic region. Often, the location unit is a zip-code or phone-exchange, as such data are often collected simply by asking customers for it at checkout.[334] Many major retail establishments have used this approach to collect customer location information.

Such data usually need to be aggregated to show sales by geographic region and distance. Normally, the regional sales data should be collected into a matrix. One column is the distance of the region from the point of sale. The second and succeeding columns consist of sales variables, such as:

- Total sales revenue for that region
- Number of unit sales for each region

[334] There are other methods as well. Special "club" cards that connect customer purchases to their location are sometimes used, although customers must normally be given some incentive to provide this information and shoulder the burden of carrying it and presenting it at checkout. Surveys, including "intercept" surveys of actual shoppers, can also be used. However, be cautious when using survey data that is contaminated by self-selection bias or poorly designated questions.

- Percentage of a retailer's sales that can be attributed to each region
- A retailer's *penetration rate* in each region, if data are known for competing retailers[335]

Geographic Analysis of Sales

Once geographic data have been acquired, the first analytical step is to explore that data along the geographic dimension. Until recent years, this spatial analysis was prohibitively expensive for most applications. Proper spatial analysis is still more difficult, and the software and data requirements more extensive, than analysis that assumes away the geographic component.[336] However, today it is possible and cost-effective in many applications.

One focus of this book is to directly incorporate such analysis. In this section we make use of GIS software and data that have a spatial component.

Example Data and Analyses

We illustrate this with standard tools of retail sales analysis and one innovation of our own. For the following analyses, we have used data adapted from actual consumer purchases of luxury automobiles in a large metropolitan area of the U.S. These analyses involved a large dataset which included the location of competing retail outlets (automobile dealers), the sales of competing brands by zip code area, and the residence location of a subset of the customers.

Drive-Time Analysis

Most consumers in industrialized countries like the U.S. drive automobiles when shopping. Even in urban areas served by mass transit, and in countries with low vehicle ownership, buses are typically the backbone of the mass transit system.

Thus, the cost in time to get to a retail outlet is best modeled by *drive time* — the time it takes to drive a vehicle to the location. Drive time is often the most important indicator of likelihood to shop at a certain retailer.

Determining Drive Time: How does one determine drive time? The only method of doing so is to analyze specific geographic components of the area, notably the road network. If there are only one or two *points* to consider, this can often be done by hand, with a map and a knowledgeable

---

335 The penetration rate refers to the percentage of a region's total sales that can be attributed to a given retailer. For example, if Television City sells 25 big-screen televisions to customers in a zip code, and all other retailers sell a total of 75 big-screen televisions to customers in that same zip code, Television City has a penetration rate of 25%.

336 Many standard financial reports have a very limited geographic component, for example, segregating sales by state or by store. Such analysis, especially when the retail outlets are distant from each other, is an example of assuming the spatial component away. The working assumption is simply that every buyer in a state (or residing around a certain store) has the same proclivity to shop at that retail outlet. As will be shown, this is almost never the case.

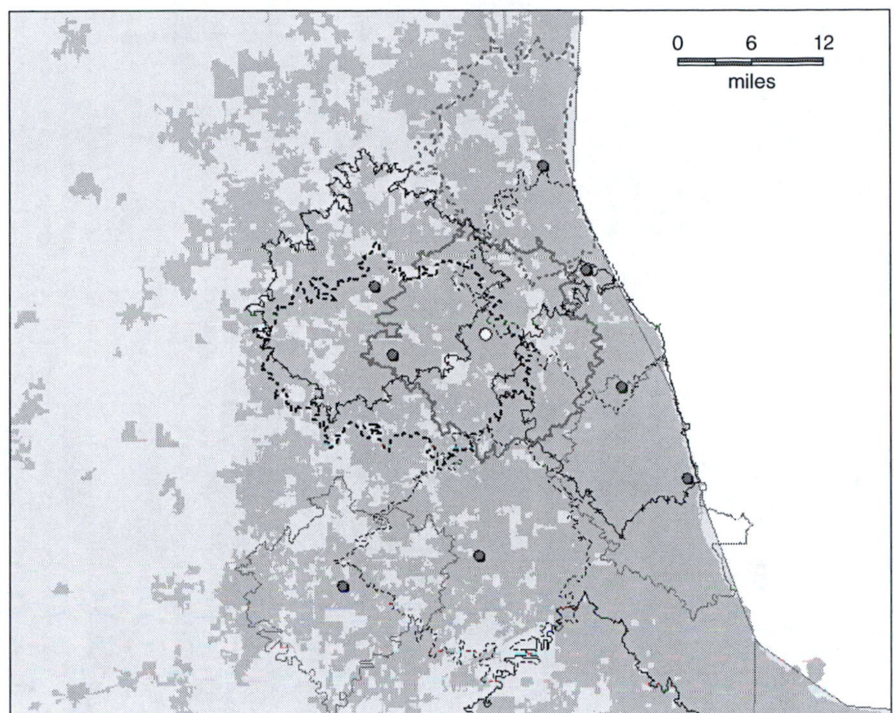

FIGURE 13-1
Drive-time analysis.

local resident.[337] If there are more than a few points to consider, then a GIS facility with adequate local data and a drive-time estimation routine is the best approach.

Example Drive-Time Analysis: An example drive-time analysis is shown in Figure 13-1, "Drive-time analysis." The irregular polygons around each point are a "drive-time area" for a specific time interval.

Equal-Distance Rings

A less sophisticated analysis, which can be done by hand on a paper map, is to draw concentric circles around a retail outlet (See Figure 13-2, "Equal-distance rings"). Using simple distance as a proxy for time introduces some distortion in the analysis for those customers that drive, walk, or take a bus using roads and streets. However, the traditional "as the

[337] By "points," we typically mean the retail outlet. However, the term is sometimes used to include customers as well.

[338] Many states have laws that allow franchised retailers in certain industries, such as automobile retailing, to protest the appointment of a same-line retailer within a certain distance from

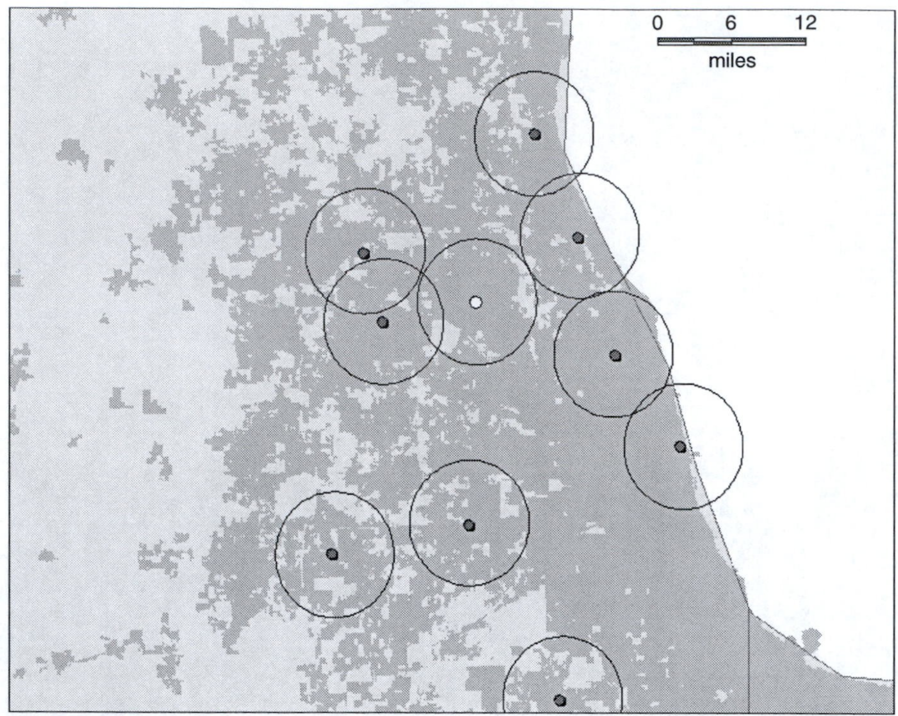

FIGURE 13-2
Equal-distance rings.

crow flies" method is a rough guide, and is often the method used in legal contracts and state laws that govern franchised retailers.[338]

"As the crow flies" rings can be an acceptable method. In circumstances where distance accurately reflects the likelihood to shop, where there are no major geographic boundaries (such as rivers) in the area, no major political boundaries, or where the distances are either quite short or quite long. In fields other than retail sales, it has broad application.[339]

Sales Patterns by Area

When data on individual purchasers are available, a much richer understanding of the sales–distance pattern can be found by plotting the customer

the location of a current retailer. Often this distance is about 10 mi, and is frequently modified to exclude retailers in a different county or other jurisdiction. States with such laws include Michigan, Wisconsin, Illinois, and Texas.

339 Some of these are: trade patterns (although typically an inferior method to shipping time), "environmental justice" analyses (see P.L. Anderson, *The Effect of Environmental Justice Policy on Business: A Report to the U.S. Council of Mayors*, Lansing, MI: Michigan Manufacturers Association, 1998), other references found at http://www.andersoneconomicgroup.com/Projects/archives/Brownfield/brown_study.htm), and analyses of air quality and emissions sources.

locations on a map. One approach is a "dot map" showing dots where customers reside. This illustrates quite effectively the actual pattern of sales. See Figure 13-3, "Sales patterns and distance."

A second approach is to collect the number of purchasers in each surrounding area, using boundaries such as zip code areas, census tracts, or municipal boundaries. This method is often used when data deficiencies allow only a partial identification of customer residences.

We present an example in Figure 13-4, "Sales patterns and distance: alternative perspective." This *thematic map* shows by different shades the total number of purchasers in each area.

Daylight Areas

Retailers operate in a competitive environment. We expect that a profitable retailer will attract new entrants to a market and that customers will travel farther to find goods they cannot find nearby. This results in competition. Price is one dimension of that competition, and location is another.

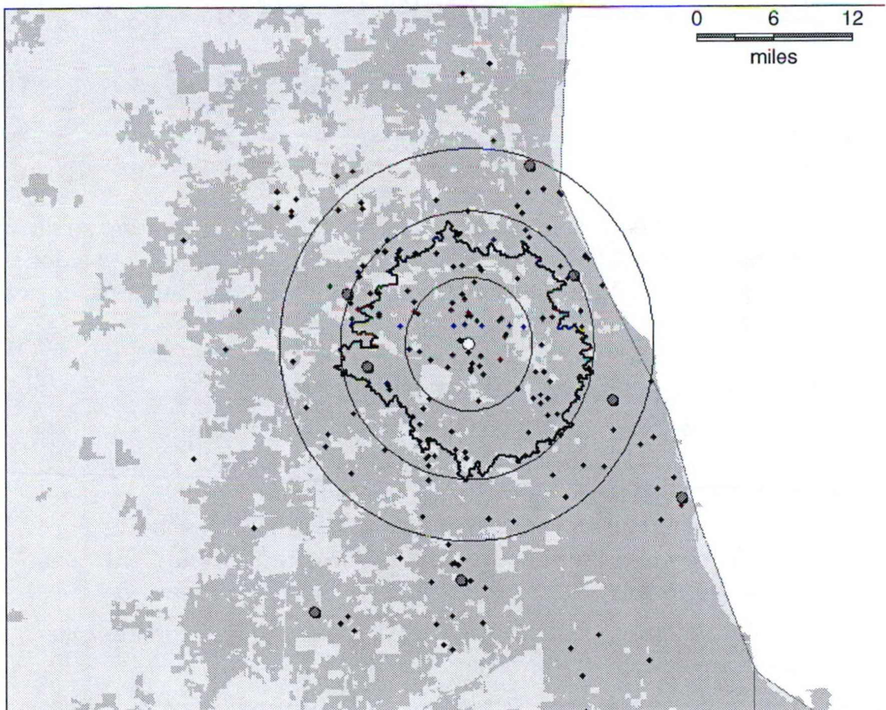

FIGURE 13-3
Sales patterns and distance.

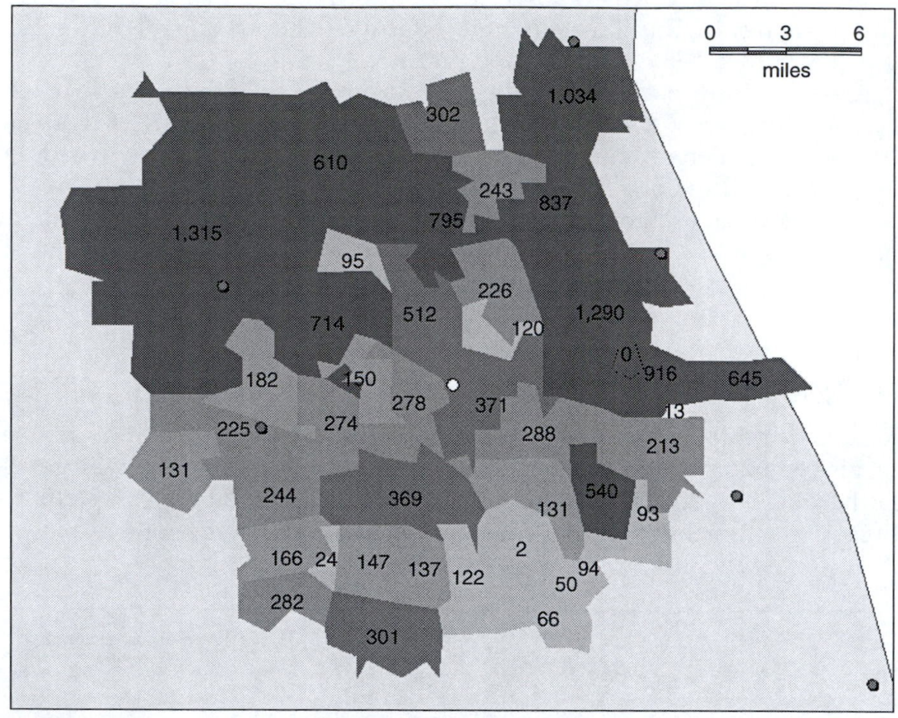

FIGURE 13-4
Sales patterns and distance: alternative perspective.

One innovation we have used in modeling competition among retailers in the same region is the identification of "daylight" areas. We define daylight areas as areas in which consumers have one, and only one, convenient retailer. The method for producing a daylight map is:

1. Identify all competing retailers in a region.
2. Create a primary trade area for each retailer. This requires analysis of the sales–distance pattern and other data, and is an important task for the analyst. These trade areas should normally be determined using the same parameters (such as drive time or distance) for each retailer. Plotting the penetration rate is an excellent visual tool for this purpose. See Figure 13-5, "Market penetration rate."
3. Show the trade areas for each retailer on a map. In many cases, these trade areas will overlap.
4. By darkening the areas that are in the primary trade areas of competing retailers, we reveal the daylight areas where an additional retailer could locate and serve a primary market area with limited competition.

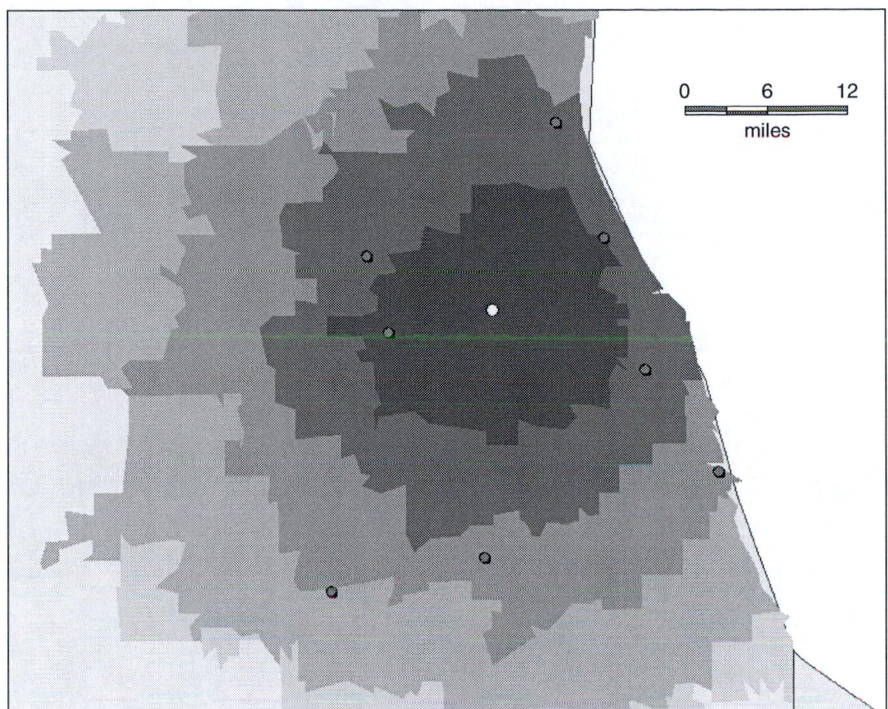

FIGURE 13-5
Market penetration rate.

5. The size of the daylight area and the number of consumers in that area provide a strong indication of the likely sales potential of the additional retailer.

See Figure 13-6, "Daylight areas."

Finding Distances

The next step in analyzing retail sales data involves measuring the distances between the retail center and the customer locations. Ideally, drive time from each customer to each retailer would be used, but functions for easily obtaining this information for large number of locations are not yet developed in GIS facilities.[339a]

In this section, we explain the best ways of determining distances. We break the processes out by the level of GIS capabilities needed to conduct the analyses and start with a discussion of the differences between individual data and data grouped by region.

[339a] Technology in this field may improve enough to develop such utilities soon.

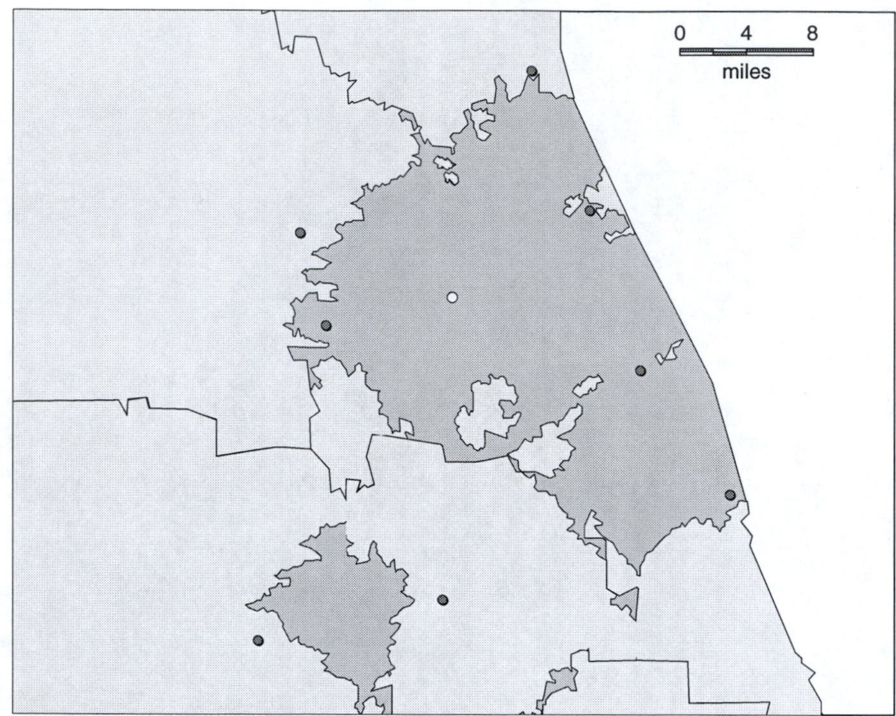

FIGURE 13-6
Daylight areas.

Regional or Individual Location Data

Determining the distances is typically done in one of two ways, depending on the available data:

1. With full individual customer addresses available, you should measure the distance between the address and the retail location. This involves the use of *geocoding* software or an outside geocoding consultant.
2. With regional sales data only, the process involves finding the geographic centroid of the regions and measuring the distances between the centroids and the retail center. This, of course, loses some information. Therefore, the results will be less precise than if a larger number of actual consumer addresses were known.

The simple centroid is, of course, not the most accurate point to use to represent buyers when sales in the region are distributed asymmetrically. (This is especially true when a centroid actually lies outside the region.) The center of concentration of consumers is the ideal measure. With many sufficiently small regions, however, errors in the distance measure due to the use of centroids may assumed to be nonsystematic. The consequent analysis may be unbiased, or have negligible bias.

Approach Using Professional GIS

The better GIS facilities allow analysts to incorporate SQL (structured query language) into the assessment of distances. After geocoding the customer locations or regional centroids and determining the geographic coordinates of the retail center, SQL capabilities can be used to add distances to the database using a single query command.

The actual implementation of this process varies between programs. However, those with knowledge of SQL and related GIS software should be able to adapt the commands to work with their system.[340]

This approach works for both regional sales and individual customer location data, resulting in a GIS database that includes distance variables. These data can then be exported into an analytical program such as MATLAB or a spreadsheet program.

Approach Using Basic GIS

If you have access to a GIS or mapping software program that does not provide SQL capabilities, you are still able to gather the distance information. This involves using the ruler tool that is available within all GIS platforms to manually measure the distances between the point of sale and the customer locations. By plotting each of the sale locations on the map and, if using regional sale data, determining the centroid of each region you can obtain much of the information that can be provided with the more sophisticated analysis described above. As you manually measure each distance, record the information in the original database. Caution should be exercised when measuring distance. While some imprecision is inevitable, users should ensure that distance from each consumer is measured in the same manner. With this in mind, when only basic GIS is available, we recommend reporting simple distances rather than driving distances or drive times where greater errors are likely to occur.

Approach Using No GIS

Without any kind of GIS platform, the process of measuring distances will take longer. However, you can often still accomplish it using one of the following methods.

[340] For example, in the GIS program MapInfo, we use the SQL Select feature to create a distance column using the query SphericalDistance(retailLat, retailLong, CentroidX(obj), CentroidY(obj),"mi") where (retailLat, retailLong) are the actual (X,Y) coordinates of the site, and obj is the GIS object of the geocoded points.

Internet mapping applications: Using mapping capabilities provided by popular Internet-based mapping engines, you can compute driving directions between two points. When the online program presents the quickest driving directions between two points, it provides a drive-time estimate. Using this tool to measure drive time between each customer location and the retail center, the analysis can proceed without the use of a GIS facility. However, this will require manually entering each of the sales into the mapping engine and individually processing each record. This can be extremely time consuming.[341] Furthermore, the accuracy of free online systems cannot be expected to be at the same level as professional systems used by trained analysis.

Paper maps: The most basic approach involves a paper map of the sales region and the use of a ruler. This process is similar to that employed in the use of basic GIS capabilities but involves manually plotting sales points on the map and manually measuring the distances. As our goal here is to increase the efficiency and precision of your analysis, we recommend using this approach only if you are unable to complete the analysis using either a GIS program or Internet-based mapping products.

Outside consultants or data vendors: If you are prepared to undertake all the analytical tasks and have the raw data on addresses, you can purchase the geocoding services from a number of outside consultants as well as data vendors. Keep in mind that this is a relatively new area of applied economics, and both the quality of the data and the knowledge of the consultants vary significantly. In particular, be certain that any outside consultant or data vendor can furnish you with the specific source of their base data and identify the results of any geocoding.[342] The best data vendors will have data updated from a variety of sources on an annual basis and will be able to explicitly identify those sources. The best consultants will have some knowledge of the industry and the available data.

Applying a Gravity Model to the Distance–Sales Analysis

Overview

In this section, we describe the steps for modeling location and sales data using a retail-sales adaptation of the social gravity model presented earlier

---

[341] Those skilled at computer programming may be able to write a soft-client application that takes addresses from a database and runs them through online program. However, this is likely to be stop-gap approach. Furthermore, the providers of these "free" programs need to be paid for their effort if they are to stay in business.

[342] Be wary of vendors who indicate broadly that their data "come from the census." The U.S. Census Bureau does not provide geocoding software and detailed, year-by-year information on all geographic locations and variables. While Census TIGER files and the separate, decennial census results are used to develop most demographic forecasts and geocoding software for the U.S. market, the data vendor should clearly state which data are decennial census enumerations of specific areas and which are forecasts or revisions made by other parties.

For additional information, see the U.S. Census Bureau Web site and, in particular, its overview of TIGER at: http://www.census.gov/geo/www/tiger/overview.html.

in this chapter. Equation (5) restates this model, where sales (S) equals a constant (β) multiplied by the inverse of distance (D) to a constant power (k).

$$S = \beta \cdot \frac{1}{D^k} \tag{5}$$

We describe below how to use past sales data to estimate the β and k constants. Once these constants are determined, the equation can describe the relationship between customer sales and distance from the retail location. We describe how to apply this equation to measure future sales, the impact of competition, or the success of a new location, beginning in "Determining Sales and Market Areas" on page 364.

Sales–Distance as a Demand Function

The sales–distance relationship is fundamentally an implicit demand function. The one argument is a price variable proxied by distance. For this reason, one should consider the same admonitions and cautions that apply when estimating regular demand functions. See Chapter 5, "Library Functions for Business Economics," especially "Estimating Demand Schedules" on page 98 and "Step Two: Estimate Demand Equation" on page 99.

Elasticity of Demand in a Gravity Model

If the distance–sales relationship is an implicit demand equation, we should be able to evaluate the price elasticity of demand at least for the portion of the price that is transformed into distance. In a well-specified model, this elasticity should illustrate rational consumer behavior. Below, we describe how to determine the elasticity of the gravity model equation.

By using standard mathematical economics and by substituting a distance variable for the price variable, we define the elasticity of sales with respect to distance using the equation:

$$\varepsilon = \frac{dS}{dD} \cdot \frac{D}{S} \tag{6}$$

where elasticity of sales with respect to distance equals the derivative of sales with respect to distance, multiplied by distance (D) divided by sales (S). We apply this by differentiating Equation (5):

$$\varepsilon = \frac{dS}{dD} \cdot \left(\frac{D}{S}\right) = -k \cdot \beta \cdot \frac{1}{D^{k+1}} \cdot \left(\frac{D}{S}\right) \tag{7}$$

Reducing the equation yields:

$$\varepsilon = -k \cdot \beta \cdot \frac{1}{D^k} \cdot \frac{1}{S} \tag{8}$$

Substituting in this equation the original equation [Equation (4)] for sales (*S*) produces:

$$\varepsilon = -k \cdot \beta \cdot \frac{1}{D^k} \cdot \frac{1}{\left(\beta \cdot \frac{1}{D^k}\right)} \tag{9}$$

which, reduced, yields the elasticity parameter for a gravity model:

$$\varepsilon = -k \tag{10}$$

We find that elasticity of demand with respect to price in a gravity model equals the negative *k* coefficient.[343]

Desirable qualities of the gravity model in practice: The following are the desirable characteristics of the gravity model:

1. The elasticity is, as expected by economic theory, negative.
2. The elasticity is constant. This should also be expected as there is no reason to expect that consumers living closer to a retail outlet would have a larger or smaller elasticity than those living farther away.[344]
3. The demand function (with respect to the distance portion of the price) can be largely parameterized by a single number, the *k* coefficient. If the retail price is assumed to be the same for residents of any area, then this captures quite well the variation in sales given the implicit change in price caused by varying the distance of the consumer's residence to the retail outlet.[345] We call this the *decay factor* as it indicates how rapidly sales "decay" as distance increases.
4. The parameter β captures the differences in scale and the differences in units in one constant. We call it the *scale factor.*

343 The variable *k* has, in our empirical work, always been a positive number, meaning that -*k* is a negative number. This is consistent with the theory presented in the first sections of the chapter.

344 In empirical work, we will often truncate the data to eliminate the very few observations from residents of faraway places.

345 In certain cases, this assumption will not hold. An example of this is where state or national borders are within driving distance of the retail outlet, and tax rates, duties, or price schedules are significantly different in the two territories. In, say, duty-free shops, the demand will be much higher from consumers who cross a national boundary.

This equation thus has very desirable practical qualities and allows us to separate the portion of the relationship that is based on arbitrary factors (such as units and population) from the part of the relationship that is the distilled distance–sales relationship.

Data Collection and Formatting

"Customer Location Data" on page 337 discusses the collection of data on sales location and the computation of distances between retail centers and customer locations. This analysis requires that the data be in one of the following two forms:

- Individual customer location data, in the form of a single-column list of distances, with each distance depicting one customer or sale.
- Regional sales data, in the form of a two-column matrix, with each row representing one sales region. One column depicts the distance of the sales region from the retail center and the second column lists the number of sales, penetration rate, or other relevant sales variables.

Place Data into Bins

In order to analyze the data using the distance–sales model described in this chapter, the data can be compiled into nonoverlapping "bins" for analysis. In this case, a bin is defined by distance from the retail center and is associated with a sales figure representing the sales at that distance.

Because regional sales data should already be in this format (with one column showing distances and the other the number of sales at that distance), the process of classifying the data into bins is only necessary for individual customer location data.

A bin's distance variable is determined by the average maximum and minimum distances of a bin. The sales number is an aggregate figure of all sales that fall within the maximum and minimum distances. For example, if your data set included sales at the distances of 1.0, 1.5, 2.3, 3.1, 5.5, 6.3, 7.5, 8.2, and 10.0 mi, dividing the data into three equidistant bins would result in a bin from 1.0 to 4.0 mi, one from 4.0 to 7.0 mi, and one from 7.0 to 10.0 mi. Four sales would be recorded in the first bin at an average distance of 2.5 mi; two sales would be recorded for the second bin at an average distance of 5.5 mi; and three sales would be recorded for the third bin at an average distance of 8.5 mi.

The bins may not overlap. Although the analysis does not require that the bins are of equal distance in width, MATLAB and other software programs will often assume equal-distance bins by default.

Steps to Create Bins

To create bins using MATLAB, do the following:

1. Open the database of distance and sales data and import it into MATLAB. Using the procedure discussed above, you may open a .dbf file in Excel and then using Excel link import the distances vector into MATLAB.
2. Once the data is in MATLAB, use the hist function to place the data in bins:

   [Sales , R] = hist(A , BINS)

   where A is the distance vector, and BINS is the number of bins.
3. The resulting variable Sales is the number of sales per bin and R is the centroid of each bin (the average bin distance). Combine the sales and centroid information into a single matrix:

   SalesMatrix = [Sales, R].
4. If there are bins with zero sales, you may wish to manually erase those bins (rows) from the matrix.

Specifying and Estimating a Sales–Distance Model

The preceding sections of this chapter demonstrated the theoretical reasons for using a social gravity model as the basis for modeling the sales–distance relationship. In most cases, the classic gravity model equation is the simplest and best choice. In other cases, a different specification may be selected that is better-suited for the data. We describe the estimation of the parameters of such models in this section; we also discuss an alternative specification, a polynomial model.

Data collection note: The data required for both specifications are the same. "Data Collection and Formatting" on page 351 describes the type of data needed and how to place the data into bins.

Step-by-Step Method

We recommend that the following steps be done after the acquisition of the data:

1. *Initial exploratory data analysis:* A good approach to this and other problems is to first explore the data. Exploratory Data Analysis (EDA) is a fruitful source of analytical techniques for this purpose. We describe techniques of exploring the data in "Geographic Analysis of Sales" on page 340. Before attempting to specify and estimate a sales–distance model, explore the spatial data patterns.

   A second, simple EDA technique is to simply plot the distance–sales relationship on an X-Y chart, with dots indicating each (sale and distance) pair. With the data already analyzed in the manner described above, a histogram of the data, broken down by distance, would provide excellent summary information.

2. *Identifying analog stores:* A second step that should be taken in many cases is to look for existing retail outlets with similar characteristics and examine their sales–distance relationships. This is sometimes called the *analog method* after William Applebaum popularized it in the 1960s.[346]
3. *Primary model specification — gravity model:* With the data prepared in bins as described above, specify and estimate the distance–sales relationship. We provide examples below.

Example I: Monte Carlo Data on Retail Sales

We first provide an example of this technique using Monte Carlo data:

1. To create this test data and then estimate it, we used the `retail_test` function which is part of the Business Economics Toolbox. It first takes in assumptions about the underlying relationship and generates the random numbers to include in the data. Second, using the assumptions and the random numbers, it creates test data. Third, using the special functions we have developed for modeling retail sales, it estimates those parameters.
2. Figure 13-7, "Underlying sales–distance relationship: gravity model," shows the test data generated to test our ability to properly identify a gravity model. The underlying relationship is shown in the top pane. The data are based on this relationship, plus some random noise, and are shown in the bottom pane.
3. Using only the test data, we estimated the relationship as shown in Figure 13-8, "Estimated sales-distance relationship." Note that the actual parameters were somewhat different from our estimated parameters, indicating that (as should be expected in all such cases) the random noise in the data obscured the precise relationship.
4. The command line output from the function is shown in the following code fragment.

[346] W. Applebaum, "Methods for determining store trade areas, market penetration and potential sales", *J. Mark. Res.*, 3, 1966. Cited in, e.g., A. Ghosh, *Retail Management* (Fort Worth, TX: Dryden, 1994), Chap. 11.

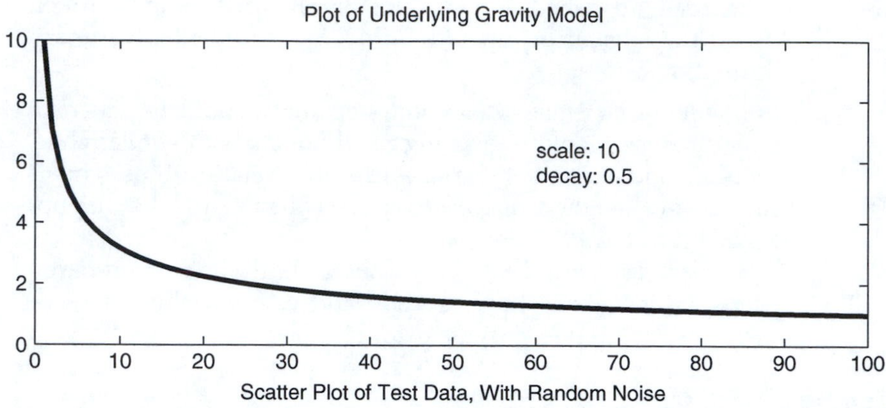

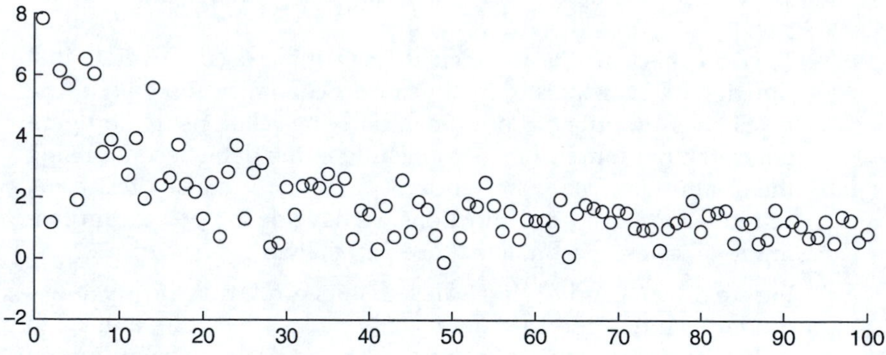

FIGURE 13-7
Underlying sales–distance relationship: gravity model.

Code Fragment 13-1. Test Gravity Model Output

```
Coefficients Used in Gravity.m:

coeff_mess =

'Estimated Parameters'

    'Scale: 7.61'

    'Decay: 0.4051'

    'using gmeval.m function'
```

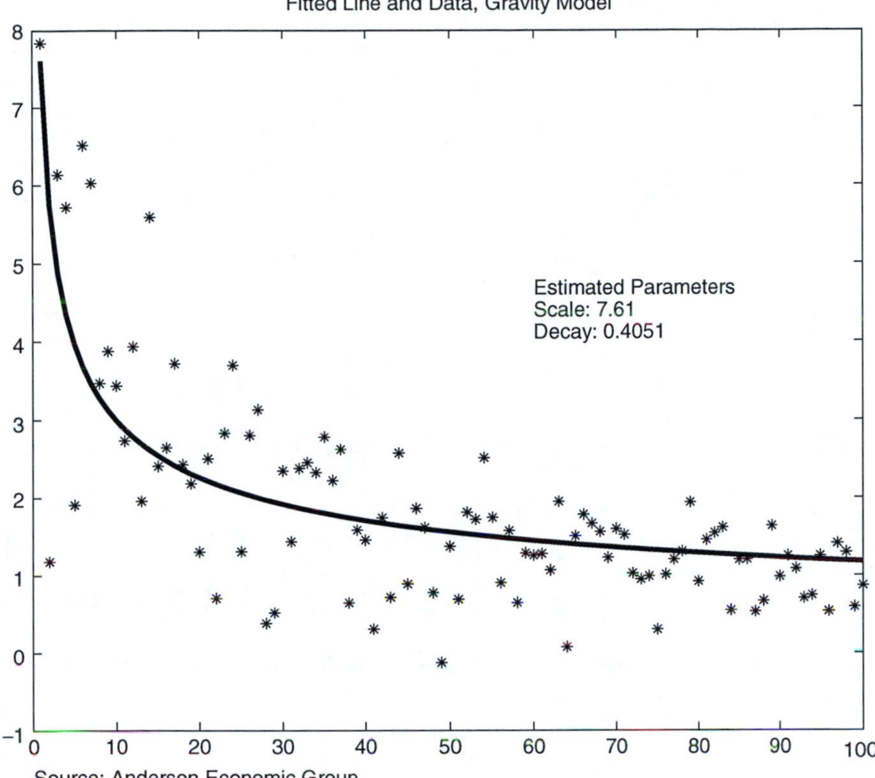

FIGURE 13-8
Estimated sales–distance relationship.

Example 2: Distance–Sales Relationships for Automobile Dealers

We modeled the distance–sales relationship for retail automobile purchasers in one of the largest metropolitan areas of the U.S. The results for two of these dealers, over multiple years, are shown in Figure 13-9, "Distance–Sales relationships for automobile dealers."[347]

Note that the relationship appears relatively stable over the years. Indeed, although total sales fluctuated during this time period, the relationship between distance and sales did not. The decay and scale parameters estimated differed between the two dealers but were fairly consistent over time for each dealer.

---

[347] This is a different geographic model, and a different segment, from the examples presented earlier. The segment in which these dealers competed was the low-price and mid-price imports.

The data were collected by zip-code area, so there is some aggregation bias due to grouping all consumers within a zip-code area into a single distance class.

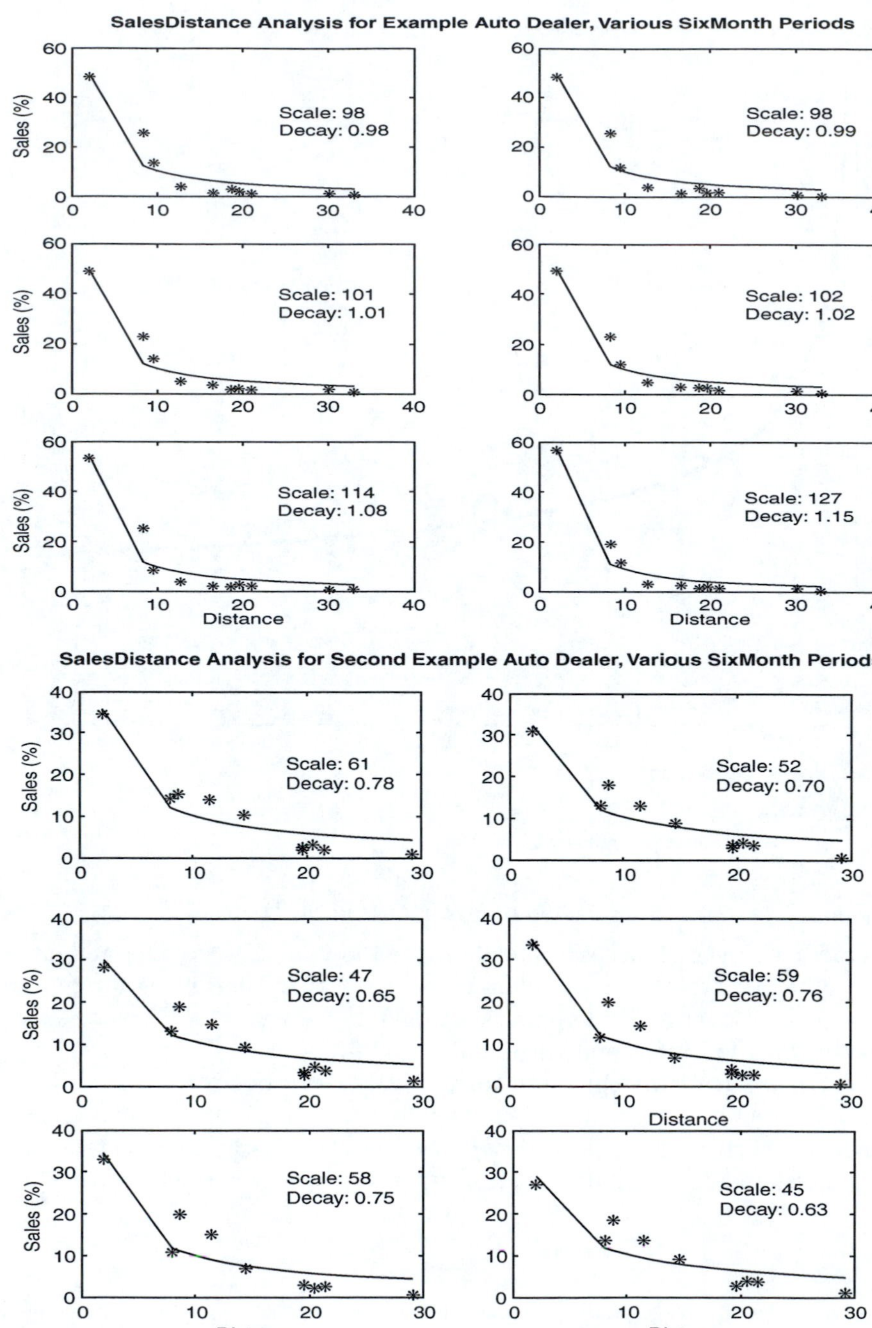

FIGURE 13-9
Distance–sales relationships for two automobile dealers.

Nonlinear Optimization and Gravity Model Estimation

The specification and estimation of a gravity model requires sophisticated software. As the model is nonlinear, a simple OLS (ordinary least squares) regression will not estimate the parameters. In this case, a software environment like MATLAB is needed.

In the examples presented above, we used the following methods:

1. We custom-programmed a gravity function, gravity, which implements the gravity model equation specified above. This function takes the data and coefficients and produces the distance–sales relationship. Note that this function will not estimate the parameters; it takes parameters as an argument and produces a distance–sales curve, given the appropriate data.
2. To estimate the parameters, we use a nonlinear optimization algorithm available within MATLAB. In particular, we make use of the gravity.m function and the nonlinear least squares estimation command nlinfit. Linear models can be directly estimated using a well-known formula that minimizes the sum of the squared errors. Nonlinear models typically require an initial guess at the correct parameters and then, iteratively, a search to find the parameters that minimize the sum of the squared residuals. The MATLAB nlinfit command uses the Gauss–Newton method to find these parameter estimates.[348] Other optimization and numerical methods are also available within MATLAB.[349]
3. To package the use of these multiple commands, we custom-programmed the gmeval function, which calls both the nonlinear least squares and gravity functions.

---

348 With this, the intellectual antecedents explored in the beginning of the chapter come full circle. The Gauss–Newton method merges the work of Isaac Newton (of gravity equation and calculus fame) and Carl Gauss (developer of the method of least squares, for which the Gauss–Markov theorem is named so). Regression (using the least squares method) was introduced by Sir Francis Galton and developed by Karl Pearson and others.

See the discussion in "Newton's Law of Gravity" on page 332 and on the following page; also, J. Stanton, "Galton, Pearson, and the Peas: a brief history of linear regression for statistics instructors," *J. Stat. Educ.*, 9(3), 2001 found at: http://www.amstat.org/publications/jse/v9n3/stanton.html. For a discussion of the history of the Gauss–Newton method, see J.C. Nash and M. Walker-Smith, *Nonlinear Parameter Estimation* (New York: Marcel Dekker, 1987); reprint 1995, found at: http://www.nashinfo.com/files/nlpechap.pdf.

349 For typical sales–distance paired data, the standard Gauss–Newton algorithm (which has been modified with Levenberg–Marquardt routines to "learn" faster) in the nlinfit function has worked well in our applied work. However, the optimization and statistical toolboxes within MATLAB provide additional options. See the MATLAB *User Guides* for the Statistics and Optimization toolboxes.

Alternative Specification: Best-Fit Polynomials

An alternative to the classic gravity model is a polynomial equation. Such an equation allows for the use of different factors and powers of those factors. Using such equations, one can often find a specification that offers the best fit for the data among a number of alternatives, including the gravity model.

Software Advances

After the sales data are plotted by distance — done during the exploratory data analysis or while examining a gravity model specification — recent versions of MATLAB make fitting polynomial equations of multiple degrees relatively simple, at least with equations involving only two variables.

Once the data are plotted in MATLAB, a figure window appears showing the analysis.[350] Within the Tools menu of this window, select Basic Fitting to open the basic fitting tool window. Using this tool, you can fit equations of different polynomials to the data very quickly.

Example: Quadratic Equation

For example, if you select the quadratic equation box, MATLAB will fit an equation in the form

$$y = p_1 \times x^2 + p_2 \times x + p_3 \tag{11}$$

to the data; if you select the cubic equation box, MATLAB will fit an equation in the form

$$y = p_1 \times x^3 + p_2 \times x^2 + p_3 \times x + p_4 \tag{12}$$

to the data (where p_1, p_2, p_3, and p_4 are constant numbers estimated by the toolbox, y is the sales variable, and x is the distance variable). If you use MATLAB to determine the best-fit equations for linear, quadratic, cubic, fourth degree, and possibly other degree polynomials, you can determine the equation that best estimates the actual sales based on distance data. To do this, return to your original data source and compare sales figures projected for each distance by the best-fit polynomials to the actual sales figures at those distances.

[350] As with other references to specific software features, these instructions may vary from version to version. In particular, the curve-fitting features within the figure window were not available in versions before R13. If you are using an older version or a different software package, you will need to directly specify and estimate the polynomial model. Other mathematical and statistical software are capable of similar procedures.

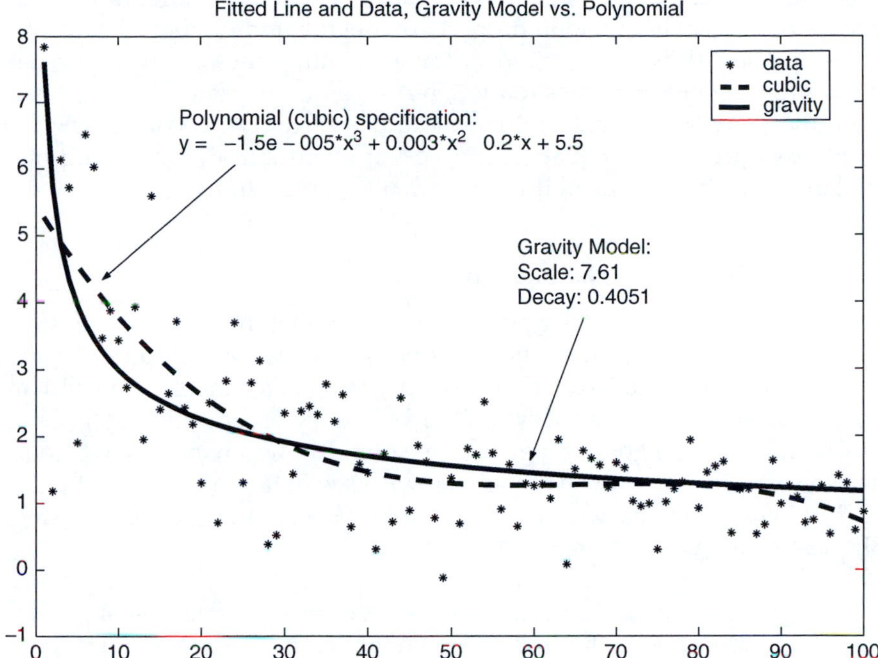

FIGURE 13-10
Gravity vs. polynomial specifications.

Comparison: Gravity and Polynomial Models

Using the same method that produced the test data and gravity model described above, we generated a new set of Monte Carlo data and then estimated the sales–distance relationship using the gravity model. Although the gravity model worked well in this case, it (of course) did not explain all the variations in the data. Could a polynomial model work better?

Figure 13-10, "Gravity vs. polynomial specifications," shows the result of this exercise. Using the MATLAB basic curve-fitting tool, we were able to quickly overlay a cubic (third degree) polynomial equation onto the data. Eyeballing the data indicates that the cubic polynomial does a better job of explaining the variations for almost all of the ranges. The little hump upward and the tail downward in the cubic curve bring it closer to the distribution of the test data.

Goodness of Fit and Understanding the Market

This raises a fundamental question about using statistical methods to estimate a behavioral relationship. Is the objective of such an exercise to get the best fit, or to get the best model?

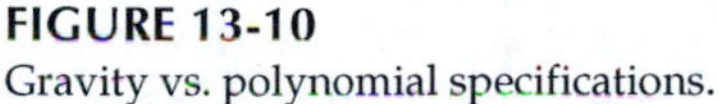

Going back to our first maxim, we clearly believe it is more important to understand the market and to properly model the underlying behavior than to get the best fit with any single data set. Getting the model right means understanding not just one situation, but an entire market.

Furthermore, keep in mind the well-known result from econometrics: the goodness of fit of any equation can be improved simply by adding more explanatory variables, even if these variables are meaningless.[351]

Weaknesses of Polynomial Equations

The main attraction for using polynomials to model the distance–sales relationship is not theoretical support; neither is it consistent empirical validation. The attractions boil down to two: (1) better fits for the data at hand and (2) the software available today easily lets you try several specifications. Of course, good fit is a good thing, so why shouldn't we abandon a rigorously derived, empirically proven model that works well for one that "fits better"?

There are problems with using polynomial equations to model sales–distance relationships. Among them are:

1. *Weak theoretical support:* There is little reason to believe, *a priori,* that consumers think about polynomial combinations of distances when they decide where to shop.[352] Remember that most consumers — and most sales clerks — use calculators to determine a simple percentage sales tax on their purchases. Will the same consumers attempt to weigh a polynomial combination of distance?
2. *Allowance of negative sales:* Consumers cannot purchase less than zero items. However, polynomial specifications can easily produce negative predicted sales for certain distances. This problem could be ameliorated by using a method that constrains the result to nonnegative territory,[353] but the use of such constraints then accentuates the other problems we identify here.
3. *Non-monotone behavior:* Consumers would not know what monotone behavior is, but they all practice it. Monotone functions have an output that constantly increases (or decreases) as the inputs increase or decrease. If consumers purchase less when they get

[351] To be precise, the goodness of fit measured by the R^2 statistic (the ratio of the explained variation to the total variation) will go up when an explanatory variable is added. For this very reason, adjusted R^2 and other measures are used.

[352] In contrast, many assumptions in economics are based on quadratic preferences, loss functions, or similar notions. In general, it seems humans operate as if they understand roughly what the square of a number is.

[353] MATLAB's extensive optimization routines allow for constrained optimization. We do not present this method in this book, but it should be considered when circumstances rule out a model that more naturally fits the data.

[353a] The elasticity is derived in "Elasticity of Demand in a Gravity Model" on page 349.

farther away, they will likely purchase even less when still farther away. A polynomial specification — even one with a "best fit"— often results in predicting inexplicable behavior. The higher the degree of polynomial, the more the curves appear in the plot of the resulting function. This not only means that more data are necessary to properly estimate the additional parameters but also that the estimated equation is likely to have unusual and hard-to-explain detours from a smooth path. If those detours fit the data better, then it may appear that the equation does a better job of explaining the data — but does it really explain the underlying market behavior?

4. *Unclear elasticity:* The gravity model presented in Equation (4) assumes a constant elasticity of sales with respect to distance.[353a] Therefore, the model assumes that the percentage change in sales, given a percentage change in distance, remains constant for all consumers and at all distances. The assumption of a constant elasticity is one which is made frequently in demand and supply behavior, and it has been thoroughly studied. Though not correct in the extreme, it has broad application in the range of circumstances faced by both buyers and sellers. On the other hand, a high-degree polynomial will often have a complicated elasticity which will vary significantly. This makes reviewing the results of an analysis difficult and makes it nearly impossible to compare demand behavior from one location to the next.[354]
5. *Intractable structural form:* The structural form of a gravity model allows for simplified output and relatively simple incorporation into other economic analyses. A polynomial form, especially a third degree or higher, could be intractable.

Conclusion: Gravity Models vs. Polynomials

We believe that the strong theoretical and empirical grounding of the gravity model supports its position as the primary tool of distance–sales analysis. However, we present polynomial specifications and acknowledge that within limited data ranges they can, in some cases, work better.

Therefore, all the same, before you use a polynomial equation, especially a cubic specification or higher, be sure you know the reason why and have considered the weaknesses described above.

---

[354] The gravity model we specify in this chapter distills the distance–sales relationship down to two parameters: scale and decay. These can be compared across multiple locations. By comparison, consider a second-degree polynomial equation that has been used in multiple locations. How many variables or parameters are in the elasticity statistics for each? How do you compare them? What if you are comparing demand estimated with a third-degree polynomial with that from a second-degree one?

Complex Models with Additional Variables

A more comprehensive model of detail attraction would make use of the information in a gravity model and consider other attributes of store attractiveness. For retail markets in which all the stores are roughly the same in size and amenities, it may be possible to ignore these other variables.[355]

However, when store attributes differ significantly (such as one offering more product choices, more services, staying open later, having more expensive goods, or other factors), the location factor may still be the most important factor affecting sales, but it will not be the only factor.

A Generalized Model

In such cases, if data are available for sales from multiple stores, and data are also available for the factors that distinguish the retail outlets, a general predictive model can be used. Such a model can take the functional form:

$$S_i = f(d, m, y) \tag{13}$$

where S_i denotes sales of store *i* in the region; *d* is a vector of distance variables, indicating the distance between store *i* and other stores; *m* is a vector of amenity variables such as size, hours, quality or price of merchandise, and atmosphere of the area surrounding the retail outlet; and *y* is a vector of data on the disposable income available to consumers in the area. We include the income variable because the data may come from retail outlets from different areas. With cross-sectional data on stores of a certain type, this equation could be specified and estimated in any number of ways.

The Huff Attractiveness Model: Similarity to Gravity Model

One specification for retail demand was created by David Huff in the 1960s. As will be shown below, it is very similar to a gravity model in its treatment of distance, and explicitly incorporates an additional variable. The additional variable is an important amenity: the size of the retail outlet.

The Huff specification defines the attractiveness *a* to consumer *i* of store *j* as:

$$a_{ij} = \frac{m_j^a}{d_{ij}^b} \tag{14}$$

---

[355] Franchised retailers, who are normally contracted to meet certain uniform standards, are a subset of retailers where this assumption is more tenable. However, even in these cases, you will see variations due to the performance of the store personnel, management practices, and other individual factors.

where m_j^a is the size of store j, the exponent a is a parameter; d_{ij}^b is the distance between consumer i and store j, and the exponent b is a parameter indicating the sensitivity of the consumer to distance.[356]

The model as defined in the equation has an unobservable variable ("attractiveness") equal to a formula with two unknown parameters. It is a model of individual choice and, therefore, could not be used with aggregate data.[356a] Thus, some transformation is necessary. To make it tractable, we first substitute the probability of purchasing goods from a certain store (which can be estimated from actual sales) for the unknowable term "attractiveness."[357]

The equation still involves a large number of parameters. In practice, the parameters a and b are often taken on the basis of past experience to be 1 and 2, respectively.[358] With these series of simplifying assumptions, the equation becomes:

$$s_{ij} = \frac{m_j}{d_{ij}^2} \tag{15}$$

In this practical Huff equation, the size variable m is the single amenity, the distance variable is given the exponent 2, and sales by consumers (or group of consumers) from one area at one store is the dependent variable.

The Similarity between Practical Huff and Gravity Equations

We have seen this equation before. It is the gravity model equation of Equation (4) and Equation (5), with the exponent on distance (the decay parameter) set to 2 and the scale parameter set to the size of the store.

Thus, the practical Huff equation is a special case of a gravity model. Indeed, using the procedure described above, in which we estimate the decay parameter directly from the data, the scale parameter is an observed indication of the amenity vector of the particular retailer.[359] In cases where size is known to be an important variable, it can be included in the specification as suggested in the general model discussion above.

---

[356] D.L. Huff, "A probabilistic analysis of shopping center trade areas", *Land Econ.*, 39, 1963; "Defining and estimating a trade area", *J. Mark.*, 28, 1964; cited in A. Ghosh, *Retail Management* (Fort Worth, TX: Dryden, 1994), Chap. 11.

[356a] With aggregate data on consumers in different areas, we may be able to estimate the necessary parameters.

[357] Economists might recognize this as using the "axiom of revealed preference." However, Huff relied upon a psychology concept known as Luce's Choice Axiom for this simplifying assumption.

For a discussion of the uses of Luce's insight and the difficulties in applying an individual choice model to aggregate behavior, see the introduction and first chapter in A. Sen and T.E. Smith, *Gravity Models of Spatial Interaction Behavior* (New York: Springer, 1995).

[358] See A. Ghosh, *Retail Management* (Fort Worth, TX: Dryden, 1994), Chap.11.

[359] Size would be only one of the likely amenities, although for different-size stores, an important one.

Determining Sales and Market Areas

Projecting Total Sales: Estimation

Assuming that we have data on sales, and a valid model for sales, we can sometimes estimate the parameters of the model using the data. The resulting model can be used to project sales in other markets, or the same market when the variables change. Before estimating sales, ensure that the subject sales location compares to the sales locations on which the data are based, in terms of market and product. It is pointless to project sales for a grocery store based on data from a car dealer, or for a high-end restaurant based on data from a fast-food establishment or even a mid-range restaurant.

Below, we describe the steps for estimating total sales using the general gravity model, assuming it has been estimated using available data. The same methodology can be used to project total sales for any of the best-fit polynomials as well.

Mathematical Derivation

We restate the standard gravity model equation where sales (S) equals a constant (β) multiplied by the inverse of distance (D) to a constant power (k):

$$S = \beta \cdot \frac{1}{D^k} \tag{16}$$

To calculate total sales, we can integrate the equation across a specified distance range:

$$Sales = \int_a^b \left(\beta \cdot \frac{1}{D^k} \right) dD \tag{17}$$

where b represents the maximum sales distance and a represents the minimum sales distance. We complete the calculations below:

$$= \beta \cdot \left(\frac{D^{-k+1}}{-k+1} \right) \Bigg|_a^b \tag{18}$$

$$= \beta \cdot \left[\frac{b^{-k+1}}{-k+1} - \frac{a^{-k+1}}{-k+1} \right] \tag{19}$$

We can then replace b with the maximum sales distance and a with the minimum sales distance from our analysis, and β and k with the estimate of the values of these parameters. We can then use this equation to project total sales across a specified region. A numerical method can also be used to project sales, especially if the distance-sales equation is different from Equation 16.

Edge Effects and the Impracticality of Zero and Infinity

The analysis of spatial data is often hampered by edge effects which are discontinuous, or unusual behavior near the edge of a region. For distance–sales analysis, we often encounter edge effects requiring some adjustment very close to the retailer and far away from the retailer.

Basic Adjustment

The derivation of the distance–sales relationship is based on a demand function that implicitly assumes a maximum sales distance of infinity (∞) and a minimum sales distance of zero. In practice, we recommend that you truncate the data to a maximum and a minimum distance.

There are three reasons for this: truncated area will almost always be in line with the actual characteristics of the market area,[360] the mathematics of estimating models is much more tractable if we slightly restrict the range, and the data are often aggregated or otherwise deficient.

We consider the rationale for the minimum- and maximum-distance truncations separately.

Minimum-Distance Truncation

If there are very few residential properties within a mile or more of the retail establishment, a minimum retail distance of zero may result in an overestimate of the total sales of a store with a primarily residential customer base, or allow a handful of observations to bias the estimated parameters of the model.

Recall our derivation of the distance–sales relationship from the fundamental microeconomics of consumer demand. Distance in the demand equation is an indicator of time and other costs of traveling to and from the retailer. Part of that time cost is the same whether the trip is across the street or across the country.[361] This *de minimis* cost is not important when the distances involved are all significant. However, ignoring it can cause problems when a few consumers are within 1 mi and others are within 5 mi, while the majority is 10 mi or more away. The first group is indeed closer than the other groups, but does it really take one fifth the time-cost for a

[360] At least until space travel becomes economical, there is an upper limit to the distance a retail customer can travel on earth. In practical applications, truncating the market area to that portion providing 90 to 95% of the sales has worked well, as this eliminates outliers without affecting the fundamental relationship.

[361] Consider the actual costs to humans of taking a shopping trip. They include making arrangements for whatever must be watched while the shopper is gone, possibly getting children or others into a vehicle, walking in from the parking lot, and other tasks. These are not cost free and are roughly the same whether the trip is 1 mi or 10.

person 1 mi away to purchase something at a retailer compared to that for a person 5 mi away?

This is further complicated by the imprecision of data. Often the analyst will have aggregate data sorted, for example, by zip-code area. In metropolitan areas, the centroid of the zip-code region may be a good indicator for zip-code areas whose boundaries are more than 5 mi away. However, the centroids may not be a good indicator when the retailer is located in the zip-code area or near its boundary. Finally, the mathematical specifications for distance–sales relationships do not work well for distances close to zero.

Maximum-Distance Truncation

The rationale for a maximum-distance truncation is quite apparent. Any good model, including a polynomial or gravity model, should estimate sales well within the range for which data are available. However, such models will not be accurate at distances where there have been relatively few sales in the past. When only a handful of sales arises from very distant customers, these could be considered outliers that should not be given equal weight with other observations.

In addition, as discussed in "Weaknesses of Polynomial Equations" on page 360, a polynomial specification can produce negative sales projections outside its intended range. With a gravity model specification and reasonably well-behaved data, the expected sales approaches zero as the distance extends outside the range of data. Hence, such models are less susceptible to distorted results because of a handful of large-distance sales observations.

Defining Market Areas

A quantitative definition of the relationship between sales and distance allows us to define trade areas based on the percentage of sales that we expect the retail center to attract from different geographic areas. Depending on the purpose of the analysis, we may decide to use a single trade area or multiple trade-area segments (e.g., primary, secondary, and tertiary market areas).

If we decide to define a single trade area for a store, we may decide that this trade area should represent the region from which 70% of customers are attracted. To determine the geographic extent of this trade area, we use the above methodology to calculate total sales and then adopt it to determine from within which distance 70% of customers are attracted. We continue to use the standard gravity model as an example, although the same approach will work for other equations.

$$(Sales \cdot 0.70) = \beta \cdot \left[\frac{b^{-k+1}}{-k+1} - \frac{a^{-k+1}}{-k+1} \right] \tag{20}$$

Equation (20) assumes that *Sales*, β, *k*, and *a* (the sales–distance minimum limit) are already determined or estimated. If we solve for *b*, we then determine the upper limit of the distance range that represents 70% of total sales. Our market area would then be defined by a distance ring of radius *b* (or drive time *b*, if distance *D* was expressed in drive time).

If we want to define multiple trade area segments for a retail location, we follow the same approach. For example, we may decide to define a primary market area that accounts for 50% of sales, as well as a secondary market area that accounts for another 30% of sales. To do this, we would define the distance-radius of the primary market area based on 50% of sales and then the distance-radius of the secondary market area based on 80% of sales (50 plus 30%).

Being able to incorporate these trade area definitions into a GIS facility is very useful for communicating the results of the analysis and for fully understanding the reach of the market area.

Estimating Sales Shifts and Cannibalization

Cannibalization refers to the loss of existing sales which occurs due to the introduction of a new product or location. For example, if BMW introduces a new automobile, some new car sales will likely be directed away from the purchase of competitive brands, and other new car sales will likely be redirected from the purchase of other BMWs. In this latter case, BMW "cannibalizes" some of its sales from its other models.

Cannibalization can also happen with the introduction of a new retail location. This is the type of sales cannibalization that we discuss in this chapter. If BMW opens a new dealership, the company will probably pull some sales away from competitive dealerships in the area. However, the new dealership will also pull sales away from other BMW dealerships. Using the sales–distance models discussed in this chapter, the amount of cannibalization can be estimated.

Cannibalization in Retail Franchises

In industries that rely on franchising or manufacturer-owned-and-operated stores, the introduction of a new retail outlet into a market already served by existing retailers selling the same brands generally has two effects:

1. Overall brand visibility and customer convenience are increased.
2. Some existing customers change their patronage from the previous retail location to the new one.

The first effect supports higher overall sales. The second cannibalizes existing sales in that the "new" sales are really sales taken from an existing retailer of the same brand.

Intelligent manufacturers are concerned about cannibalization and avoid it when possible.[362] However, given the divergent interests between the franchisor and franchisee and the trade-off between increased brand visibility and cannibalization, there are many decisions that rely heavily on an estimate of the amount of truly new sales in an area, compared with the amount of existing sales that are merely cannibalized. The techniques in this chapter provide a method to estimate these.

Required Data

Because such assessments deal with shifts in sales between retail centers, as well as changes in total sales, the analyses require both total-sales and penetration-rate data. Typically, due to the requirements of the analysis, regional sales data are used instead of individual customer data.

The quality and availability of data at this level differs significantly. In some industries and markets, data are available on the sales of each retailer by region. This occurs more often for major purchase items such as automobiles and in heavily regulated industries such as telecommunications. In such cases, both total sales and penetration rate data can be derived for each region.

Many times, however, some data must be derived using the selected model. This approach assumes that all retailers in the analysis face the same market behavior. Therefore, the analysis is more accurate if the different retail locations compare closely in terms of product selection, accessibility, and other factors that attract customers.

To derive the required data by region, use the selected model to project sales for each retailer by geographic region. The sales figures of the different retailers can be aggregated and adjusted to provide total sales by region, and then the retailer penetration rates can be estimated. We describe methods for these tasks in the following sections.

---

362 For example, Blockbuster's 2001 SEC 10-K report indicates that they select store locations in "markets [that] are most likely to offer growth opportunities with minimal cannibalization of our existing stores." Blockbuster 10-K, page 5; found at the EDGAR filings available at: http://www.sec.gov.

Changes in Total Product Sales

Brand visibility and customer convenience correlate with the proximity of a potential consumer to the point of sale. If a new retail location is closer to a consumer than existing retail locations, an increase in overall brand visibility and convenience for that consumer will probably increase total sales of the product in the area.

We suggest the following steps to determine changes in total product sales in a region when a new competitor enters the market:

1. For each region in the analysis, measure the distance between the region's centroid and the closest point of sale *before* the introduction of the new retail center.
2. For each region in the analysis, measure the distance between the region's centroid and the closest point of sale *after* the introduction of the new retail center. The distance will change only if the new retail center becomes the closest point of sale to the region.
3. Determine the percentage change in distance after the introduction of the new retail center. Equation (21) calculates the percentage change in distance (*PCD*). D_1 is the distance to the closest point of sale before the introduction of the new retail center and D_2 is the distance to the closest point of sale after the introduction of the new retail center.

$$PCD = \frac{D_2 - D_1}{D_1} \tag{21}$$

4. Determine the percentage increase in sales for each region which results from the introduction of the new retail center. Equation (22) estimates the percentage change in sales (*PCS*) based on the product of the percentage change in distance (*PCD*) and the elasticity of sales with respect to distance (ε). Equation (23) defines elasticity as in "Elasticity of Demand in a Gravity Model" on page 349.

$$PCS = PCD \cdot \varepsilon \tag{22}$$

$$\varepsilon = \frac{dS}{dD} \cdot \frac{D}{S} \tag{23}$$

Elasticity differs between models. In some models, it fluctuates with different quantities or at different distances. However, the standard gravity model has an elasticity that remains constant at all levels of sale and distance. Because of this, using the standard gravity model can simplify the analysis.

Equation (24) shows how to determine the percentage change in sales for a gravity model analysis, using the constant variable -*k* as elasticity.

$$PCS = PCD \cdot (-k) \tag{24}$$

5. Calculate total change in sales for each region that results from the introduction of a new retail center. To do this, apply the percent shift in sales to the total sales figures for each region.

$$NewSales = PCS \cdot TotalSales \tag{25}$$

$$NewTotalSales = NewSales + TotalSales \tag{26}$$

After calculating the increase in total sales by region, we can sum the regional figures to estimate the total new product sales that are created as a result of increased exposure and convenience.

Measuring Changes in Sales by Retailer

The following describes how to estimate for each retailer in the market the changes in sales which result from the introduction of a new facility. This approach weights the future likely cannibalization of existing retail locations, based on their existing level of sales in a region.

1. Using the methodology described above, collect data for each region on (1) the sales by retailer before the introduction of the new point of sale.
2. Estimate sales by region for the new retail center using the gravity model or best-fit polynomial chosen for the analysis.
3. Estimate the unadjusted sum of all sales by adding together the estimated or actual sales information for all retailers in a region, including the estimated sales figures for the new retail center.[363]
4. Determine the penetration rate for each point of sale after the introduction of the new retail center. To do this, divide the actual or

[363] Note that new total sales for a region is not equal to the sum of the new retail center sales and the total sales for a region before the new retail center entered the market. This is because some of the sales that the new retail center attracts will be cannibalized from the sales at other locations.

projected sales of each retailer by the unadjusted sum of all sales determined in the previous step. Equation (27) calculates the penetration rate for a retail center (PR_1) by dividing its estimated retailer sales (RS_1) by the unadjusted sum of all sales (*SumSales*). The retail center's total sales are determined using the selected model or by actual data.

$$PR_1 = \frac{RS_1}{SumSales} \tag{27}$$

5. Estimate sales for each retailer after the introduction of the new retail location. Although we use the sum of all sales figures to calculate the penetration rate, the estimated sales for each retailer ($NewRS_1$) is determined by multiplying the penetration rate (PR_1) by the total sales after the introduction of the new retail center (*NewTotalSales*).

$$NewRS_1 = PR_1 \cdot NewTotalSales \tag{28}$$

6. Determine the changes in retailer sales by region by subtracting retailer sales prior to the introduction of the new sales location from the retailer sales after the introduction of the new location.

Measuring Changes to Sales

The results of the above analyses can then be used to compare (1) new product sales and (2) sales shifted from other retailer locations, which result from the opening of a new retail center.

The total new sales that result from the opening of a new location and the method of calculation were discussed in "Changes in Total Product Sales" on page 369. The total increase in product sales throughout the market represents the genuine new sale creation that results. However, the difference between new total sales and the original total sales in the market will most likely be less than the total sales projected for a new retail location in "Measuring Changes in Sales by Retailer" on page 370. This difference is accounted for by cannibalization.

To measure overall cannibalization of other retailer sales, subtract genuine new sale creation from the estimated total sales of the new store. For example, if a new retail center is estimated to sell 100 items and the total sales in the market area are only expected to increase by 60 items, then the new retail center causes the cannibalization of 40 items away from other stores.

To measure the effect of a new retail location on the sales of other retail locations, measure the difference between the total sales of each retail location before and after the introduction of the new point of sale. These figures were calculated in "Measuring Changes in Sales by Retailer" on page 370. For each retailer, the total change in sales across all regions shows the loss of sales to the new store. This loss represents the cannibalization of that retailer's sales.

14

Applications for Manufacturing

Statistical Quality Control

Over the past three decades, the use of statistical quality control or "statistical process control" in manufacturing has increased dramatically. With the growth of international trade, objective measurements of quality — and the processes that ensure that quality is a central objective of production — have become particularly important.

This book deals with the economics of business, and an essential part of that economics is the ability to consistently produce high-quality products. Statistical methods of testing for quality are growing in importance, and a basic knowledge of this field will be helpful to many readers of this book.

In this chapter, we present first the classic statistical process control methods, and then introduce the newer Taguchi methods.

Statistics and Statistical Process Control

The core statistical applications used in statistical process control (SPC) are the same as those used in many other applications of statistics, including those in business economics and finance. MATLAB has a set of routines that accomplishes the basic calculations used in most statistical process control applications. This chapter briefly covers such SPC applications.

Purposes, Innovations, and Limitations of This Chapter

The purpose of this chapter is to present SPC to an audience that is intensely interested in the future earnings of a firm, but may be lacking in knowledge about an important tool used to ensure those earnings. As the tool relies on the same fundamental economic skills used in other chapters of this book, such knowledge could be rapidly acquired.

While this chapter presents basic SPC routines, it is not a thorough reference to SPC. A practitioner will want to review other references.

While the routines shown here are relatively straightforward, we suggest two innovations to those working in manufacturing:

- First, consider bringing the capability of customized SPC routines to facilities in different locations using a Web-server approach. Using a centralized facility to perform the complex tasks and communicating the relatively small amount of information among the facilities using the Internet are basic principles of "thin-client" computing. As the data involved in most SPC routines are small in size (in terms of bandwidth), it could be communicated with little difficulty.
- Second, consider customizing the routines for specific production processes, incorporating past data, recording new data, checking data before analysis, and other improvements. Such customization brings benefits that can outweigh the costs of using a programming environment such as MATLAB instead of a packaged SPC software product.

History and Uses of SPC

Much of the development of SPC originated in the work of Walter A. Shewhart at Bell Labs in the 1920s. Shewhart developed simple routines that — without the extensive calculations made possible by calculators and computers that would not be available for another half-century — provided manufacturing engineers an excellent tool to monitor the consistency of production. He called this approach *economic control* of the quality of manufactured products.[364]

It is a fitting testament to the man that the same techniques are in use today.

Principles of SPC

What is a Process?

A process is any series of machines, people, facilities, logistics, or other activities that take inputs and produce a certain product.[365] We say a process is "in control" when it consistently produces the product within the predicted range of variation.

This definition implies that SPC requires disciplines to define the product, measure it, *and* define the process that produces the product.

---

364 W.A. Shewhart, *Economic Control of Quality of Manufactured Product* (New York: D. van Nostrand, 1931); cited in C.P. Quesenberry, *SPC Methods for Quality Improvement* (New York: John Wiley & Sons, 1997).

365 This is sometimes called the "5 m&e" definition, for men, methods, machines, materials, measurements, and environment. See Quesenberry, *SPC Methods for Quality Improvement*, Chapter 4.

What Is Quality?

The term "quality" is somewhat amorphous in business. However, the SPC discipline requires firm operational definitions of product quality. An operational definition of quality generally requires one or more specific measurements or other operational checks.[366] If "quality" cannot be defined and measured quantitatively, then the SPC routines cannot be used.

Principles of SPC

There are two principal uses of the classic SPC routines: to recognize when a process begins to go out of control, and to aid the adjustment of a process.

These two purposes fit SPC neatly into the Plan–Do–Check–Act management paradigm, which was established by the American quality expert Edward Deming and popularized first in Japan after World War II.[367] SPC is not, in and of itself, a management paradigm. However, organizations that value product quality highly can adopt an SPC protocol that fits neatly into their management paradigm.

The Classic Shewhart Charts

The principles of SPC are more important than the actual statistics. For our discussion assume that a business is following those principles; meaning that there is an operational definition of quality based on actual measurements; that we have identified the process involved; and that we have taken measurements of the products. Using these measurement data, we can produce the classic Shewhart charts.

Control Charts — Xbar: Using measurements from samples taken at regular intervals, calculate the mean measurement from each sample.[367a] Plot these over time. Given past data on these measurements, we know what the mean and standard deviation should be. Control lines on the plot show the UCL (upper control line) and LCL (lower control line), indicating the expected variation in this measurement. In some instances, a specification line is also plotted, indicating the range beyond which the measurement is beyond specifications.[368] An example, generated by the simple program below, is Figure 14-1, "Xbar chart."

---

[366] Quesenberry, *SPC Methods for Quality Improvement*, Chapter 1; citing W. Edward Deming, *Out of the Crisis* (Cambridge MA: MIT Press, 1986).

[367] Quesenberry, *SPC Methods for Quality Improvement*, Chapter 1.

[367a] Because the sample mean of a distribution is often symbolized as $\overline{X}$, these are known as "X bar charts."

[368] "Out of Spec" does not mean "out of control." *Control* is based on *predictable* variation. A process that is "in control" will occasionally produce, due to statistical variation, outliers. Furthermore, specifications may be set by those unfamiliar with the variation in manufacturing processes, or may be substantially looser or tighter than the predicted variation.

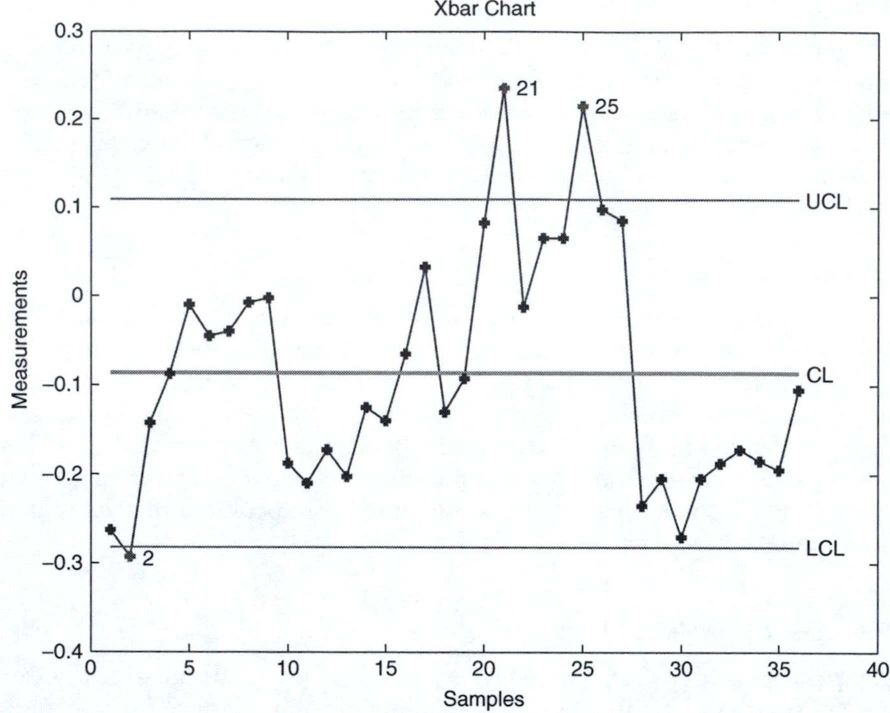

FIGURE 14-1
Xbar chart.

Control Charts — S Charts: An S chart plots the standard deviation of the samples taken at regular intevals. Such a chart visually indicates when a process starts to exhibit increasing variation and, thereby, threatens to go out of control. Using this chart in conjunction with the Xbar chart provides a well-balanced view of the quality of the manufacturing output. See Figure 14-2, "S chart."

Capability Chart: A capability analysis uses information on variation in past measurements together with proposed specifications and statistical distributions. With this information, assuming the process remains in control, we can predict the share of parts that will be outside the specification. See Figure 14-3, "Capability chart."

These figures, produced on a workstation, provide a supervisor on the assembly line strong information about the quality of the parts being produced.

A Simple SPC Program

We provide, in the Appendix to this chapter, a simple MATLAB program that creates data simulating measurements of goods produced in a

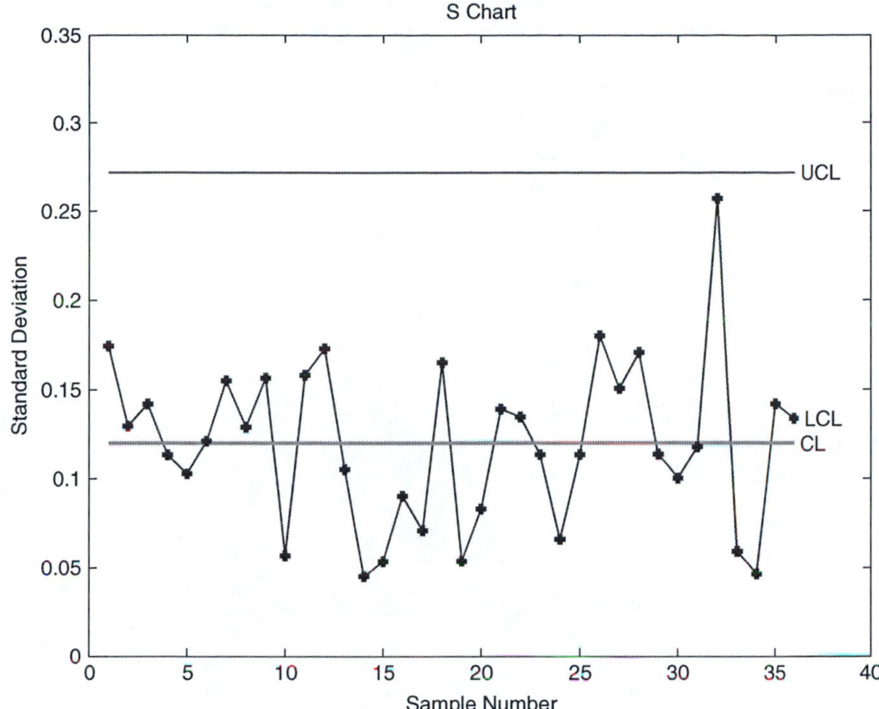

FIGURE 14-2
S chart.

manufacturing facility. It then examines those data using standard SPC tests. The program will also produce the classic Schewhart graphs shown here.

Cautions on the Use of SPC

The warnings against "garbage-in, garbage-out" are especially important when using statistical programs in which the data may be gathered, input, and analyzed by different individuals. Even when used by trained individuals, SPC routines can provide misleading information when the process analyzed is different from one that matches the underlying assumptions.

For example, the capability plots rely on an assumption that the underlying sample will be random and distribution of measurements normal. If the measurement protocol does not ensure random samples, the first assumption is violated. If the process itself does not produce a normal distribution of results, then the second assumption is violated. If such data is input, the routines will produce neat, precise statistics and pretty figures — showing the wrong information!

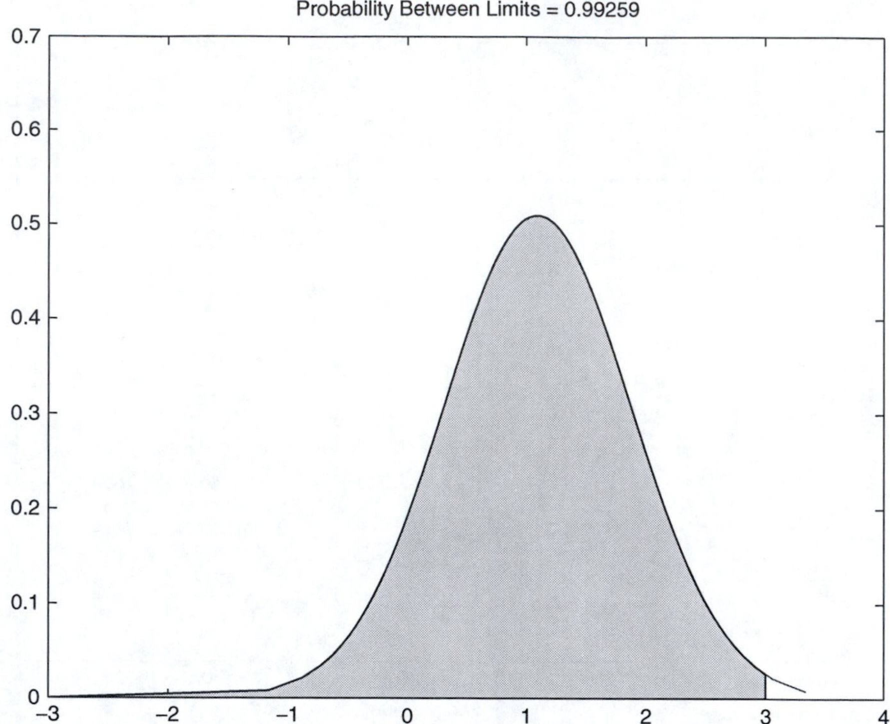

FIGURE 14-3
Capability chart.

A "Thin-Client" Approach

Consider now the first innovation we suggest: bringing the power of customized SPC routines to various locations. You could do this by sending an engineer to each location. You could also do this by having someone record data and transmit it back via telephone, mail, or e-mail. Either approach invites delay and expense. It also creates the likelihood that the persons recording the data, analyzing the data, and receiving the analysis are all different and fail to communicate with each other about relevant factors.

However, using a thin-client approach, the supervisor running the production process can receive immediate analysis of the SPC data. This achieves the central goal of SPC — enabling those running a production process to catch small changes in manufacturing output before they become quality problems. It also allows for informal or formal discussions among those who take the measurements and those who analyze the data. Indeed, it is possible that these would be the same people.

SPC Using MATLAB Web Server

Creating a thin-client SPC program could be accomplished in the following steps:

1. Create and test SPC routines appropriate for your manufacturing facility.
2. Create and test appropriate test procedures.
3. Port the routines to a Web server running MATLAB and test the application.
4. Create a data exchange protocol for process engineers to input the data collected on the assembly line into the SPC routine. This could be simple HTML pages or a more complex XML implementation.
5. Review the graphical output and information needed, and incorporate those features into the Web pages that report the results of the analysis. Graphical output could be produced in the standard JPEG or GIF format, or in the new SVG format.

We discuss the specifics of creating a MATLAB Web server application of this type in Chapter 16, under "Thin-Client Computing" on page 406.

Customizing the Application

If a straightforward SPC analysis was all that was desired, specific SPC packages could probably accomplish the same goal with less expense. However, in other cases, the advantages of using MATLAB would outweigh the additional cost.

You could customize this application to fit specific production processes and goals, using any of the following methods:

1. Selecting specific SPC routines, based on the control process selected for the specific production process
2. Customizing the graphical output, for example, by adding additional text boxes, subplotting multiple plots, or providing more informative titles and annotations
3. Specifying the tolerances allowed, sample sizes, and other standards
4. Catching poor data or likely measurement errors

5. Recording the data or tests either by an automatic diary entry or through a user-selected routine
6. Providing historical data in graphical form
7. Creating a custom graphical user interface

Extending Beyond SPC: Taguchi Methods

The classic approach described above is the foundation of much of statistical process control. It is simple, robust, and requires relatively little advanced statistical analysis. The fact that these methods were largely developed a half-century ago, before computers could assist with the calculations, reveals the importance of the contributions of Shewhart and others.

A major advance beyond the classic methods was developed by the Japanese engineer Genichi Taguchi in the 1980's.[368a] The methods that resulted are known as "Taguchi methods," and are based on both statistics and critical insights into the role quality plays in the profitability of an enterprise. Because Taguchi methods look at the entire enterprise, rather than at the statistics of a production process, it can be considered from a management perspective that is accessible to a business economist. From this perspective, we summarize the key insights as follows:

- Customers become increasingly dissatisfied with products as they vary more frequently from the standard they expect.
- Variations arise not just from the production processes themselves, but also from other areas of the environment in which a product is produced, including human and managerial factors. Taguchi called these "noise factors."
- The variation in a process creates costs for the organization. These costs include not only the costs of returned or defective products, but also the additional managerial, production, and technical resources that must be devoted to the problem. These costs increase with variation.
- The cost of these variations can be expressed as a *loss function*. Such a loss function is a much richer formulation of the actual costs of imperfect processes than the classic approach, which relies largely on a discrete step function in which a process is either in control or out of control, and a product is either within specifications or outside specifications.

[368a] Genichi Taguchi, Online Quality Control During Production (Tokyo, Japan: Japanese Standards Association, 1981). Other Taguchi methods resources area available from the American Supplier Institute, whose web site is: http://www.amsup.com.

- In many situations, the cost of the variation (the loss function) can be modeled as a constant multiplied by the square of the deviations from the desired objective.[368b]
- This "quadratic loss" function allows an organization to estimate the value of reducing variation in a proces and to make appropriate management decisions.

The economic reasoning behind Taguchi methods is, in our opinion, more important than the statistical innovations.[368c] We view Taguchi's crowning insight as the treatment of "quality management" decisions not as solely engineering decisions, but as business management decisions. Such decisions are tractable from a classic microeconomic perspective and have implications for the profitability of the firm, the value of brands and intangible assets associated with a product, and the value of an enterprise.

Appendix

Code Fragment 14-1. An SPC Demonstration Routine

```
%    SQCdemo
%    statistical quality control example routine
%    from MATLAB function descriptions and instructions
%    Edited PLA Nov 14 2001; calls MATLAB Statistical
%    Toolbox

% Part I: Shewhart Control Charts

% These are the Shewhart charts developed at Bell
% labs in the 1920's.

%--load sample "parts" data, with samples of four
% measurements taken 36 times

load parts;
```

[368b]See, e.g., Charles Quesenberry, *SPC Methods for Quality Control* (New York: John Wiley & Sons, 1997), Section 9.9.

[368c]That is not to minimize the importance of the statistical innovations Taguchi introduced, such as the use of "orthogonal arrays" to reduce the number of examinations that must be conducted to evaluate a process. These are outside the scope of this book, but the use of Taguchi methods by business managers and business analysts is not.

```
whos;

runout;

% then create Shewhart control chart (x-bar, or
% sample means);

% then Shewhart standard deviation chart (similar to
% R or range charts);

figure(1),

xbarplot(runout);

figure(2),

schart(runout);

%   Part II: Capability Charts

% A "capability plot" shows probability of a sample
% measurement occurring within a specified range,

% assuming a normal distribution.

%   Create data: normal with mean 1 and sd 1, 30X1
%   vector

data = normrnd(1,1,30,1);

%   Calculate and create figure

figure(3),

p = capaplot(data,[-3 3])

% output of p = 0.9784 means that

% The probability of a new observation being within
% specs is 97.84%

%   MATLAB Resources

%   ------------------------------------------------

%   type "help stats" and look at SPC for info.

%   see also "SPC" in stat toolbox user manual.
```

15

Fuzzy Logic Business Applications

Human Logic and Fuzzy Thinking

Human beings invented logic and use it every day. However, most important decisions rely on something more than the type of rigorous logic taught in math and rhetoric classes. We are required to make decisions in a world full of partial information and imperfect data. This requires a logic that takes into account information expressed by imperfect human beings concerning events about which they have less than complete knowledge.

The basics of textbook logic courses is "Boolean" logic, based on data that can take two forms: "yes" and "no." Such logic is essential in many situations. "If the traffic light is red, stop" is an excellent rule to follow when driving. Boolean logic does quite well in such situations. However, what happens when a traffic light is green or yellow? A green light quite far in the distance may provide a trained driver with enough information to know that he should slow down, as the light is likely to change before he gets there. Another driver, facing a yellow light, may wish to consider the speed, the distance to the light, the traffic conditions, and the weather, before deciding whether to stop or go.

An Extension of Boolean Logic

Decisions of this form cannot be made easily with Boolean logic. An extension of Boolean logic, known commonly as "fuzzy logic," has been developed to deal with these situations. Such logic includes the yes/no situations for which Boolean logic is quite adequate, and extends this thinking into situations where the data are not confined to neat, yes/no information.

I believe the term "fuzzy" is unfortunate, as it implies that there is something unrigorous or even illogical about such logic. In fact, there is nothing fuzzy about fuzzy logic. It requires very careful, rigorous, and logical thinking to implement correctly. When done properly, it gives *exactly the same answer as Boolean logic* in those situations where the required yes/no answers are available, and also gives sound answers to the many situations where the data cannot fit into yes/no categories.

Probability Statements

Many business situations boil down to probability statements rather than yes/no answers. The stock market, for example, is a market filled with participants making risk-adjusted discount decisions with imperfect information. Most decision rules of investors are best modeled as probability statements.[369] Insurance underwriting is obviously an evaluation of probabilities. So is investing in real estate, choosing which tickets to purchase in a theatre or baseball season, choosing a restaurant, voting in an election, choosing to pursue a new business venture, closing down a money-losing subsidiary, and deciding to keep fishing when the sun goes down and the mosquitoes come out.

Fuzzy logic uses rules based on probabilistic thinking for those occasions when yes/no data is not available, not complete, or represents only a fraction of the available information about an important problem.

Example: Selection of Investments

For example, when selecting investments, one person may wish to concentrate on companies with a good record of consistent earnings. What, then, is the definition of "good?" How could one use fuzzy logic to evaluate this question, and is such a method superior to Boolean logic?

Steps in Boolean Logic Approach

A rigorous examination of this question would involve the following steps:

1. Examine the earnings of a universe of stocks.
2. Calculate mean, standard deviation, and other descriptive statistics.
3. Using statistical inference, describe the underlying distribution of earnings based on the information you have. You might describe

[369] For example, an investor will first evaluate his or her own risk profile (a probability exercise), then choose markets that reflect that risk profile (a second probability exercise), and then pick individual securities or mutual funds within those markets (a third probability exercise). Even investors who are convinced that they cannot beat the market are making probability-based decisions when they simply hand over money to a portfolio advisor or invest in a large mutual fund. We explicity incorporate uncertainty in our discussion of investment in Chapters 10, 11 and 12.

more than one distribution, such as earnings over the past 1- and 5-year horizons.

4. Make an assumption about whether past performance will predict future performance, with or without bias.[370]
5. Select, using your own preferences, the area of the distribution of returns that you consider "good."
6. Define "good" returns with a metric that uses available data.

Such a practice is standard in financial portfolio theory. Even if one had never heard of fuzzy logic, the basic elements of such logic are present in this exercise: defining conversational terms by using probabilistic and statistical measures, mixing information from multiple criteria, and returning to a straightforward quantifiable decision based on this information.

Indeed, without using such logic, it would be difficult or impossible to do this exercise.[371]

Sorting Stocks: Boolean Approach

Once you have completed the exercise above, you must still sort stocks into categories. A Boolean logic approach to this would use a strict cutoff, which has the effect of defining all stocks as "good" or "not good." If "good" is defined as "returns better than 6.2% per year for the past five years," it means a stock with a cumulative return of 30% is not "good," and furthermore is just as "not good" as a stock that has declined in value. Similarly, a stock with a cumulative return of 100% during the period is "good," but so is a stock that has 6.2% returns every year.

Sorting Stocks: Fuzzy Logic Approach

A fuzzy logic approach would define a probability distribution of stocks, based on the same 6.2% judgment, and also incorporating other information. Suppose your overall stock-picking approach allowed certain stocks to be selected even if they had one or two bad years but had other outstanding characteristics. Suppose also that you valued stocks that had very high returns (say, better than 10% per year) more than those that just had good

---

[370] This critical step is often overlooked or buried in a legal disclaimer. By predicting with bias, we mean that past performance will systematically understate or overstate future performance. For broad, well-established parts of investment markets, most analysts commonly assume that past performance is an unbiased predictor of future performance. It is not clear, particularly when the selection of the past period is a judgment call, that this is the case.

[371] In particular, the definition of "good" is inherently judgmental, even when using rigorous statistical techniques. Even a seemingly objective definition such as "more than one standard deviation above the mean" carries with it many subjective, human judgment factors.

returns of 7%. A fuzzy logic approach would enable you to capture this information and use it.

Equivalence of Boolean and Fuzzy Logic

You may observe that, for the subset of stocks that fit into the obvious categories of "good" and "bad" stocks, both the Boolean approach and fuzzy logic approach provide the same answer. This illustrates a central attraction of fuzzy logic: it is equivalent to Boolean logic, when the data can be converted into the yes/no categories essential to Boolean logic.

Fuzzy logic should be considered an extension of Boolean logic, not a separate system. The same premises and measurements should produce the same answer using both systems, when the proper data are available. When only mixed data are available, the fuzzy approach should produce an answer that fits the logic used in describing the system in Boolean terms.

History of Fuzzy Logic

Human logic is as old as humanity. The classic texts of ancient Greece, as epitomized by the Socratic dialogues, were exercises in logic. In recent centuries, a rigorous mathematical logic was established and codified as Boolean logic.

However, humans exercise, as previously observed, a kind of logic that handles information which does not fit neatly into the yes/no structure of Boolean logic. The modern study of this began with the pioneering work of Lofti A. Zadeh in what he called "fuzzy sets" in 1965, which was extended to a system of fuzzy logic in the 1970s.[372] The first fuzzy logic controller was documented by Mamdani and Assilian in 1975.[373] Since then, the use of fuzzy logic has expanded greatly.

---

[372] Lofti Zadeh, Fuzzy sets, *Inform. Contr.* 8, 338–353, 1965; Zadeh dates the use of fuzzy logic to a 1973 paper entitled "Outline of a new approach to the analysis of complex systems and decision processes," *IEEE Transactions on Systems, Man, and Cybernetics*, 3 (1), 28–44, January 1973. Biographical information on Zadeh can be found at the Berkeley Initiative on Soft Computing Web site at: http://www.cs.berkeley.edu/~zadeh/pripub.html.

[373] E.H. Mamdani and S. Assilian, "An experiment in linguistic synthesis with a fuzzy logic controller", *International Journal of Man-Machine Studies*, 7 (1), 1–13, 1975.

Current Uses and Names of Fuzzy Logic

While fuzzy logic is used widely, the term is not. Perhaps burdened by the unfortunate "fuzzy" adjective, application developers and users of this technology often characterize it as "artificial intelligence," "expert systems," or "neuro-fuzzy systems." These are all efforts to incorporate subjective, partial, or qualitative information into rigorous decisions rules, and are related or based on the basic insights of fuzzy sets.

References on Fuzzy Logic

For readers of this chapter, a well-written introduction to the field is the *Fuzzy Logic Toolbox Users Guide* published by The MathWorks.[374] There are a number of other Web sites that provide overviews.[375] Unfortunately, very few references on its use in business economics or finance exist, although there are some publications that deal with similar applications (often using different terms such as *expert systems*) or discuss applications that are crude versions of fuzzy logic systems, such as rudimentary credit scoring.[376]

An Income Tax Audit Predictor

Example: Income Tax Audit Predictor

As a demonstration of the power of this approach consider the question of whose income tax returns will be audited in the IRS. The U.S. Internal Revenue Service, like most tax authorities, cannot audit all tax returns. It must, therefore, select a portion of those returns to audit. Generally, taxing authorities like to maximize the effect of those audits, by concentrating on those returns where an audit will be most likely to increase revenue or improve compliance.

The IRS system for selecting returns is a closely guarded secret, and we do not claim to have any confidential information on their system.[377] However, we constructed our own predictive model of the IRS system — using publicly available information, including the IRS's own pronouncements about the focus

---

[374] Portions of this are available on the Mathworks Web site at: http//www.mathworks.com.

[375] For example, the *Fuzzy Logic for Just Plain Folks* online book at: http://www.fuzzy-logic.com; and the resources at http://www.austinlinks.com/Fuzzy/ and at the Berkeley Initiative on Soft Computing, at: http://www.cs.berkeley.edu/~zadeh/.

[376] A credit scoring method involves giving points for specific attributes and then adding up the points to determine whether a borrower is qualified for a certain loan. Simple scoring involves the quantification of partial and subjective information, but it misses some of the logic and probability distribution features of true fuzzy logic.

We describe a credit risk fuzzy system in "A Credit Risk Fuzzy Logic Simulation Model" on page 397.

of its enforcement and compliance efforts, as well as statistics on the characteristics of returns that are audited.

Purpose of the Predictor

The primary purpose of this exercise is to demonstrate how the techniques of fuzzy logic can be used to model complex business problems. An example of such a task is the decision about which tax returns to audit. We believe citizens have both a moral and legal obligation to pay their income taxes — whether they believe they will be audited or not.[378] However, we observe that humans have resisted paying taxes for all recorded history.[379] This creates the impetus for auditing tax returns on a selective basis.

History and Structure of the Predictor

The predictor was one of the first fuzzy logic routines ever presented on the Internet. From just before April 15, 2000, through much of 2001, it was run on a MATLAB Web server accessible to millions of taxpayers.[380] Taxpayers across the country were able to anonymously test it with real or imagined data.[381]

---

[377] One IRS statement on the subject [FS-97-5] reads:

"The IRS examines (audits) tax returns to verify that the tax reported is correct. It is not an attempt to increase the tax. Selecting a return for examination does not suggest that the taxpayer has either made an error or been dishonest. In fact, many people may be surprised to know that many examinations result in a refund to the taxpayer or acceptance of the return without change, in addition to those that result in an assessment of additional tax ...

"HOW RETURNS ARE SELECTED FOR EXAMINATION

"The Headquarters office in Washington has program management responsibility, but cannot initiate or oversee the day-to-day conduct of examinations ....

"Computer Scoring — Most returns are selected for examination on the basis of computer scoring. A computer program called the Discriminant Function System (DIF) numerically scores every individual and some corporate tax returns after processing. It rates the potential for change, based on past IRS experience with similar returns.

"IRS personnel screen the highest-scoring returns, selecting some for audit and identifying the items on these returns that are most likely to need review."

[378] In support of this assertion, we cite: among the moral authorities, the statement recorded in the New Testament to "render unto Caesar what is Caesar's;" among the legal authorities, the Constitution and laws of the U.S. and all its states.

[379] See, e.g., C. Adams, *For Good and Evil: The Impact of Taxes on the Course of Civilization*, (Lanham, MD: Madison, 1993). The Adams history book and the incentives faced by taxing authorities are discussed in this book in Chapter 11, "Business Valuation and Damages Estimation," under "Approaches to Valuation" on page 251.

For other discussions of taxation, see also Chapter 10, "Applications for Business," under "Iterative Market Value and Tax Model" on page 226; and Chapter 8, "Tax Revenue and Tax Policy."

[380] The Web site was http://www.bizclimate.com, which was sponsored by Anderson Economic Group. At the time of this writing the Web-based audit predictor is no longer being provided on this site.

[381] We hasten to note that no taxpayer names or other data were recorded, and no followup was done to determine whether the taxpayers were, in fact, audited.

TABLE 15-1

Questions for Audit Predictor

Variable	Question	Response Range
Income	What is your annual wage and salary income, in dollars?	0-150,000
Flags	How many of the following items are claimed on your income tax return? (A list of deductions or credits known to be under special IRS scrutiny are then provided)	0-10
Deductions	Did you claim the standard deduction, itemize typical deductions like mortgage interest, or itemize miscellaneous deductions like employee business expenses?	0, 0.5, 1
Accuracy	On a scale of 1 to 10, with 10 being meticulous accuracy, how careful are you in preparing and documenting your return?	1-10

The Web site briefly introduced the topic and allowed visitors to run their own prediction test by answering four relatively simple questions. The questions posed are summarized in Table 15-1, "Questions for Audit Predictor." These questions provided information on income, the number of "audit flag" items on the return, whether deductions were itemized, and a self-evaluation of accuracy.

The answer to each question was converted to a numeric value assigned to a variable and submitted to the fuzzy logic program running on a MATLAB Web server. The fuzzy logic routine then provided the answer to the question "How likely is this return to be audited?"[382]

The answer, in return, was printed on a Web page for viewers, along with an appropriate disclaimer indicating that this was not a substitute for properly submitting their tax returns.

The Rules of the Audit Predictor

Fuzzy logic requires very rigorous rules, although they are expressed in the form of probability statements that define concepts such as "good" and "poor." These rules are implemented by *membership functions,* which are probability statements. The combination of the rules and the membership functions form an *inference system.* A fuzzy logic inference system is sometimes abbreviated as FIS.

Such an inference system is similar to Boolean logic, except that Boolean logic requires a yes/no determination. Fuzzy logic allows for probability statements that produce the identical result for strong yes/no situations but translate gray areas into probabilities.

382 The "how likely" question was answered by comparing the audit likelihood of the hypothesized return to that of the median return, meaning that a "high" likelihood should be understood as "higher than the median." The actual likelihood of audit, in a mathematical sense, varies from year to year.

We describe such an inference system, using the audit predictor as an example, in the following sections.

Audit Predictor Rules

This section states the rules which are the basis for the income tax audit predictor fuzzy inference system. The first rule is:

1. If (income is low) and (flags is no_flags) and (deductions is standard) and (accuracy is not poor) then (audit_risk is low).

This incorporates the first observation about tax filers and the likelihood of audits: if your income is low, and there is nothing unusual in your return, and you make no mistakes on your return, then the likelihood of an audit is low.

This rule is true, but it is inadequate by itself. We therefore summarize our other information about the factors that predict audit likelihood in addition rules. These are stated in Code Fragment 15-1 on page 390.

What is "low"?

Many of the rules use terms to describe the likelihood of an audit in terms like "low" or "high."

It is the translation of concepts like "low" into mathematical statements that is the basis for the "fuzzification" in fuzzy logic. Rather than use hard-edged rules (e.g., income below $10,000 is low), we make a *probability statement,* using a defined distribution. These form membership functions. Such a membership function for low income would indicate that, of course, a person with income of $5,000 would be considered low income, but a person with income of $11,000 would likely be low income as well.

The graphical illustrations included in the next section illustrate the implementation of this research.

Code Fragment 15-1. Audit Predictor Rules

```
%    --------------

%    Now print rules in workspace

rules_text = showrule(fismat)


rules_text =
```

```
1. If (income is low) and (flags is no_flags) and
(deductions is standard) and (accuracy is not poor)
then (audit_risk is low) (1)

2. If (income is low) and (flags is no_flags) and
(deductions is standard) and (accuracy is
compulsive) then (audit_risk is low) (1)

3. If (income is low) and (flags is no_flags) and
(accuracy is poor) then (audit_risk is medium) (1)

4. If (income is low) and (flags is not no_flags)
then (audit_risk is high) (1)

5. If (income is middle) and (flags is no_flags) and
(deductions is standard) and (accuracy is not poor)
then (audit_risk is low) (1)

6. If (income is middle) and (flags is no_flags) and
(deductions is itemized) and (accuracy is not poor)
then (audit_risk is medium) (1)

7. If (income is middle) and (flags is one_or_two)
and (deductions is itemized) then (audit_risk is
medium) (1)

8. If (income is middle) and (flags is
three_or_more) then (audit_risk is high) (1)

9. If (income is middle) and (flags is one_or_two)
and (deductions is miscellaneous) and (accuracy is
poor) then (audit_risk is high) (1)

10. If (income is high) and (flags is no_flags) and
(deductions is itemized) and (accuracy is not poor)
then (audit_risk is medium) (1)

11. If (income is high) and (deductions is
miscellaneous) and (accuracy is not compulsive) then
(audit_risk is high) (1)

12. If (income is high) and (flags is three_or_more)
and (deductions is miscellaneous) then (audit_risk
is high) (1)

%   ---------------
```

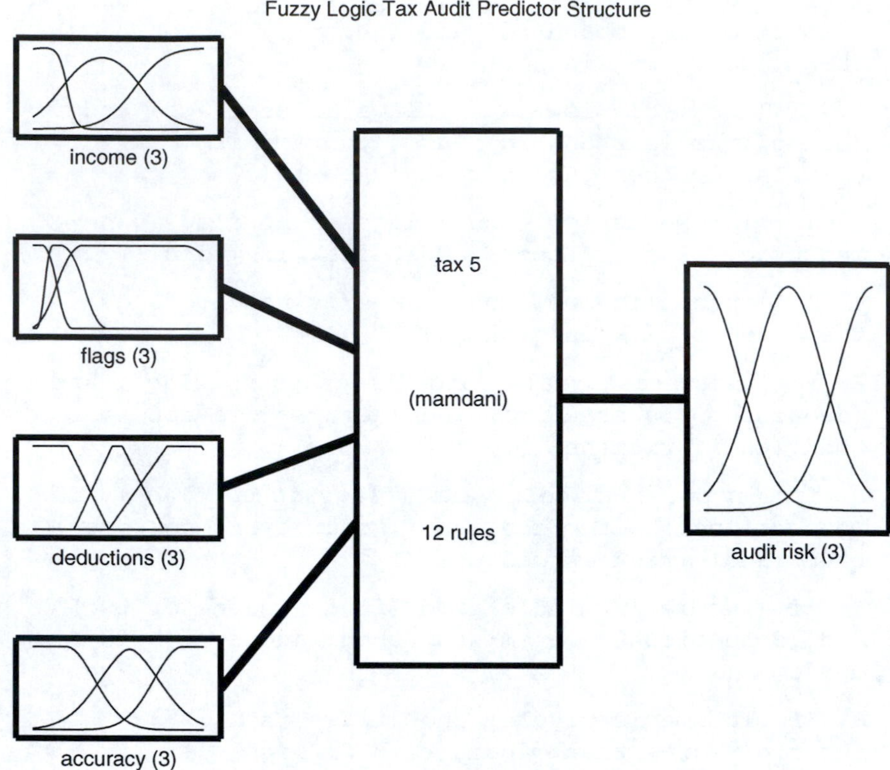

FIGURE 15-1
Graphical display of audit predictor.

Graphical Illustration

The following figures illustrate this inference system:

1. Figure 15-1, "Graphical display of audit predictor," illustrates the entire structure of the inference system. Note that the number of rules is shown, as are the membership functions on the left. From the inputs and the rules, the inference system provides an output measurement which is shown on the right.

2. Figure 15-2, "Income membership functions," shows the membership function for income. As the rules often refer to income ("if income is high, ... then ... "), this membership function translates a dollar figure of income into a conceptual description such as high or low. The membership function for "middle income" is centered in the range of $40,000 to $73,000 in annual income for the year 2000.

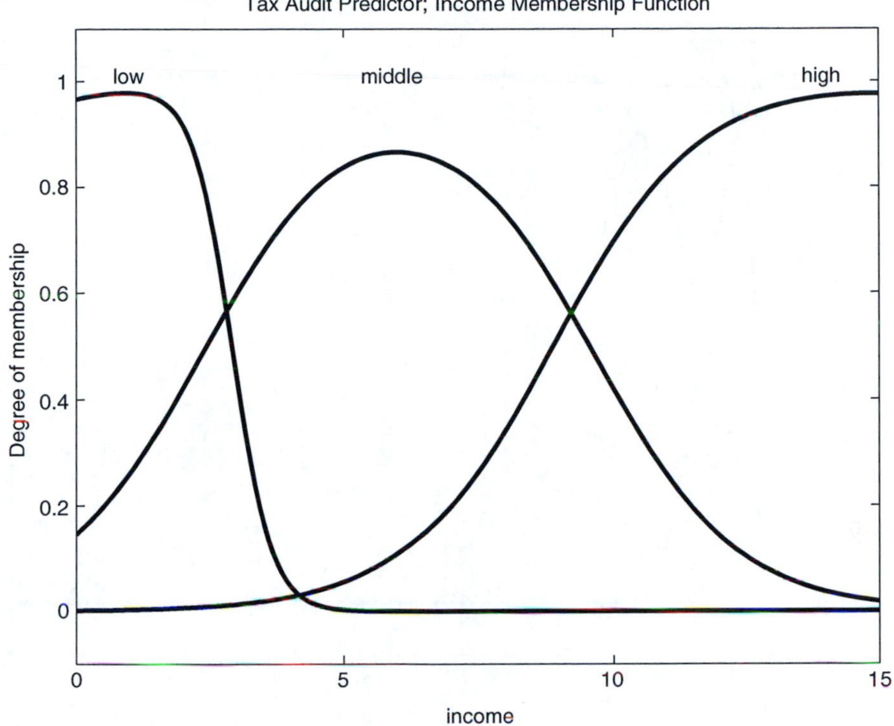

FIGURE 15-2
Income membership functions.

3. Figure 15-3, "Audit flag membership function," shows the membership function for audit flags.
4. Figure 15-4, "Surface of rules," provides a 3-D view of the likelihood of an audit, based on income and the number of flags. Note that the actual function is 4-D, but only 3 dimensions can be shown on this surface graph.
5. Figure 15-5, "Ruleviewer for audit risk FIS," provides a comprehensive graphical illustration of the effect of the various inputs, membership functions, and rules on the resulting value from the FIS. In this graphic, the inputs used are an annual income of $75,000, 5 audit flags, itemized deductions, and mediocre accuracy. These are shown in the first four columns of distribution graphics in the figure. As there are 12 rules, each input is evaluated in the 12 rows.[383]

[383] Some rules may not involve some inputs. In this case, the distribution box is blank. You can confirm that rule no. 8 does not involve two of the inputs, and therefore two boxes on row 8 are blank.

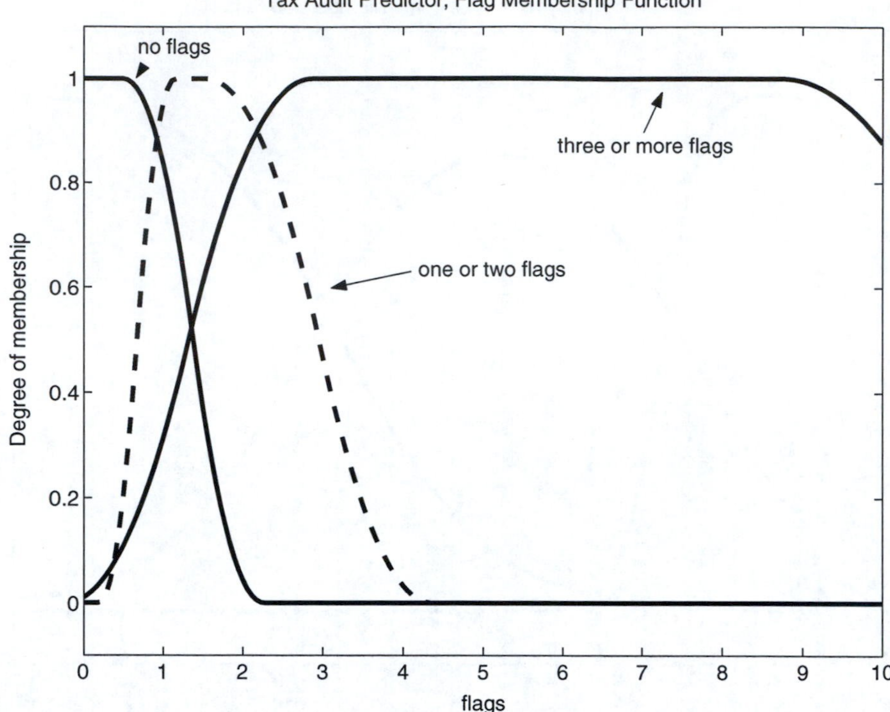

FIGURE 15-3
Audit flag membership function.

The audit predictor FIS evaluates such characteristics as resulting in a high probability of an audit. This is consistent with common sense; a person who makes mistakes on his return, and whose return contains a number of audit flags, should be audited more frequently than the typical taxpayer.

What cannot be shown in this figure is that the ruleviewer in the MATLAB toolbox is interactive, allowing you to see how the FIS evaluates multiple scenarios quite quickly.

Running a Fuzzy Logic System

This chapter provides a very quick overview of fuzzy logic and an example of how such a system has been implemented in the past. The program "Fuzzy Logic Audit Risk," shown in Code Fragment 15-3, on page 403 in the Appendix to this chapter is a guide to running an entire fuzzy system.

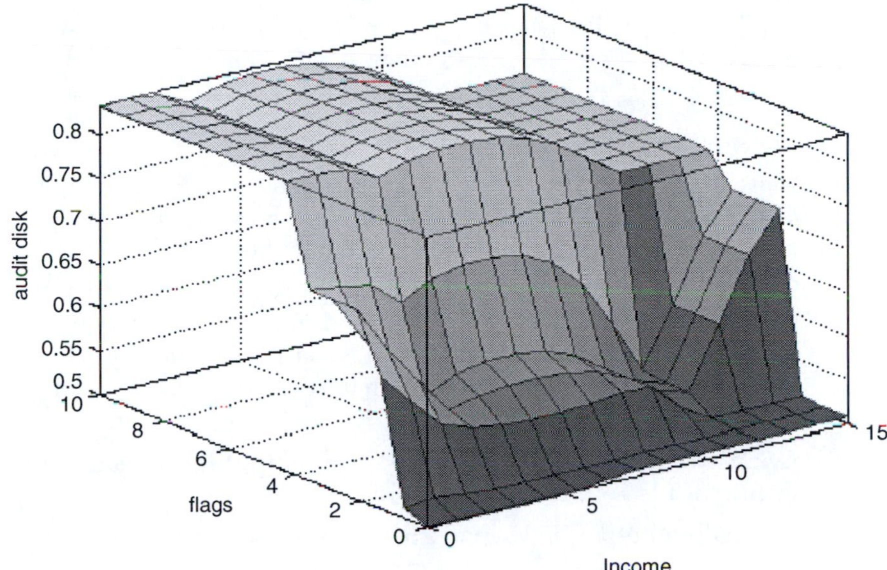

FIGURE 15-4
Surface of rules.

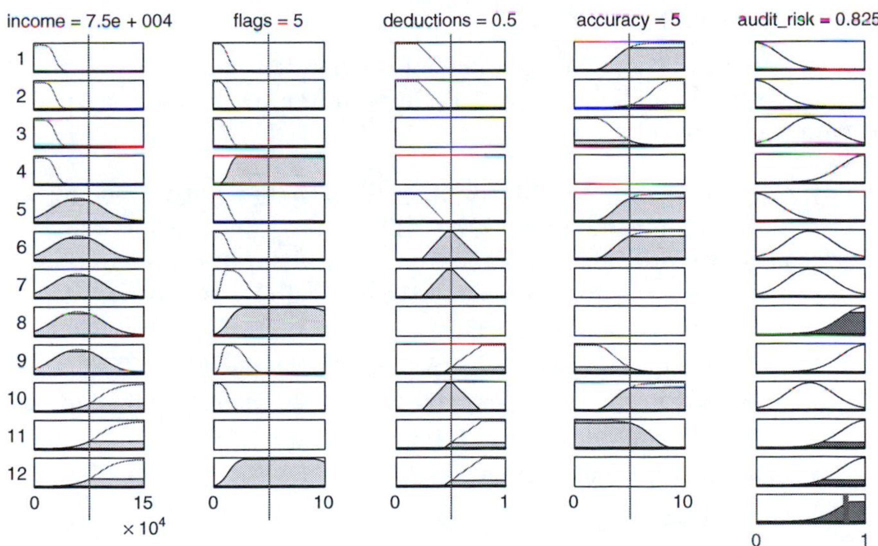

FIGURE 15-5
Ruleviewer for audit risk FIS.

In particular, the program:

1. Reads in an FIS file. This FIS file must be created separately, using a tool included with MATLAB'S Fuzzy Logic Toolbox.
2. Prints the rules in the workspace.
3. Plots the membership functions for key inputs.
4. Calls an ingenious rule viewer that provides a visual explanation of the fuzzy logic at work, given different inputs. This is a particularly effective tool in the MATLAB Fuzzy Logic Toolbox.
5. Generates test inputs for the FIS. Testing the FIS is an important step, particularly since the rules and membership functions incorporate subjective information. The script we created to illustrate this approach generates random information to use in testing the system.
6. Evaluates the test data and produces an output for it, using the test input file.
7. Graphs the output and reports on the analysis in the command window.

We recommend following a similar approach when implementing other fuzzy logic models.

Variation: Simulation Models

One variation discussed in the following section is to utilize a fuzzy logic component in a simulation model. For a Simulink model of this type, the recommended procedure is:

1. Initialize the model, either with an initialization block or by manually running an initialization callback program. Such a callback may also print the rules, plot the membership function, or open a ruleviewer. This step should include the first several steps outlined above.
2. The testing tasks should be accomplished outside the general initialization routine unless you want a simple test to be part of every use.
3. Enter any custom variables.
4. Simulate the model.
5. Graph or otherwise report the output.

A Credit Risk Fuzzy Logic Simulation Model

Another common business decision, involving subjective as well as objective information, is the evaluation of credit risk. This task is so important that "credit scoring" is now an established industry.[384]

We developed a fuzzy logic credit risk algorithm that is similar to the audit risk model presented above. However, we implemented it in a Simulink simulation model. This model demonstrates how a fuzzy logic inference system can be used even in simulation models.

Using a Simple Credit Risk FIS

Consider the role of a lender confronted with numerous business loan applications. She wishes to quickly analyze a number of applications that have already passed an initial review. We assume that, as a result of this prior review, we will only examine firms that have been historically profitable, going concerns with satisfactory debt repayment histories.

While an actual examination of business credit risk would look at more variables, for this inference system we have simplified the evaluation to two measures of a firm's ability to pay debt: its return on assets (ROA) and its coverage ratio (earnings as a multiple of debt coverage). We take current and historic information necessary to calculate these measures and collect them for multiple applications in one data file.

We also summarize the collected wisdom of experienced loan officers in a credit risk FIS. This system includes membership functions for ROA and coverage ratios, quantifying judgments such as "good" or "poor." We then summarize our underwriting standards in a set of rules that tells us when the desirability of lending money to the applicant is high, low, or uncertain. Such rules would include statements such as "if ROA is good, and coverage is adequate, then desirability is high."

We then create a simulation model to feed the information into the FIS and provide other graphing and reporting functions.[385] This will allow us to take the entire data file representing multiple applicants and analyze them all using consistent standards.

---

[384] One of the industry leaders is Fair Isaac, whose "FICO scores" are commonly used in consumer lending. Fair Isaac now sells copies of their ratings, along with other information, directly to consumers and offers some information on their scoring methodology on their Web site at: http//www.myfico.com.

[385] There are some advantages to creating this in a simulation model, notably the ability to visually present the model. However, the same analysis could be performed in a MATLAB environment without the simulation model.

Structure of the Credit Risk Model

The Credit Risk model in Figure 15-6, "Credit risk simulation model," is structured as follows:

1. The model is initialized with variables for assets, earnings on those assets, and debt. For testing, random information is included to generate multiple scenarios. The same initialization file reads in a credit risk FIS which will be used in a fuzzy logic controller block in the model.
2. The earnings are calculated in an earnings subsystem, using the variables created when the model was initialized. The costs of debt are calculated in another subsystem.
3. Using these variables, the model calculates ROA and coverage ratios for each scenario.
4. This information is then given to a fuzzy logic controller. The credit risk FIS, referenced by the controller, rates each case as it is generated.
5. The resulting scores are collected and reported.

Although we have not shown them here, the membership functions, rules, and other attributes of this model can be illustrated as they have been in the section "An Income Tax Audit Predictor" on page 387.

Extensions of This Model

This simple model could be extended greatly, for example by:

1. Using the numeric score from the fuzzy logic controller to directly control the disposition of an application
2. Including more extensive underwriting processes for those applications that fail to pass the first FIS
3. Creating a second stage requiring additional information or analysis
4. Increasing the number and sophistication of the membership functions and rules

Conclusion: Credit Risk Simulation Model

The credit risk model demonstrates the following:

1. Fuzzy logic is a powerful tool, and similar analyses are already being performed every day, although they may not be called fuzzy logic.
2. Fuzzy logic can be used directly in a simulation model.

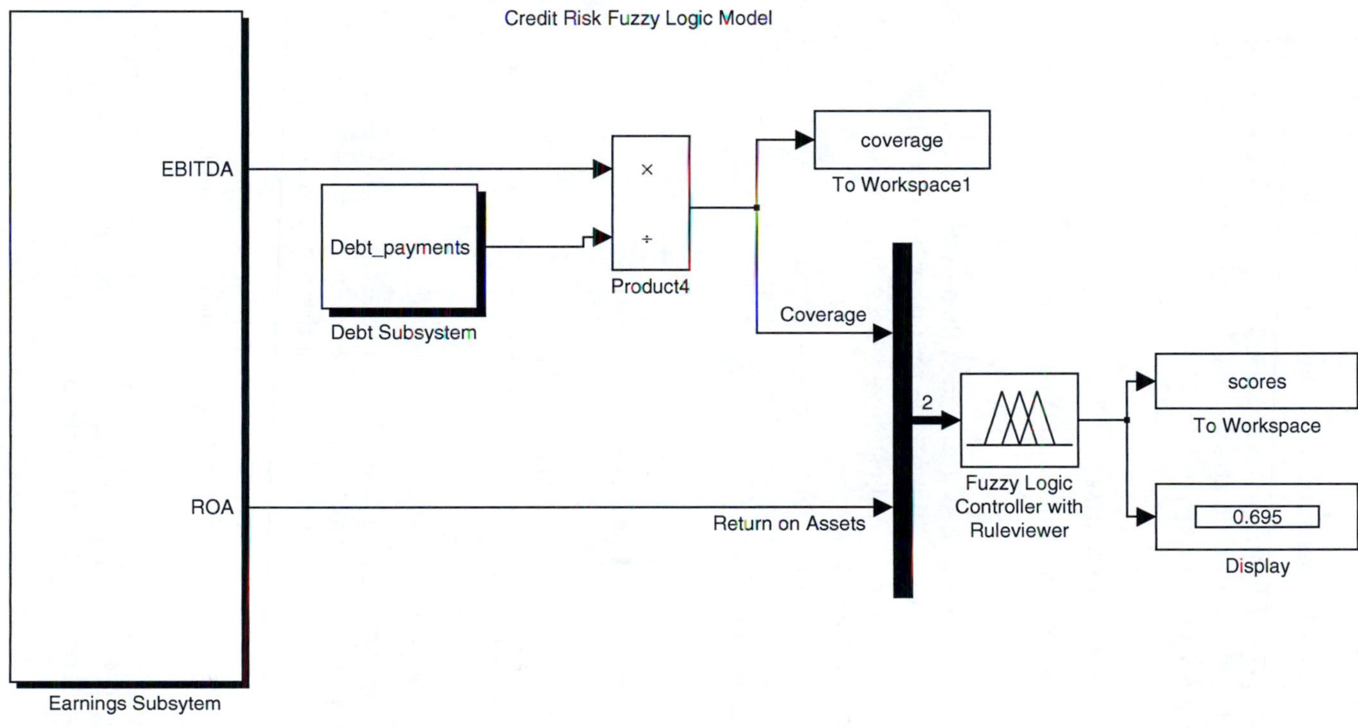

FIGURE 15-6
Credit risk simulation model.

Implementation Issues and Troubleshooting Tips

Fuzzy logic is a new field, and there are relatively few references on its use. However, based on our experience, we have identified implementation issues and troubleshooting tips.[386]

Fuzzy Logic Models in MATLAB

There are a number of specific points to keep in mind when creating a model using fuzzy logic:

1. The inputs should be clearly identified in the FIS and preferably named in a way that is logical for the model as a whole.
2. The FIS can be created using a special editor which is invoked by the command fuzzy.
3. You must read in the FIS before it can be used. One way to accomplish this is through an initialization script given in the code fragment below.

Code Fragment 15-2. Read FIS Command

```
fismat = readfis('myfis');
```

Fuzzy Logic Models in Simulink

Fuzzy logic can also be incorporated into a Simulink model, as demonstrated above. There are special blocks for membership functions and fuzzy logic controllers. The following troubleshooting tips apply to such models:

1. The FIS must be "read" into the workspace, using the command readfis. This is often done in an initialization routine, before the model is simulated.
2. If you receive an error message indicating that the FIS you have selected is unknown, make sure that the FIS has been read into the workspace. For example, if the readfis command shown in

---

386 Of course, these tips may apply only to specific versions of the Fuzzy Logic Toolbox. If the recommended action does not work exactly in the version you are using, try to find a similar action. Using the Help function or checking the manual for the toolbox are always good ideas.

the example code fragment 15-2 had been issued, the parameter for the fuzzy logic controller would be fismat, not myfis.

3. The inputs can be fed by a command-line script or into a Simulink model. In an m-file, the syntax is described in Code Fragment 15-2, "Read FIS Command" on page 400.
4. Carefully check the orientation of the inputs.
5. Look under the mask of the fuzzy logic controller as well as simply clicking it to set the parameter.[386a]
6. Check to see that the proper FIS name is included in all the blocks (the animation and the controller blocks), drilling down twice for the controller. Do not assume that the parameter will pass through to the lower levels; there may be a bug preventing this.
7. When all else fails, delete the block. Break the library link for a new block from the Simulink library and import it into the model. Go through all the above steps systematically as you hook up the new block.

Appendix

Code Fragment 15-3. Fuzzy Logic Audit Risk

```
%   Fuzzy Logic Example:  Risk of Audit%
%   (c) 2001-2003, Anderson Economic Group
%   Use with license only.
%   August 2003, ver 3
%   Uses: Fuzzy Logic Toolbox, generate_tax_input.m
echo on;
%   ------------------------------------------
%   Fuzzy Logic Audit Predictor
%      Patrick L. Anderson
```

[386a] If you have the block with the ruleviewer animation, you may need to look under the mask twice in order to get down to the actual FIS controller block. You enter the parameter first by clicking on the controller block, then put it into the animation block, and then look under the controller block mask again (for a total of three times) to set the parameter.

```
%     AEG LLC, Lansing Michigan

%   -------------------------------------------

%   Read in Fuzzy Inference System (fis).

%   Note that FIS must already be created; that the
%  file must be in the path;

%  that the name of the file must be in single
%  quotes.

%

fismat = readfis('tax5.fis')

%   Note that fuzzy inference system is a MATLAB
%  structure.

%   Examine the items within the structure.

fismat.name

fismat.input

fismat.input.name

fismat.rule

%   ---------------

%   Now print rules in workspace

rules_text = showrule(fismat)

%   ---------------

%   Now use specific tools to plot parts of FIS

figure(1), plotfis(fismat);

title('Fuzzy Logic Tax Audit Predictor Structure',
'fontweight', 'bold');

text(0.4,-0.2, ['Printed by AEG, ' datestr(now, 1)],
'units', 'normalized', 'fontangle', 'italic');

figure(2), plotmf(fismat, 'input', 1);

title('Tax Audit Predictor; Income MF',
'fontweight', 'bold');
```

```
% text(0.0,-0.2, ['Printed by AEG, ' datestr(now,
1)], 'units', 'normalized', 'fontangle', 'italic');

figure(3), plotmf(fismat, 'input', 2)

title('Tax Audit Predictor; Flag MF', 'fontweight',
'bold');

% text(0.0,-0.2, ['Printed by AEG, ' datestr(now,
1)], 'units', 'normalized', 'fontangle', 'italic');

%   ----------------------------------------------

%   Now use GUI to see rules

ruleview(fismat)

echo off;

%-----------------------------

%   Test script

%

%   Create input vector of many cases, using random
%   variables; display

generate_tax_input

%   evaluate input vector

rating = evalfis(test_input, fismat);

% Display numerical and verbal assessments;

disp('Gross income, audit flags, deductions,
accuracy inputs are:');

disp(test_input);

disp('Audit risk index on a 0 to 1 scale (1 is
highest): ');

disp(rating);

%   Create bar chart of ratings of sample tax returns

figure, bar(rating);
```

```
title('Ratings of test tax returns', 'fontweight',
'bold');

ylabel('Rating (higher number is higher risk)');

text(0.0, -0.1, ['Printed by AEG, ' datestr(now,
1)], 'units', 'normalized', 'fontangle', 'italic');

%   ---------------

%   Check to see what is in the workspace now.

whos

%   end of script
```

16

Bringing Analytic Power to the Internet

Static and Dynamic Data on the Web

The use of the publicly available communications network known as the Internet has exploded in the past decade, causing tremendous changes in the interaction among individuals in business, government, and private life. However, despite the very real changes wrought by the Internet, the content on the vast majority of Web pages is quite static — offering information that is selected by the individual viewer but not analyzed or customized for that viewer.[387]

Of course, the advent of widely available static information is a huge jump forward. There is no need to apologize for making high-quality information available on static pages, if that information is useful to the viewer. However, there is more that can be done, and the growth of database-driven Web sites, cascading style sheets, and user-customized content is offering dynamic content to more and more users. We consider this evolution in the content and functionality of the Internet to be important and divide its growth into roughly three stages, as shown in Table 16-1.

In this chapter, we suggest a few methods in which dynamic information can be created through a customized analysis and then transmitted quickly to a viewer through the Internet.

The Third Wave: Complex Analysis

While the common Internet functions are impressive enough, they leave many desirable tasks unfulfilled. Economists, scientists, businesspeople, and

[387] The legions of Internet users that feel their pages are truly dynamic should consider a few additional facts before e-mailing me their critiques of this statement. First, the vast majority of Web pages consist of static text, static links, and static image file references. Second, many pages with features that appear to change — those with banner ads and animated images — are really just fancier versions of static pages. (Most movie-type image files are another example.) Third, even relatively simple calculators (such as ones that can calculate sales taxes and shipping rates for an e-commerce retailer) are still lacking in most Web sites.

There are very impressive, truly dynamic Web sites out there, and we acknowledge a few of them later, but they are still in the minority.

TABLE 16-1

Waves of Internet Functionality

Stage	Character
First Wave	Internet used as dumb bulletin board and as communication tool. (Bulletin Boards, e-mail, static World Wide Web pages)
Second Wave	Internet used as interactive tool for simple transactions in commerce. (dot.com retailing, online auctions, simple calculators)
Third Wave	Internet used as interactive tool for complex data interchange and analysis (trade exchanges, interactive software updates, complex analysis)

consumers would often like something more than the simple calculators and search functions. Consider the following example, both drawn from retail sectors that have been on the cutting edge of development for consumer-oriented websites.

Example 1: Financial Industry

A lending institution might first establish an e-brochure, posting current lending rates and contact information on its site. Most lending institutions have already done this. This would be typical of a first-wave Internet application.

The same institution might establish mortgage calculators and even rudimentary qualification routines. These applications would use simple routines to calculate the monthly payment on loans of different principal, term, and rate, as well as provide some logic-based functions that determine whether borrowers of certain income levels qualify for loans. This is now done on some Web sites created for lending institutions. These second-wave applications greatly improve the user experience for a consumer and also save time and money for the lender.

A third-wave application would go beyond this and perform complex analyses and interactive data exchange. It may, for example, provide analysis of prepayment on a portfolio of mortgages or create a credit evaluation routine using fuzzy logic. It may consider the tax advantages of different mortgage options. It may also supply a user with a customized data source document that could be used to apply for a mortgage at one or more sites. It could even display a richer graphical representation of their options.

Such complex applications require more analytic power than can normally be found on Web sites. As of this writing, only a few lending institutions provide such tools, and even those provide just a subset of what could be built today.[388] Using the tools described in this book, however, such power can be provided to visitors to an Internet Web site.

---

[388] For example, the e-loans.com site provides an analysis of tax-benefits associated with various lending options. On the other hand, none of the sites visited recently provides an XML document that could be used to exchange information among multiple lenders. (We acknowledge there are competitive and security considerations here, as well as technology ones.) None provide the type of graphics that could be generated for the user with the SVG functionality described later in this book.

Example 2: Automotive Retail

A second example is advertising for automobiles. Years ago, auto manufacturers created Web sites with pictures and text describing the vehicles. These then evolved to include interactive features allowing a Web site visitor to "build your own vehicle" by selecting options and colors. The Web application would dutifully produce a nice list of options and even generate a realistic-looking image of a vehicle with the chosen color.

These were excellent first- and second-wave applications. Recently, some sites have moved into third-wave applications which provide a much richer experience. Databases connected to the Web application not only allow a build-your-own feature but also allow visitors to select competing vehicles. The application then dynamically builds a comparison analysis for the user.

It is interesting to note that the most interactive, rich sites are often sponsored by intermediaries rather than the car companies themselves.[389] These organizations provide the service free to consumers but gain revenue from referral charges and advertising.

Key Advances

Advances in technology now allow complex tasks to be performed on behalf of Internet users. The advances we will discuss are primarily not hardware improvements (such as faster CPUs, graphic cards, or storage systems) or even application software used on a stand-alone basis. The key advances are:

1. The emergence of a powerful data standard allowing rich data sets to be identified, stored, and communicated.
2. The emergence of engines that will take user input and perform calculations, database searches, and other tasks, and then produce information useful to the user.
3. The incorporation of databases into many Web applications.
4. The growing use of broadband connections.[390]

Information in This Chapter

This chapter describes methods to exploit three of these advances, in particular:

[389] These include MSN Autos, Edmunds, and Kelley Blue Book.

[390] While not discussed in this book, the growing use of broadband can be seen as a reflection of data-rich usage and a reduction in the price of using the Internet for complex tasks. However, the business difficulties — including bankruptcies — of a number of firms that built broadband capacity in the 1990s and early 2000s are an example of how easily expectations can exceed reality in this world.

1. The powerful data standard known as extensible markup language (XML); how it allows data to be located, described, and exchanged over the Internet; and the emergence of a promising graphics format — scalable vector graphics (SVG) — based on that data standard. See "The Coming XML Standard" on page 415, and "Scalable Vector Graphics" on page 419.
2. How MATLAB can be used on Web servers, bringing sophisticated analytical power to the Internet. See "MATLAB on the Web" on page 410.
3. Briefly, the potential use of databases in applications using MATLAB.

The Network as Platform

Choosing the Platform: Windows vs. the Web

The Microsoft Windows environment has dramatically extended the reach of personal computing. It features a point-and-click graphical interface which was originally developed by Apple and Xerox. Windows is the dominant operating system for desktop personal computers in the world, but it is not the dominant platform on the Internet.

Unix and its variants, including Sun Microsystem's Solaris, run most servers that power the Internet, as well as most enterprise-scale operations. Such operations rely more heavily on scalability and reliability, features which generally have been stronger in the Unix OS than in the Microsoft OS. The newer Linux OS, developed largely as freeware, also runs many of the largest — and some of the smallest — server operations around the world.[391]

One beauty of the Internet is its ability to create cross-platform applications in which the design and the host of the application itself do not matter to the ultimate user. This allows the developer to create the application on the best, or best available, server environment and provide it to people running laptops, desktops, terminals, and other computers or similar appliances around the world. While earlier claims about the Internet were clearly overblown (it did not "change everything", and did not even revolutionize business), it did usher in an area in which "the network" was the most important platform.

[391] Both hard-core Linux aficionados and programmers around the world would encourage me to note that GNU/Linux and many other programs distributed as freeware are actually subject to the GNU public license. The GNU public license allows others to use and distribute source code for software and generally excludes any warranty by the creator. See http://www.linux.org; for copies of the license, write to the Free Software Foundation, 59 Temple Place, Suite 330, Boston, MA 02111, or see the GNU project Web site at: http://www.gnu.org.

Controlling the Desktop

Typically, the only interface necessary between the user and the developer across the Internet is a Web browser. Because for many people the browser is the desktop from which they access most of their information, the race to control that desktop is a serious one.

The most popular Web browsers have been developed by Netscape and Microsoft, building on the earlier work of the MOSAIC group at the University of Illinois. The Microsoft Internet Explorer browser today has the largest market share, although that market share was built largely by bundling the browser with Microsoft OS.[392] However, there are many Web browsers available to those who choose not to rely upon the dominant providers in this market. These include Mozilla, which is an open source browser that is designed to be cross-platform.[393] Over the next several years, we can expect more innovation in this market and a continuing battle to control the desktop.

Calculations, Databases, and Dynamic Pages

There are many ways for Web developers to provide dynamic content on their sites. These include:

1. Integration of databases, which allows the Web server to access large amounts of data and then provide customized information for specific users.
2. Server-side applications.
3. Client-side applications, such as Java applets, that run on the Web client.

All these can provide dynamic content and will continue to be developed in the future. These approaches will often be superior to one using MATLAB Webserver for most tasks. However, for some intense analytical tasks, a mathematical model run on a Webserver may be the best approach, and we consider it next.

---

[392] This was one of the claims in the *U.S. vs. Microsoft* anti-trust case which culminated in a number of orders to Microsoft and a settlement of claims. The "Findings of Fact"in Civil Action No. 98-1232 and 98-1233 of the U.S. District Court for the District of Columbia (released in November 1999) provide a detailed listing as well as useful technological explication. See the U.S. Department of Justice, Anti-Trust Division, at: http://www.usdoj.gov.

[393] Available at http://www.mozilla.org. Others are also available; users should select their browser with as much care as they do any other software.

MATLAB on the Web

The MATLAB environment provides, for certain niche applications, powerful advantages over other applications described above. While clearly not in the stage of development of popular Web applications today, it can nevertheless produce dynamic content that would be difficult — or impossible — to produce using more popular methods.

Many of the applications that MATLAB can provide on a workstation, including advanced computational and graphical capabilities, can be provided to a Web browser in a faraway location. The MATLAB Web server is a separately licensed product that calls on MATLAB to perform calculations on the data it provides, and then takes the results of those calculations, (as well as a graphic) and serves it back to a visiting Web browser.

This offers to interested and motivated people the ability to dramatically enhance the sophistication of the dynamic content offered on their Web sites.

Design Considerations for Web Server Applications

The developer must consider a number of differences in the look and feel of an application that will be delivered over the Internet, rather than a desktop environment. These include the following:

A Difference in Windows

In this case, the "windows" we are talking about have little to do with Microsoft, but a lot to do with functionality. In most desktop operating systems, one or more windows can be up at one time.

This powerful feature is lost when you provide an application over the Web.[393a]

While you can force a Web browser to open a new window, such a window will generally be an independently functioning browser, and will not have the same connection on what is happening in the other browser window. Thus, the developer must generally deliver separate, independent views.

393a This is an overview of current technology, and the researcher should expect development to continue in this area.

Other Differences

Typing and Pointing

Keep in mind that the desktop environment allows for a large variety of inputs and corrections in multiple programs. A Web server environment allows a much more limited set of inputs. Therefore, an application requiring extensive interaction may not work well in a Web server model.

This limitation is a challenge for developers. Not everything works well on the Web. People are quite reasonably frustrated by what they perceive as repetitive requests for the same information, inexplicable rejections of the information they submit, and time delays. Consider carefully your customer before assuming they will prefer — or even endure — a Web server application that causes inconveniences.

Callbacks

In a workstation environment, you will be able to manually select and even edit different script and function files, and execute them in whatever order you choose. In a Web server environment, no editing of the callbacks is allowed and the order of execution is limited by the interface.

Graphical User Interfaces

MATLAB allows for extensive customization of GUIs in its workstation environment. Web server environments require a new GUI in the form of a Hyper Text Markup Language (HTML) page. HTML is much easier to program than MATLAB GUIs, but is not designed as a language for use in programs.[394]

Therefore, you can assume that a very pretty interface can be developed for a Web server application, but its functionality will be limited.

Bandwidth and Load Time

Even a slow workstation will run most MATLAB applications very quickly. However, even a rudimentary Web server application will involve a noticeable delay. Some of this delay will be in sending the instructions back and forth through the intranet or Internet. Some will involve the Web server itself.

Text and Graphical Output

Output from MATLAB in a workstation environment can be voluminous and quite varied. Within a Web server environment, however, you will be limited to specified graphics and text. You can produce almost any graphic that can

---

[394] See "HTML and its Limitations" on page 414.

be produced by MATLAB, but this graphic must then be converted to an image format that can be displayed on a Web page. This is a relatively expensive process (in computer and network resources), and therefore a Web server environment is practically limited in the amount and size of graphics it can provide the user. We discuss graphic formats later in this chapter.

Development Expertise and Environment Stability

MATLAB has been around for a long time and has a huge base of users in multiple continents. Web server applications of all kinds have been around a short time. Furthermore, only a very small fraction of MATLAB users and programmers have created Web server applications.

That does not mean you should wait another decade. For some applications, the potential to provide sophisticated analyses through a Web server environment would be a breakthrough worth much extra effort.

Setting Up the Web Server Model

The MATLAB Web server toolbox comes with a *User Guide* and other references, and we do not repeat the information in those references here. However, we recommend a user take the following steps before beginning a Web server application:

1. Develop a MATLAB application on a work station.
2. Restrict the inputs to a small number.
3. Put the inputs in a primitive GUI or a dialogue box.
4. Test the model with multiple users, using the primitive GUI or dialogue box.
5. Review the design considerations described above for a Web server application before porting the application to a web server.

After these steps have been completed, you will have the basic information necessary to determine whether the environment is suitable.

Planning for the Future

We provide useful cautions about Web server models as well as indicate their promise. However, we should also note that the practical limitations of Web server models today could be reduced over the next decade, given the rapid expansion of network and Internet usage and the compelling economics of thin client computing. Therefore, when considering whether to make the jump to a Web server environment, it may make sense to be cautiously optimistic about future capabilities.

Text and Data Formats on the Web

Graphics Formats

Before discussing the best use of graphics in a Web application, we must discuss image file formats. Such information will be useful to those interested in simply publishing the results of their analysis, as well as to those interested in Web servers.

Most graphics on the Web today are in one of the following formats:

1. Picture files, which are normally described by the acronyms that are also used as file extensions: JPEG, RGB, PPM, GIF, and PNG. These represent an image by a large number of dots (pixels) and are known as raster formats. Raster files vary in the amount of data they contain, and some formats use compression algorithms to pack more information into a smaller file. The number of dots determines the amount of information, and therefore all raster formats suffer from data loss deficiencies. Despite their deficiencies, raster format files such as JPEGs and GIFs are the most common image files used today on the Web.
2. Picture files such as metafiles (.wmf, .emf) and bitmaps (.bmp). These are also raster formats and are often used in low-resolution applications such as document files produced by standard word processing programs.
3. Other raster formats typically used where high resolution is desired, such as tagged image format files, or TIFFs.
4. True print files such as PostScript (.ps), encapsulated PostScript (.eps), and Adobe Acrobat (.pdf) files, which contain the instructions for a printer. These are vector format files which describe the images in terms of shapes (such as lines, curves, and characters) using mathematical equations and specified fonts. Print files are typically provided for download from a Web site rather than being embedded in a Web page itself.
5. A newer vector format called Scalable Vector Graphics (.svg), which is described later.

Further information on these formats can be acquired from the MATLAB Help information and from texts like *Graphics and GUIs with MATLAB*.[395]

Publishing Your Results on the Internet or an Intranet

Today, publishing the results of an analysis in a medium that is widely available is easier than ever before. With the advent of the World Wide Web and electronic publishing tools, economists can now communicate large amounts of information including graphics, data, and text to a vast audience.

Unfortunately, getting people to actually find read, and understand your results may not be so easy. The potential obstacles include:

1. You may have contractual, ethical, or other restrictions that preclude sharing your results with others. This is especially likely in cases involving confidential business data, strategy recommendations, or findings that are not favorable to at least one party.
2. Your analysis may be quite technical or incompletely summarized for a nontechnical audience.
3. You are a great economist but have not quite become familiar with the World Wide Web.
4. You love surfing the Internet but have never actually created a Web page.
5. Your results are great, you are happy to post them on a Web site, but prospective readers cannot find your work.

This chapter cannot solve all these problems, but can help you present your information in a format that can be transmitted easily to others who can find your work on an intranet or the Internet.

HTML and its Limitations

The HTML language has enabled vast amounts of information to be conveniently transmitted over the Web. However, HTML has built-in deficiencies that render it incapable of effectively sharing data. An understanding of the history and limitations of HTML is vital to the effective design of Web-enabled applications, and we summarize the topic briefly in Chapter 4, "Limitations of HTML" on page 67, and "History of HTML; Other Formats; XML" on page 84. In the following section, we discuss how a forthcoming standard addresses these deficiencies.

[395] Patrick Marchand and O. Thomas Holland, *Graphics and GUIs with MATLAB*, 3rd ed. (Boca Raton, FL: CRC Press, 2003).

The Coming XML Standard

The growth of data exchanged over the Internet has revealed the inadequacies of the common HTML standard. HTML was a revolutionary concept, and without it we would not have the World Wide Web as we know it today. HTML is a language designed to *display* information.

We describe how HTML descended from an earlier, very comprehensive language called Standard Generalized Markup Language (SGML) in Chapter 4 under "History of HTML; Other Formats; XML" on page 84. From that same root comes XML. This language is already in wide use and is an important standard to consider when developing Web applications using data.[396]

Structured Data and the Document Object Model

We described earlier in this book the advantages of organizing data and recommended the use of structures to contain data in a hierarchy.[397] The same principles suggest that hierarchical structures be used when possible to organize data in other environments.

XML was designed to preserve structure, and the widespread use of XML in transferring data via the Internet suggests that we consider whether the structure of XML can be used when transferring data from or into an analytical software environment. There has been tremendous work across the world in creating protocols to allow people to create documents containing data and transfer them to others.

Much of this work has been organized by the World Wide Web Consortium (W3C) which has produced standards for use worldwide.[398] One of these standards is the Document Object Model (DOM), which establishes a format for placing information in a tree-like structure.[399] The DOM standard was adopted as a W3C recommendation in 1998 and is used along with the XML language in a growing number of industries and countries.[400]

---

396 There are numerous references on XML programming. However, to understand how XML can be used in business, we recommend the excellent book by Kevin Dick, *XML: A Manager's Guide*, 2nd. ed. (Reading, MA: Addison-Wesley, 2002).

397 See Chapter 4, under "Using Structures to Organize Data" on page 74.

398 The W3C Web site is at: http://www.w3.org.

399 The standard itself (DOM Level 1, Specification, October 1998) states:

"This specification defines the Document Object Model Level 1, a platform- and language-neutral interface that allows programs and scripts to dynamically access and update the content, structure, and style of documents. The Document Object Model provides a standard set of objects for representing HTML and XML documents, a standard model of how these objects can be combined, and a standard interface for accessing and manipulating them."

This and other specifications and information are available at: http://www.w3.org/DOM/Activity.

400 The relationship between the XML language and the DOM standard is summarized by the W3C as follows:

"W3C's Document Object Model (DOM) is a standard Application Prgramming Interface (API) to the structure of documents; it aims to make it easy for programmers to access components and to delete, add, or edit their content, attributes, and style. In essence, the DOM

The use of the XML language and the DOM standard for organizing information can assist economists in some of their analysis and publishing tasks. However, its true importance lies in the potential for sharing information. We discuss this with particular focus below.

MATLAB and XML

MATLAB already uses XML for Help pages. You can see the files in .xml format by browsing through the Help directory.[401] Thus, when you use MATLAB Help, you are already using XML to help organize your work and exchange information. The fact that appears to be seamless is part of its beauty.

MATLAB in recent versions contains a small set of commands that allow file input and output of XML-formatted information. These include a limited XML *parser* (program to identify the information in an XML file), along with commands that serialize XML files and transform them using stylesheets.[402] These features of MATLAB are not well documented but are already in use.

Example 1: Generating Help Files

For example, the command xslt instructs MATLAB to transform information in one XML-format file into a separate format as specified in a transformation stylesheet. The example in Code Fragment 16-1, "Using XML to Generate Help Files," illustrates how this can be used with MATLAB's native info.xml files to generate, in a Web browser, help files.

---

makes it possible for programmers to write applications which work properly on all browsers and servers and on all platforms ….

"In 1996, the W3C released the first version of the Extensible Markup Language (XML) specification. XML has been used to create a variety of applications: editors, repositories, databases, B2B systems Web content, network protocols, etc."

"This specification provides the syntax and the grammar of XML. However, an API is needed in order for the user to access the XML content and structure in an application … The DOM is a generic tree-based API for XML."

401 This direction applies at least to the Windows version of Release 13 and prerelease 14; different installations and later versions may not be the same.

402 In R13, the commands include:
xmlread, a parser,
xmlwrite, a serializer, and
xslt, an XML file transformer. See the MATLAB Help information under "file input-output."

Code Fragment 16-1. Using XML to Generate Help Files

```
xslt info.xml info.xsl info.html -web

ans =

file:///D:/models/info.html
```

Example 2: Creating Custom Information Files

One of the ways MATLAB organizes information on separate toolboxes is through the use of info.xml files. One of these is referenced in the code fragment just described. Another use is to create your own information files.

In current MATLAB versions, the info.xml file for each toolbox can be accessed by the MATLAB Launchpad with icons denoting various components, including Help information, specific tools, and Web sites with related information.[403]

We have done this in Code Fragment 16-2, " Info.xml file for Economics Toolbox," on page 418. This portion of the file lists, in XML format, information and icons on a subset of the tools available to users of this toolbox.

Web Services; Other XML Tools Available

"Web Services" is a generic term for applications that use the World Wide Web to access data and perform some manipulation or reformatting of that data. The SOAP protocol is one XML-based standard that MATLAB can use in the Web services applications. MATLAB sends requests to another server, and then handles the responses. Web services can be relatively simple programs that accomplish much.

There are also many other XML tools available that have been programmed by other authors. Some of these can be accessed from the file exchange area of the Web site maintained by The MathWorks.

Conclusion: Using XML in MATLAB

XML and its variants are rapidly becoming a world standard for exchanging information, and the structured nature of the files provides significant advantages for business economics. There are already rudimenatary XML tools available in recent MATLAB versions, and we expect the XML functionality in MATLAB, both in native versions and in tools programmed by the user community, to grow significantly.

We recommend those creating applications for the future that will require the exchange of data to carefully consider the XML standard.

---

[403] The launchpad is an optional view within the MATLAB desktop, which appears on the screen when you start the program. If it does not appear on your screen, try the desktop command.

Code Fragment 16-2. Info.xml File for Economics Toolbox

```
<productinfo>

<MATLABrelease>13+</MATLABrelease>

<name>Economics Toolbox</name>

<type>Toolbox</type>

<icon>D://models/tools/economics/
economics_toolbox.jpg</icon>

<list>

<listitem>

<label>Test Plots</label>

<callback>testplot</callback>

<icon>D://models/tools/economics/
economics_toolbox.jpg</icon>

</listitem>

<listitem>

<label>Help--not currently implemented</label>

<callback>doc econ_toolbox_help/</callback>

<icon>D://models/tools/economics/
economics_toolbox.jpg</icon>

</listitem>

<listitem>

<label>Publisher Page (Web)</label>

<callback>web http://www.crcpress.com -browser;

</callback>

<icon>$toolbox/MATLAB/icons/webicon.gif</icon>

</listitem>

</list>

</productinfo>
```

Scalable Vector Graphics

SVG and Traditional Graphics

We have described the various image formats, noting that they fall into two camps: vector formats that describe the shapes by means of equations, and bitmap or raster images in which the image is composed of dots.

SVGs, as the name implies, is a vector format that is designed to be scalable. There are other vector formats that can be scaled, but SVG has additional advantages that provide great promise. The SVG specification was approved by the W3C Consortium in 2001 and is now in use worldwide.[404]

XML and SVG

The advantages of SVG arise from the combination of the vector format of the images and the XML format of the data file itself. By putting the data in a format that provides a structure for both data and metadata, and which has a standardized language that is largely platform-independent, the developers of the SVG format have created an impressive jump forward in the usability and functionality of image files.

Because the XML structure of the file allows for metadata and other elements beyond image information alone, the format allows data to be a little less dumb.[405]

Extending beyond Static Display

This combination of attributes allows for images to go beyond static display and become truly interactive. By interactive, we mean graphics that allow the user to probe the image and find out additional information based on the user's needs. An animated image, in and of itself, is not interactive. A simple SVG image — for example, of a bar graph — can be made interactive by allowing the user to "see" the numeric data for each bar by moving his or her mouse over it. More complex images such as maps can be made interactive by providing multiple layers of information, which the viewer can select.

Note that because all the information is in the graphic file, this does not require any specialized software or calls to a database or server. Currently, there are free SVG viewer utilities available, and some have been integrated into Web browsers and readers of other file formats.[406]

---

404 For a discussion of the SVG standard, see the W3C site at: http://www.w3.org/Graphics/SVG.

405 We recognize that all data are inanimate and therefore, by definition, dumb. The users of the data, however, are a different story! Therefore, a graphics format that allows users to see more about the data is in some sense "smart."

406 For example, Adobe Corporation provides a free viewer that integrates into popular Web browsers, as well as integrating an SVG viewer into its Adobe Acrobat Reader. Other viewers are available from Corel and Apache. Viewers and utilities are also available from other vendors and as shareware or freeware.

Business Applications for SVG Images

Our experience with this technology resulted in the first known commercial Web application using interactive SVG images. This interactive SVG image of telephone exchange areas appeared on the Anderson Economic Group Web site in October 2000. It allowed the user with a standard Web browser to zoom in, zoom out, and click on telephone exchange areas to find out more information about service costs.

Since then we have produced similar interactive images showing election results in Michigan and across the nation, telecommunication service areas, student performance data by school district in Pennsylvania and Michigan, development of coastline areas in Virginia, tourist and business information for visitors to cities, green power generation facilities in Michigan, and other similar applications.

Using SVG Applications Today

MATLAB in past versions has not included a native SVG print utility. However, there are a number of options available for incorporating the SVG images in your work, including:[407]

1. Purchase an SVG print utility that can be used in MATLAB and other programs, such as SVGmaker, or a toolkit such as Apache Batik.
2. Purchase an image editing program that can translate among various file types; these include Adobe Illustrator, Corel Draw, and The Graphics Connection.
3. License a suite of SVG tools which allow for extensive interactivity for Web users and programming and data support for the provider. The pioneer product in this field is SmartMaps, now offered by Universal Map.[408]

The available options are growing, and we recommend that users considering applications that will generate images for use on the Web consider the advantages of this graphic format.

---

[407] The software list here is illustrative, not exhaustive, and the author makes no warranty about their performance in specific applications.

[408] Universal Map purchased the distribution rights to the software that was used to develop the first commercial SVG application noted above. Their Web site is http:// www.universal-map.com. Anderson Economic Group no longer offers support for SVG appllications, but some examples may be found at their Website at: http://www.andersoneconomicgroup.com. For information on the other products listed above, see the Websites for the supplier of each product. As this field is rapidly developing, interested users should search for newer products as well.

17

Graphics and Other Topics

Purpose

In other places in this book, we describe methods of acquiring data and analyzing it properly. In this section, we discuss *communicating* the results of that analysis and, in particular, *visualizing* information through graphics. In doing so, we address two problems:

- First, the world is awash in good analysis that is communicated poorly. Sometimes this is the result of poor communication skills (broadened here to include poor data visualization skills) or software with limited ability to present data visually.
- Second, and more frustrating, the news media is filled with analysis that is communicated in a way that exaggerates, distorts, confuses, hides, or simply invents data. In most cases, this communication orginates from individuals and organizations with enough resources to have done a better job.

We first address the latter issue, the principles of effective and ethical visualization of data.

Tufte Principles

The principles of good graphic illustration of quantitative information have been defined and beautifully documented by the contemporary statistician and social scientist Edward R. Tufte. Beginning with the landmark study *The Visual Display of Quantitative Information* in 1983 and continuing through his works *Envisioning Information* and *Visual Explanations*, Tufte has almost single-handedly codified the proper use of graphics for the modern reader.

In addition, he has illustrated numerous examples of outrageously poor design, misleading graphics, and downright misrepresentation of data. Often, these misrepresentations and grotesqueries have occurred in prestigious publications, well-known newspapers, and high-circulation newsweeklies.

Tufte's works summarize the origin of good graphical design, going back to the pioneering work of William Playfair in 18th-century England and the classic illustrations of 19th-century French engineer Charles Minard. He also documents the contributions of contemporary innovators such as John Tukey and Herman Chernoff, as well as the outrageous and commonplace violations of good taste, good sense, and good design.

We summarize below some of Tufte's principles, gathered from the pages of *The Visual Display of Quantitative Information.*[409]

Graphical Excellence

Tufte defines graphical excellence as follows:

- Graphical excellence is the well-designed presentation of interesting data — a matter of substance, statistics, and design.
- Graphical excellence consists of complex ideas communicated with clarity, precision, and efficiency.
- Graphical excellence is that which gives to the viewer the greatest number of ideas in the shortest time with the least ink and the smallest space.
- Graphical excellence is nearly always multivariate and requires telling the truth about the data.

These statements are so easy to read, they almost trip too easily through the mind. However, a few examples of extremely poor graphical design illustrate how essential they are.

Distraction and Chartjunk

Tufte documents numerous cases of what he calls "chart junk" (also now written as "chartjunk"). Chartjunk are graphical artifacts that add nothing to the understanding of the data and, instead, often take the attention away from the underlying information. A simple perusal of contemporary periodicals produces numerous examples of chartjunk. These flourishes are often introduced as an artistic effort to enhance the reader's interest in the underlying data, or even to offer additional explanation. In fact, chartjunk insults the intelligence of the reader and distracts the reader from a proper understanding of the underlying information.

---

[409] Edward R. Tufte, *The Visual Display of Quantitative Information* (Cheshire, CT: Graphics Press, 1983), including p. 51. Tufte's books and example graphics are available on the Web site: http://www.edwardtufte.com.

Distortion and the Lie Factor

One of the most common distortions introduced by poor graphical design arises from the distortion of scale or the misuse of perspective in graphics. Perhaps the most common is the use of 3-D graphical objects to denote 1-D data. The ease of use of many computer programs designed to produce graphics unfortunately has resulted in a large number of bogus 3-D bar charts, pie charts, and other graphical expressions.

In addition, bogus perspective is produced by making pieces of landscape provide graphical information about the underlying data. These are almost always misrepresentations of the data since very few data series — except actual geographic information — are appropriately visualized through the use of the perspective used by the landscape artist.

Tufte was so concerned about such distortion that he created an index to quantify it, which he called the "lie factor." He defined it as the following:

$$\text{lie factor} = \frac{\text{size of data in graphic}}{\text{size of effect in data}} \tag{1}$$

A lie factor close to one means that the reader's perception of the data is close to the actual data. Tufte then proceeds to document, using official U.S. government reports, National Science Foundation publications, exhibits from the *New York Times, Time, The Washington Post*, and other publications, lie factors that easily ranged upwards of nine.[410]

Attracting special attention was the "skyrocketing government spending" chart, for which Tufte provided examples from as far back as 1786. These charts usually exaggerate the actual growth of government expenditures by reporting them in nominal terms. For economists, this type of distortion is so well known that most quantities are properly deflated in presentation by economists. However, similar time series data are presented in the news media that are undeflated for price changes. To further distort reality, some are in bogus 3-D bar charts, or even as 3-D representations of objects, that grossly exaggerate the actual change.

Example: Distortion of Data

The following 3-D bar chart is typical of a style seen in the financial press, as well as in glossy corporate annual reports and bulletins of government agencies. In this case, it shows profits and sales for a fictitious company. See Figure 17-1, "3-D graphic of sales and income data."

The bars in the chart illustrate how both sales and income for the company have been rising consistently. What visual impression would an investor in the company draw? In all likelihood, an investor would see consistent growth in both measures of performance and may gain the additional impression that the profitability of the firm — the ratio of income to sales — had remained fairly constant.

---

[410] *The Visual Display of Quantitative Information*, pp. 57–63.

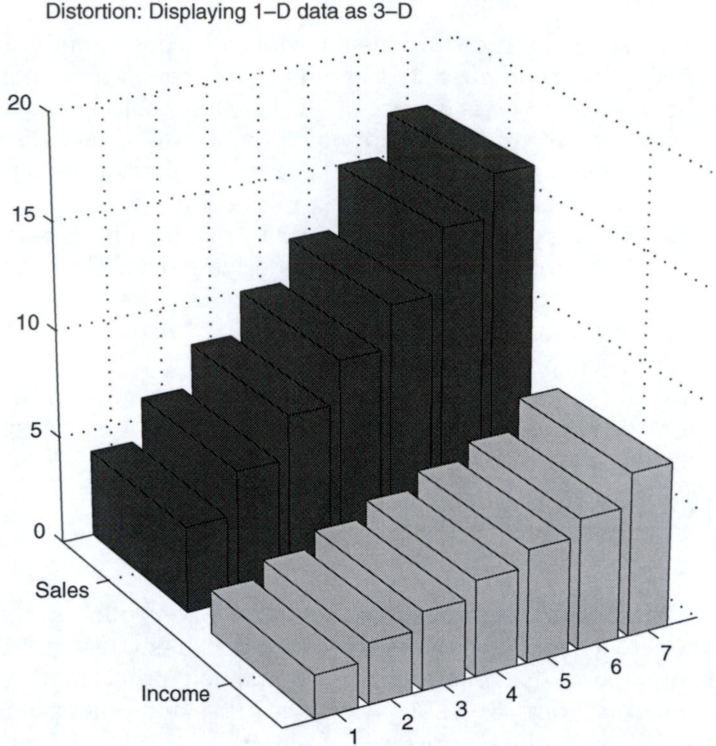

FIGURE 17-1
3-D graphic of sales and income data.

Now let us look at the underlying data without the distortion of the 3-D chart. See the much simpler Figure 17-2, "Proper display of underlying data."

This graphic provides a much richer understanding of the data and does so without distortion. In the top subplot, we see both sales and income plotted. Yes, they both increase — but note that sales are increasing a lot faster than income. The critical profitability ratio is shown in the bottom subplot. The profit margin for this firm actually dropped steadily, and quite sharply, for most of the time period! It recovered somewhat in the last year.

Look again at the 3-D graphic (Figure 17-1, "3-D graphic of sales and income data"). Can you see the profitability ratio change?

Calculating the Size of the Distortion Effect

Using Tufte's lie factor ratio, we can actually calculate a measure of distortion. In the case of the bogus 3-D bars, one distortion occurs from visualizing the *volume* of the 3-D bars, while the actual data is only 1-D scalars. Furthermore, a viewer is often comparing a number of data points (such as the

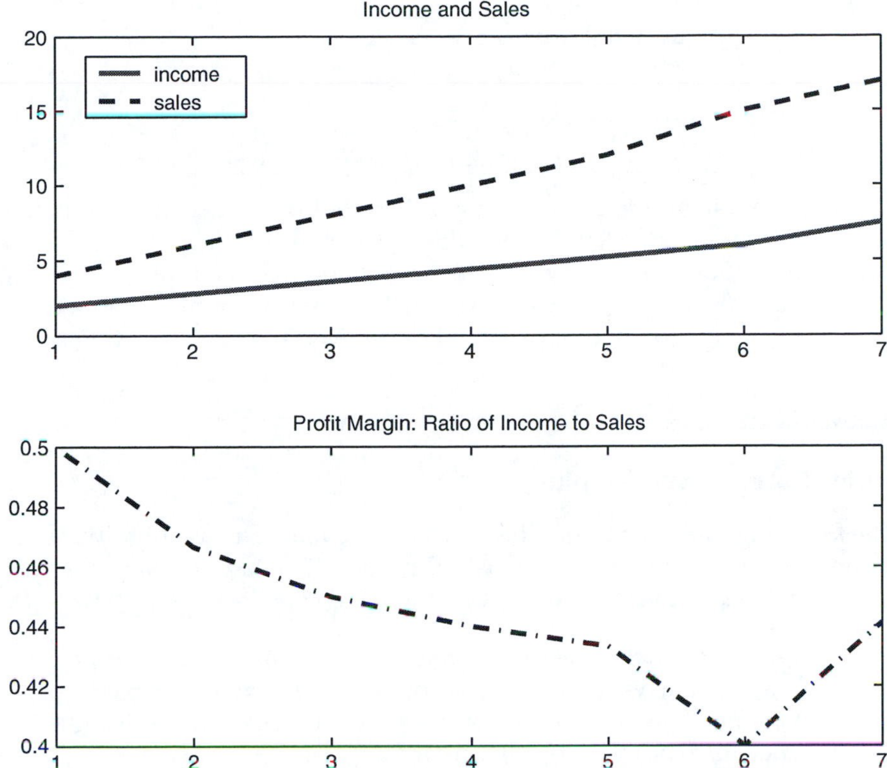

FIGURE 17-2
Proper display of underlying data.

relative size of profits and sales in the first period and the last period and the ratio of the two measures during the entire period).

The comparison of the underlying data is difficult for the viewer who only sees the volume. How is it possible to estimate how much distortion is introduced?

In the script bogusplots.m that generates the data and these charts, we compare the visual impression with the underlying data.[410a] As the visual impression is of volume, we actually measured the size of the bars for the last period, scaled them to sales = 100, and calculated the apparent volumes.

The reader can translate effectively the height of the highest bar, and the 3-D chart creates little distortion for this single datum. However, the trend

[410a] This script file is part of the Business Economics Toolbox made available to purchasers of this book. See Chapter 1, "How to Use This Book," at "Accessing the Business Economics Toolbox" on page 8. Note that the "lie factor" ratio assumes that the reader does not completely adjust for distortion introduced in the graphic. Some readers probably adjust the proportions mentally, others do not. Note also that we allowed one distorting element in the graphic — bogus perspective — but did not allow others, such as drawing the width of the bars proportional to their height.

TABLE 17-1

Actual and Visual Impression of Data

	Income	Sales	Profitability Ratio
Actual data	44.1176	100.0000	0.4412
Visual impression	14.7059	91.1765	0.1613

and relative magnitude of the data are distorted or concealed. As the data in the table show, the viewer sees sales in the 3-D chart about the same as that in the actual data but is given the impression that the profitability ratio is less than half of what it actually is. Furthermore, as discussed above, the trend and change in trend of profitability are well hidden in the 3-D chart.

Rules for Effective Graphics

The following rules summarize the principles we follow in graphics. These can be applied not only to the mathematical graphics but also to maps, thematic maps, renderings of landscapes or structures, and combinations of these types.

1. *Let the data do the talking.* This is the most important rule; pretty graphics that violate the data are often worse than no graphics at all. Graphics that distort or misrepresent the data, even if the distortion appears to be created by the computer program, violate this rule.
2. *Highlight what is important.* All graphics separate out a few dimensions of a multidimensional story. Make sure that the form of the graphic, the colors, and the amount of additional information (including labeling, artwork, details, and other markings) support the important points rather than detract from them. Illustrate the key points with contrast but never allow a visual impression that is contrary to the underlying data.
3. *Make the display as simple and relaxing as possible.* There is no reason to clutter up your graphics with chartjunk, unnecessary colors, bogus 3-D scale, landscape perspective, huge text, jarring fonts, or other distractions. The more actual information a viewer can glean in the first glance the better.

These rules are much shorter than those offered by Tufte, and we recommend that frequent creators of graphics look at his classic text for guidance.[411]

[411] There are a few suggestions that Tufte makes that we find difficult or impractical to follow. For example, Tufte notes that a portion of the ink used in a bar chart is wasted because each bar (a 2-D, 4-sided object) only communicates one dimension. While Tufte is correct on this point, there is no practical way to create a bar chart without providing width to the bars. This is a small area of disagreement, however, and we nonetheless recommend reviewing the section on "Data-Ink Maximization" in *The Visual Display of Quantitative Information*.

Methods of Specific Plots

With the instruction to produce clear, undistorted graphics strong in our mind, we now turn to the methods of producing those graphics in MATLAB.

Handle Graphics in MATLAB

One of the most powerful features of MATLAB is its Handle Graphics, which allows a nearly unlimited number of graphical possibilities, each of which can be programmed meticulously. This offers careful users enormous capabilities to present the results of their analysis.

The disadvantage of this powerful feature is the demand it places on users to correctly specify a large number of options whenever they create a graphic. If you have been frustrated by the inability of more common software packages to correctly illustrate data, then you will welcome the power of MATLAB's Handle Graphics. If, on the other hand, you are satisfied with the rudimentary, preprogrammed options available in most spreadsheet programs, you will probably find MATLAB graphics much too tedious.

We intend this book to focus on analysis rather than graphics. However, we illustrate our analyses using rules and techniques that we also wish to present to our readers. We cover some of those techniques and rules, as well as some suggestions for power and ease of use, in this chapter.

References on Graphics in MATLAB

For those needing assistance in using graphics in MATLAB, we recommend the following:

1. The "Graphics" sections of the MATLAB Help documents, which can be installed along with the MATLAB program files.
2. The book *Graphics and GUIs with MATLAB* by Patrick Marchand and O. Thomas Holland.[412]
3. The testplot file in the Business Economics Toolbox that accompanies this book.[413]

---

412 We have the third edition, published by CRC Press in Boca Raton, FL, in 2003.
One general suggestion from these authors is to set the default graphics settings when you start your session using an m-file that specifies these settings.

413 This script file generates sample data, and from this data it generates a set of graphics. It includes a handy collection of a number of special techniques and common graphics-building instructions in one place. See Chapter 1, "How to Use This Book," under "Accessing the Economics Toolbox" on page 8.

The Testplot Function

To capture a large number of methods used to properly set up and then customize graphics in MATLAB, we created a master graphics function called testplot. This function, which is part of the Business Economics Toolbox, creates sample data (some generated with random information, some from actual economic data), plots the data using a variety of graphic types, and then adjusts and annotates the graphics.

The testplot file is a function that takes no argument and returns a cell array of figure handles. This allows a user to reference particular figures if necessary. The creation of this file as a function allows for a very clean design in which the main sequence is quite short. Most of the work is done by subfunctions which generate the data and the graphics individually. This modular design also keeps the data from one graphic from clogging up the workspace when you are working on other things.

Sections of testplot are reprinted in the appendix to this chapter.

You will note that the code used to create different graphics varies considerably; one reason for this variation is to capture many different techniques.

Line Graphs

We start with a simple line graph. Figure 17-3, "Line graph," illustrates the sine curve between one and six. This graph provides a template for a simple line graph. The code to generate the underlying data and draw the graphic is in Code Fragment 17-1, "Testplot: Line Chart."

There are a few innovations in this graphic worth mentioning:

- The adjustment of the axes within the figure allows more space around the plot. The ability to do this will become more important when we add additional content to a graphic.
- Although not shown here, the code names the figure window (which is visible on the computer screen) and also tags the figure object. The former makes it easier to work with multiple figures at the same time and the latter allows for directly addressing the figure should you wish to further customize or reprint it.
- The code also adds a print date at the bottom of the graphic, which adjusts to the current date. The ability to annotate figures is quite useful for keeping track of multiple projects, analyses, and outputs.

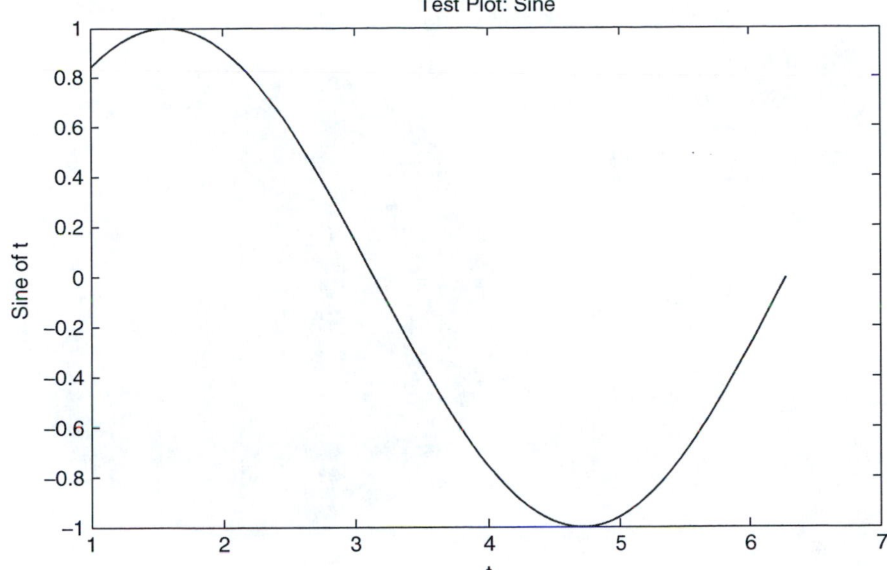

FIGURE 17-3
Line graph.

Pie Charts

Pie charts are useful in displaying the share of certain quantities that are produced or consumed by separate parties. Pie charts are easy to create in MATLAB and other programs, but there are certain idiosyncrasies to them that are worth noting. In particular, pie charts by their nature cannot display the magnitude of the total commodity. Therefore, one must be careful to provide the reader ample information on the total amount from which the "slices" are cut. See Figure 17-4, "Pie chart."

In this case, the chart is used to summarize a very complex analysis involving multiple taxes. The data come from an analysis of state, local, and federal income taxes, as well as payroll taxes such as Social Security and Medicare deducted according to the Federal Insurance Contributions Act (FICA).[414] Payroll taxes consume a much larger share of total labor compensation than many people realize, partially because a good portion is hidden from the employee.[415]

---

414 Patrick L. Anderson, *A Hand Up for Michigan Workers: Creating a State Earned Income Tax Credit, Lansing*, MI, Michigan Catholic Conference, 2002. Available at: http:// www.micatholicconference.org and http://www.andersoneconomicgroup.com.

415 One half of the Social Security and Medicare taxes and all of the unemployment insurance taxes are collected from the employer. Most employers do not report these amounts on the employee's paystubs. As discussed in the report cited above, economists generally agree that employees actually pay these taxes, even though they are collected from employers.

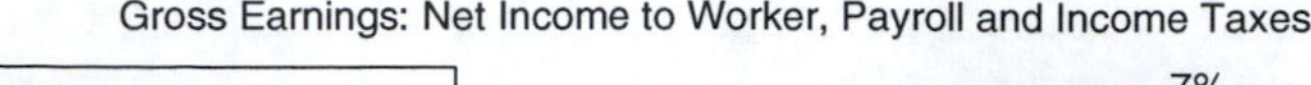

Gross Earnings: Net Income to Worker, Payroll and Income Taxes

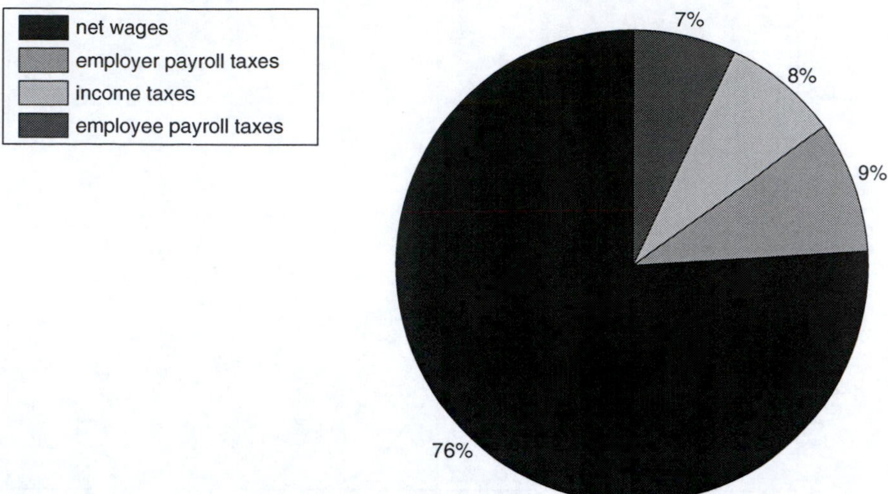

Source: Anderson, A Michigan EITC, 2002
Base Data: IRS, Michigan, CCH Master Tax Guide
Single filer, $20,000 wages, Michigan Income Tax Payer, 2002

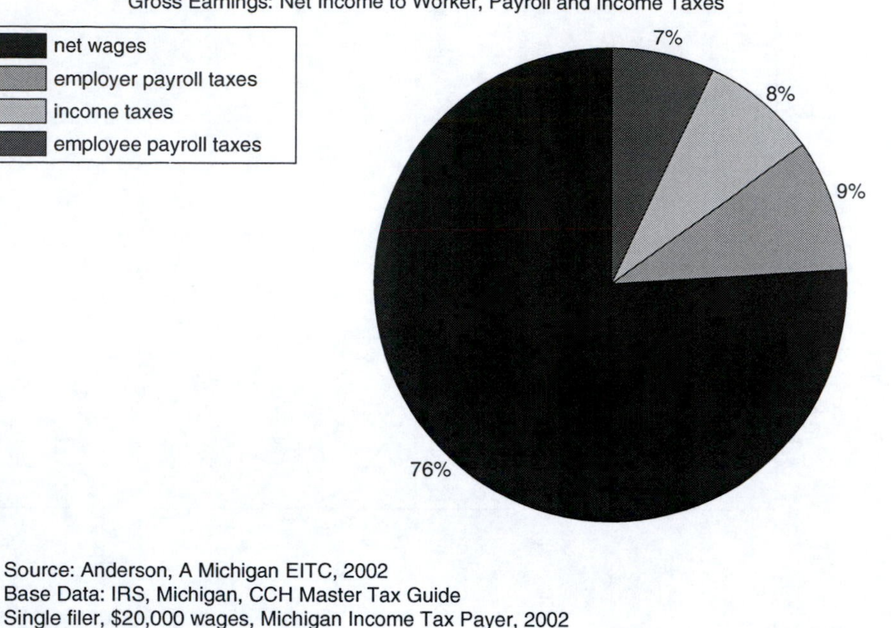

FIGURE 17-4
Pie chart.

We have customized the graphic in the following ways:

- There are two axes, with the left one created largely to allow more space for the sizable legend.
- There is extensive text indicating the source of the data and the status of the taxpayer for which it was generated. Without this information, the pie chart is largely useless because the viewer would not be able to tell the total amount of income, nor the filing status of the taxpayer.

The code fragment in "Testplot: Pie Chart" on page 442 generates the pie chart. The code alone, however, will not properly align the title and legend. So before printing, we moved the legend over (using the mouse) to a good position. In MATLAB, this can be done without opening up the property editor for the figure. To move the title, however, the figure had to be opened for editing. This is easy to do.[416]

---

[416] The straightforward way is to select the arrow on the figure's menu bar. Another method is to use the Edit | Figure Property selection on the menu bar. Version 7 of MATLAB (release 14) has more extensive graphics customization capability right from the figure window.

Bar Charts

Bar charts are more complex than they appear as there are a number of ways to display information using bars. Bar charts can be stacked, grouped, and made horizontal or vertical. Furthermore, as illustrated earlier, they are one of the more commonly abused forms of graphics. See Figure 17-5, "Bar chart."

This graphic, which displays actual data on school finance in the State of Michigan, has been extensively customized to include the following effects:

- The first and second sets of bars are colored differently. The different colors separate visually the drivers of educational costs (such as price and enrollment changes) from the actual change in revenue.
- In order to separate these graphics, two separate plots were superimposed in the same figure.
- Text strings are placed on the bars to show the numeric values of the changes in prices, enrollment, and taxes.[416a]
- The labeling is custom-programmed at the bottom of each bar.
- A multi-line title, a specific label for the Y-axis, and text strings noting the sources of the data and the author of the plot are included on the graphic itself.
- Although not shown here, the code for the plot sets the position properly for the figure, determining its size and position on the computer screen of the user.

The code for the plot is contained in Code Fragment 17-5, "Testplot: Bar Graph," on page 444.

Does This Graphic Follow Principles of Graphical Excellence?

Note that the example uses nearly an entire page to display the data, and that the last bar — which displays an increase of over 100% in property taxes for capital purposes during a 7-year period — towers over the other bars. The admonitions from the earlier section on "Tufte Principles" in this chapter should still be fresh, including those about distortion, exaggeration, misrepresenting scale, and the addition of chartjunk. Does the graphic follow the principles of graphical excellence or violate them?

Consider the following:

- The sources of the base data are named on the graphic, and the text includes a full reference to the work in which it originated.[416b]

[416a] Tufte would object, in principle, to this redundancy.

[416b] The original text in which this appeared was Patrick L. Anderson and Scott D. Watkins, *The Charter School Funding Gap*, Lansing, Michigan, Michigan Chamber of Commerce, 2003.

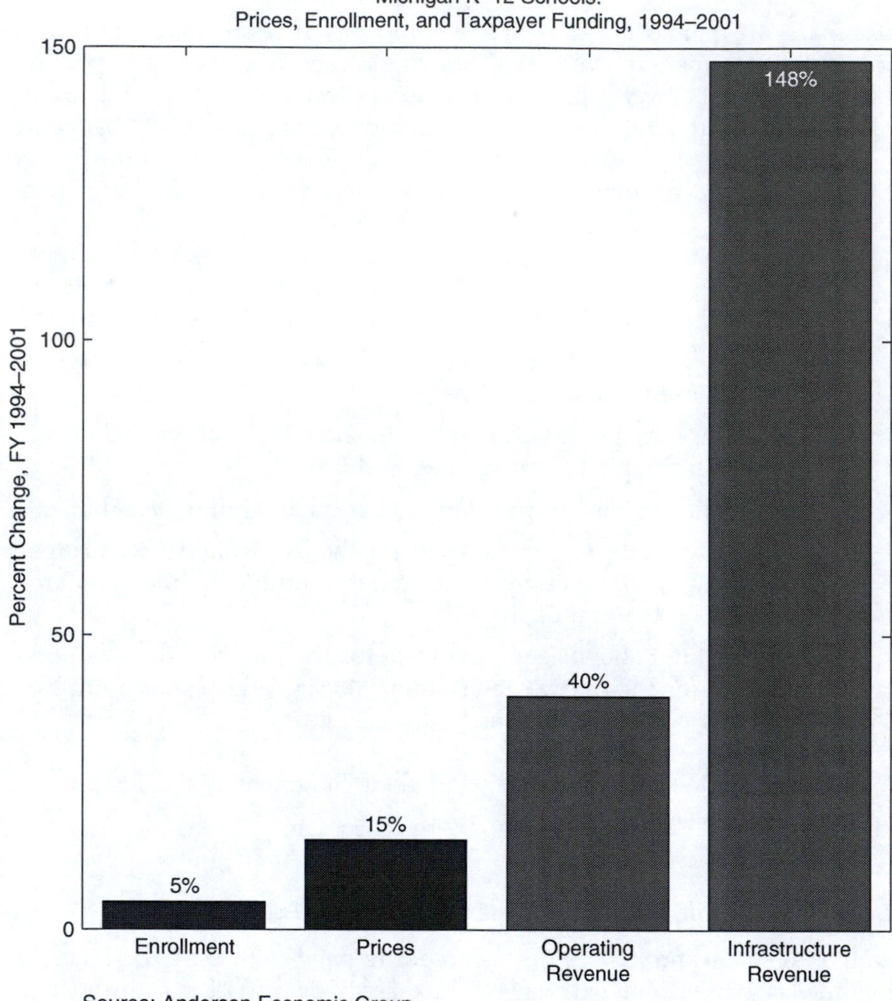

FIGURE 17-5
Bar chart.

- The common misuse of nominal values is avoided because changes in prices are explicitly included for comparison. Indeed, a not-so-obvious comparison variable, school enrollment, is also explicitly included.
- The use of color (shown as gray-scale shades in this text) adds to the reader's understanding because the comparison variables (price and enrollment changes) are shown in a different color than the variables for increases in revenue.

- The additional information on the chart, i.e., source information, lengthy title, and explicit labeling of the percent change at the top of each bar, adds to the understanding of the reader. These additions reduce the data-ink ratio, but with a purpose. Nothing is superfluous chartjunk. With that said, one could remove the date and the labels at the top of each bar without causing much damage to the graphic. These are judgment calls, as are the details in the title and source information.

Most important, the graphic does not distort the data or leave the reader with the wrong impression. The difference between the enormous growth in tax revenue for the school system in question and the underlying change in prices and enrollment does not just *look* enormous. It *is* enormous. A good graphic conveys the truth.

Subplots

Figure 17-6 illustrates a time series analysis graph template with actual data, the line of best fit, and the errors. The three related variables are displayed in subplots which are stacked one above the other.

Subplots are a powerful feature, especially when the indexes to the plots (in this case, the data on the X-axis) are the same across multiple plots. This allows the reader to see related variables visually aligned on the same page. The code to generate this graphic is in Code Fragment 17-6, "Test Plot: Subplots," on page 447.

Adjusting Axes

Another powerful feature of MATLAB graphics is the ability to adjust the properties of the figure, axes, lines, text, and other objects, each of which has a "handle" that can be used to address it. Figure 17-7, "Growth chart," illustrates a line graph showing growth over 12 months, including some random information. It also includes an extensive explanation of the function used to generate the data, which is displayed alongside the plot. The code to generate this graphic includes a number of innovations that you may find useful in customizing your graphics, including:

- The use and placement of two axes within the figure, of which one is reserved for text and displays no data.
- The replacement of the labels on the X-axis (the "XTickLabel" property) with words, in this case, alternating months.
- The use of random information, which will change the graph every time the code is run.

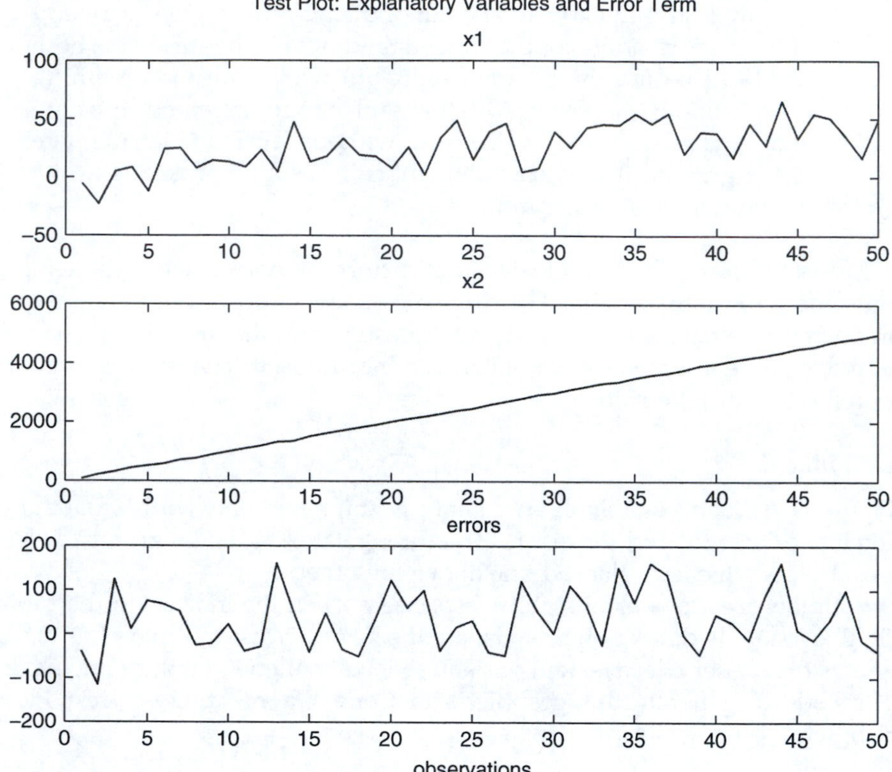

FIGURE 17-6
Time series analysis.

Generating Test Graphics

The appendix reprints the main sequence of the testplot function, as well as longer sections for specific graphics.

We provide these examples to demonstrate how to:

- Produce different types of graphics
- Customize and annotate graphics
- Identify the figure and figure window
- Align, scale, and position graphics
- Make other modifications to standard charts, which improve the understanding of the viewer

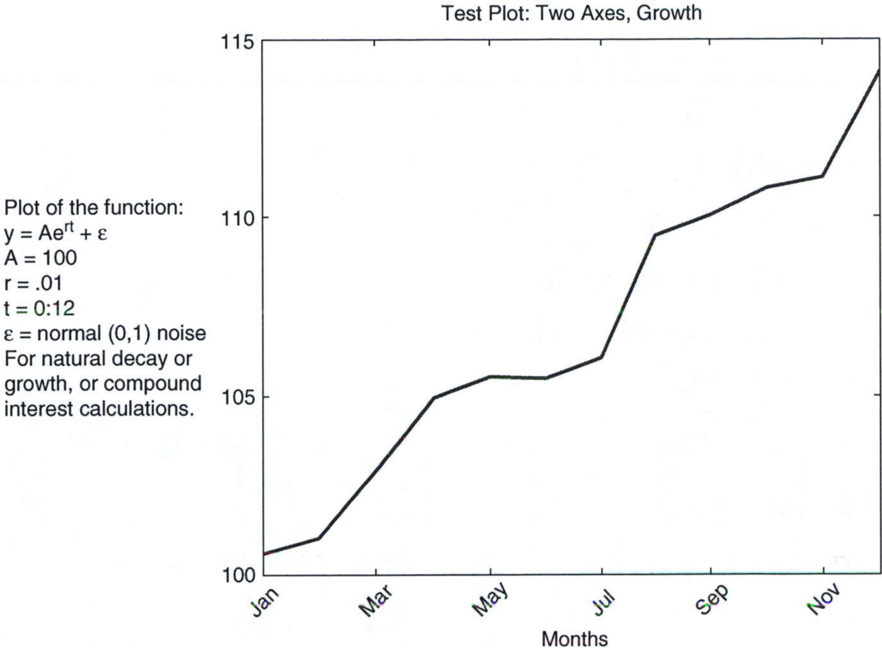

FIGURE 17-7
Growth chart.

The scripts shown also include commands to generate test data or to recreate actual data.[417] The commands that generate test data are useful if you wish to try other graphics or run test calculations on the data presented in these graphics.

The testplot function purposely uses different methods for the various examples and is designed to demonstrate technique rather than to execute quickly.

Appendix

Code Fragment 17-1. Testplot: Line Chart

```
%  first data set and plot; adjusts axes size, text
%  annotation; set
```

[417] Where actual data have been used, we have attempted to provide brief source information.

```
% figure and window title

t = 1:pi/100:2*pi;

% first plot
figure(20),
plot(t, sin(t));

% Reduce the size of the axis within the figure.
% (See reference section at bottom of m-file)
pos = get(gca,'Position');

set(gca,'Position',[pos(1), .2, pos(3) .65])

% labeling info
printdate2 = ['printed on: ', datestr(now, 26), '.'];

% set figure window title (visible) and figure
% property tag (invisible)
tagtest = ['Test Plot: Sine']; % Insert name on
% this line
set(gcf, 'Name', tagtest, 'Tag', tagtest);

% set title and axes labels
Title(tagtest, 'fontweight', 'bold');
ylabel('Sine of t');
xlabel('t');

% insert labeling info into current axis
text(0, -.12, printdate2, 'units', 'normalized',
'fontangle', 'italic');
% note: trial-and-error used to place text
% correctly
```

Code Fragment 17-2. Testplot Function

```
function handles = testplot
%  Testplot function; generates example plots
%  for use in testing, and as templates.
%  Annotated line, pie, and bar charts generated,
%  with text strings
%  and adjusted axes. "Handles" is a cell array of
%  figure handles.
%
%  PLA June 5, 2003 version h
%  (c) 2003 Anderson Economic Group,
%  part of Business Economics Toolbox
%  Use by license only.
%  ---main sequence-----------------------------------
%  reset error
reseterror
%  create cell array for figure handles
handles = cell(1,5);
%  first plot
f1 = sineplot
handles{1} = f1;
%  second plot
f2 = eitcpieplot
handles{2} = f2;
%  third plot---check base data
f3 = schoolbarplot
handles{3} = f3;
%  fourth plot
```

```
f4 = regsubplots
handles{4} = f4;
%  fifth plot
f5 = growthplot
handles{5} = f5;
disp('Test plots completed.')
disp('Cell array of figure handles produced.')
%  -------end main sequence------------------------
```

Code Fragment 17-3. Testplot: Adjusting Axes

```
%  ----------------------------------------------
function f5 = growthplot
%  ----------------------------------------------
%  fifth example
%  Adjusts position of axes to give more room; rotates
%  graphics objects
%  (See solution 5375); Uses two axes to allow
%  extensive text outside
%  the graph. See position reference at bottom of
%  script.
%  Generate some test data.  Assume that the X-axis
%  represents time,
%  y-axis is value. Equation is exponential (natural
%  log) growth or decay.
%  Noise is added (random component).
t = 0:12;
randn('state',sum(100*clock));  %  resets random
% number generator with time
noise = randn([1, length(t)]);
y = 100*exp(0.01*t) + noise;
```

```
%  Descriptive String
str(1) = {'Plot of the function:'};
str(2) = {' y = A{\ite}^{\itr{\itt}} + \epsilon'};
str(3) = {'With the values:'};
str(3) = {' A = 100'};
str(4) = {' \itr = .01'};
str(5) = {' t = 0:12'};
str(6) = {' \epsilon = normal(0,1) noise'};
str(7) = {'For natural decay or '};
str(8) = {' growth, or compound '};
str(9) = {' interest calculations.'};

%  Create figure
figure(25);

%  Set up two axes
h1 = axes('Position',[0 0 1 1],'Visible','off');
h2 = axes('Position',[.35 .2 .61 .7]);

%  Plot the data on one axes; write the descriptive
%  string in another
axes(h2);
plot(t,y,'-', 'linewidth', 2);

set(gcf,'CurrentAxes',h1);  %  equivalent to
%"axes(h1)"
text(.025,.6,str,'FontSize',12)

%  ---Labeling and Titling-----
%  set figure window title (visible), axis title
%  (visible) and figure property tag (invisible)
tagtest5 = ['Test Plot: Two Axes, Growth'];
%  Insert name on this line
set(gcf, 'Name', tagtest5, 'Tag', tagtest5);
```

```
% set title and axes labels
axes(h2);
Title(tagtest5, 'fontweight', 'bold');
% xlabel('test');
% ylabel('test');

% labeling info
printdate5{1} = ['Printed on: ', datestr(now, 26),
'.'];
printdate5{2} = ['Testplots by Anderson Economic
Group LLC.'];

% insert labeling info into current axis (set as
% left axis)
axes(h1);
text(0.05, 0.05, printdate5, 'units', 'normalized',
'fontangle', 'italic');

% ------Reset X-Tick and Rotate--------------------
% Set the X-Tick locations so that every other
% month is labeled.

Xt = 1:2:11;
Xl = [1 12];

axes(h2);
set(gca,'XTick',Xt,'XLim',Xl);

% Add the months as tick labels.
months = ['Jan';
          'Feb';
          'Mar';
          'Apr';
          'May';
          'Jun';
```

```
          'Jul';
          'Aug';
          'Sep';
          'Oct';
          'Nov';
          'Dec'];
ax = axis;     %  Current axis limits
axis(axis);     %  Fix the axis limits
Yl = ax(3:4);  %  Y-axis limits
%  Place the text labels (note: this is not the
% XTickLabel property.)
tl = text(Xt,Yl(1)*ones(1,length(Xt)),months
(1:2:12,:));
set(tl,'HorizontalAlignment','right','VerticalAlig-
nment','top', ...
     'Rotation',45);
%  Remove the default labels
set(gca,'XTickLabel','')
%  -------X-axis Label Placement---------
xal = ['Months'];   %  X-Axis Label text
%  Get the Extent of each text object.  This
%  loop is unavoidable.
for i = 1:length(tl)
  ext(i,:) = get(tl(i),'Extent');
end
%  Determine the lowest point.  The X-label will be
%  placed so that the top is aligned with this point.
LowYPoint = min(ext(:,2));
```

```
% Place the axis label at this point
XMidPoint = Xl(1)+abs(diff(Xl))/2;
tl = text(XMidPoint,LowYPoint, xal, ...

        'VerticalAlignment','top', ...
        'HorizontalAlignment','center');
```

Code Fragment 17-4. Testplot: Pie Chart

```
% ------------------------------------------------
function f2 = eitcpieplot
% ------------------------------------------------
% second data set and plot: pie chart
% Data from Patrick L. Anderson, "A Hand Up: A
% Michigan Earned Income
% Tax Credit," Lansing, MI: Michigan Catholic
% Conference, 2002; see
% http://www.andersoneconomicgroup.com
% Data: share of labor costs taken by payroll and
income taxes
taxshares = [76 9 8 7];
labels = {'net wages', 'employer payroll taxes',
'income taxes', 'employee payroll taxes'};

% plot
f2= figure,

% Set up two axes
h1 = axes('Position',[0 0 1 1],'Visible','off');
h2 = axes('Position',[.35 .2 .61 .7]);
% Plot the data on one axes; write the descriptive
% string in another
```

```
axes(h2);

pie(taxshares);

legend(h2, labels, 2);      % puts label in axes h2;
% positions legend to left

                            % must put legend in axes
% with data

axes(h1);

datanote(1) = {'Source: Anderson, A Michigan EITC,
2002'};

datanote(2) = {'Base Data: IRS, Michigan, CCH Master
Tax Guide'};

datanote(3) = {'Single filer, $20,000 wages,
Michigan Income Tax Payer, 2002'};

text(0.06, 0.16, datanote, 'units', 'normalized');

%  labeling info

printdate2 = ['printed on: ',  datestr(now, 26),  '.'];

%  set figure window title (visible), axis title
% (visible) and figure property tag (invisible)

tagtest2 = ['Gross Earnings: Net Income to Worker,
Payroll and Income Taxes'];    %  Insert name on this
% line

set(f2, 'Name', tagtest2, 'Tag', tagtest2);

%  set title

axes(h2);

Title(tagtest2, 'fontname', 'times new roman',
'fontweight', 'bold', ...

    'HorizontalAlignment', 'center', 'position',
[0.25 1.04 0]);

%  note: position acquired through hand moving, then
%  copying vector

%  from property editor. "Supertitle" or similar
%  function would also work.
```

```
% insert labeling info into current axis
text(0.6, -.11, printdate2, 'units', 'normalized',
'fontangle', 'italic');
```

Code Fragment 17-5. Testplot: Bar Graph

```
% -----------------------------------------------
function f3 = schoolbarplot
% -----------------------------------------------
% third data set and plot; bar graph; set position
% for figure; change
% labels
% Data: 1994-2001 (from Patrick L. Anderson, et
% al., "Charter School Funding Gap";
% Michigan Chamber of Commerce;
% Anderson Economic Group, 2003. See
% http://www.andersoneconomicgroup.com

% Data: 1994-2001
ch_enroll = 4.7;                % enrollment change;
% MI DoE
ch_price = 15.2;                % 109.74/95.29; GDP
% implicit price deflators,
% 1994-2001
% source: Federal Reserve Bank of St. Louis:
ch_op = 39.6;                   % souce: MI DoE
% Source: 2000 Economic Report of Governor,
% State Tax Commission, Table
% A-39; additional STC data
ch_debt = 147.8;                % source: MI State
% Tax Commission
% includes debt, bsm, sf millage;
```

```
% Michigan economic report of the
% governor(Table A-39 in 2000 report);
% Office of Revenue and Tax Analysis

change_vector = [ch_enroll ch_price ch_op ch_debt];
change_vector1 = [ch_enroll ch_price 0 0];
change_vector2 = [0 0 ch_op ch_debt];

% figure
f3 = figure;

% set figure window title (visible), axis title
% (visible) and figure property tag (invisible)
tagtest3 = ['Test Plot: Bar Chart'];    % Insert name
% on this line
set(gcf, 'Name', tagtest3, 'Tag', tagtest3);

% plot
bar(change_vector1,'b');
hold on;
bar(change_vector2,'r');
hold off;

% set title and axes labels

bar_title = {'Michigan K-12 Schools:', 'Prices,
Enrollment, and Taxpayer Funding, 1994-2001'};
title(bar_title,'fontweight','bold','fontname',
'times new roman','fontsize',14);

Ylabel('Percent Change, FY 1994-2001',
'fontweight','bold','fontname','times new
roman','fontsize',10);

% Replace labels for bars
labels = {'Enrollment', 'Prices', 'Operating
Revenue', 'Infrastructure Revenue'};    % Note: use
cell array, not string vector
```

```
set(gca, 'Xticklabel', labels);

%  labeling info (note units for text command are in
% pixels)

text(0, -8,'Source: Anderson Economic
Group','fontangle','italic','fontname','times new
roman','fontsize',10);

text(0, -12,'Base Data: MDoE; Michigan STC; FRB St.
Louis','fontangle','italic','fontname','times new
roman','fontsize',10);

%  labels on bars---from variables

text(0.9, 3, num2str(ch_enroll),'fontweight','bold',
'fontname','times new roman','fontsize',12);

text(1.9, 13,num2str(ch_price),'fontweight','bold',
'fontname','times new roman','fontsize',12);

text(2.9, 37.5,num2str(ch_op),'fontweight','bold',
'fontname','times new roman','fontsize',12);

text(3.9, 145,num2str(ch_debt),'fontweight','bold',
'fontname','times new roman','fontsize',12);

%  set figure position property (to display on
%  screen)

%  note that the units could be set to normalized,
%  and position settings adjusted, for display on

%  different size screens

set(gcf, 'position', [35  35  757  758], 'units',
'pixels');

%  insert labeling info into current axis (note
%  units are normalized)

printdate2 = ['printed on: ',  datestr(now, 26),  '.'];

printdate3 = {'AEG Testplots', printdate2};

text(.8, -.07, printdate3, 'units', 'normalized',
'fontangle','italic','fontname','times new
roman','fontsize',10);
```

Code Fragment 17-6. Testplot: Subplots

```
% ---------------------------------------------------
function f4 = regsubplots
% ---------------------------------------------------
% fourth data set and set of sub-plots
% (from beta_test.m; for use with glmfit)

% Create random variables

e1  = random('Normal', 0, 15, 50, 1 )
e2  = random('Normal', 0, 25, 50, 1 )
e  = random('Normal', 50, 75, 50, 1 )

I =  repmat(1, 50,1);
x1 = [1:1:50]' + e1;
x2 = [100:100:5000]' + e2;
X = [ x1 x2]

b0 = 0
b1 = 5
b2 = -5

% Create Regression Equation

y = b0*I + b1*x1 + b2*x2 + e;

% Plot variables

figure(24),
subplot(3,1,1)
plot(x1)
title('x1', 'fontweight', 'bold')
% get axis number for supertitle placement below
h1 = gca;

subplot(3,1,2)
```

```
plot(x2)

title('x2', 'fontweight', 'bold')

subplot(3,1,3)

plot(e)

title('errors', 'fontweight', 'bold')

xlabel('observations')

%  supertitle

axes(h1)

text(.26, 1.25, 'Test Plot: Explanatory Variables
and Error Term', 'units', 'normalized', 'fontangle',
'italic')

%  set figure window title (visible) and figure
property tag (invisible)

tagtest4 = ['Test Plot: Subplots'];    % Insert name
%  on this line

set(gcf, 'Name', tagtest4, 'Tag', tagtest4);

%  set title and axes labels

%  Title(tagtest4, 'fontweight', 'bold');   % Don't
%  use here because "supertitle" necessary for subplots
```

Appendix A: Troubleshooting

Introduction to Troubleshooting

The following short sections provide tips on troubleshooting common problems that occur in the practical use of MATLAB and Simulink. As Simulink is built on MATLAB, problems that appear in a Simulink environment can sometimes be traced to problems within the base MATLAB environment.

Troubleshooting MATLAB

Many problems in MATLAB can be resolved by looking at the MATLAB Help menu. This should be the first resource for understanding the commands, functions, and environment. Our experience indicates, however, that those using MATLAB for business economics will eventually run into the following problems, for which we provide troubleshooting suggestions below:

1. Path command problems: These often evidence themselves as "command not found" error messages.
2. Data type problems: These include difficulty using data types such as strings, cells, and structures.
3. Incorrect MATLAB file references: These can stem from improper installations, moving files, or unexplained computer problems.

In addition, problems with Excel Link are covered in Appendix B. Finally, these troubleshooting tips are based on MATLAB versions 6, 6.5, and 7. Always check the help information for the specific version you are using.

Path Command

The following troubleshooting steps may be helpful:

1. Always confirm your path when editing a troublesome script or function file. Sometimes more than one copy of a file exists in the path, and MATLAB executes the first one in its path.

2. Check to ensure that you are using the correct version of startup, and rename all others. The startup script that is stored in the \work directory is executed by MATLAB upon startup — unless the path specifies another directory in which a startup.m file exists.
3. Use the File | Set Path command to review and set the path. We recommend putting the \startup directory first, followed by the general MATLAB program directories, and then the directories in which you store your models and custom programs.
4. We strongly advise against putting your custom work in the same folders as program files. Use the work directory or a user-created model directory to store all your custom files. Set your path to add those folders within that directory that you are currently using.

Data Type Mismatches

MATLAB allows a number of data types, including matrices (the fundamental data type used in the MATLAB environment and the most natural to understand, share, and communicate), cell arrays, strings (character arrays), and structures.

These different data types give MATLAB great power. However, they also allow for easily overlooked errors to bedevil script and function files. Below, we indicate a number of error messages which, at their root, may indicate a data type mismatch. We then suggest ways to fix these problems.

Error Messages

The following error messages may indicate a data-type mismatch:

1. "Horzcat does not work;" or "all matrices must be same size."
2. "Conversion to char from cell not possible."
3. "Field in structure does not exist."
4. "Empty cell" or "nonexistent field."

Suggested Troubleshooting

1. Clear the workspace before running the function or script, and carefully note the types and sizes of any variables required to run the file.
2. Use the debugger to set "breakpoints" in the function or script. When the function stops at that point, check the workspace to see the size and type of all the variables.
3. Try manually duplicating the commands.

4. Be mindful of misspelled names and capitalization.
5. Be careful of the size and orientation of all matrices.
6. Note the difference between the protected workspace in which a function operates and the general workspace.
7. If you use varargin or nargin in a function, manually check (using the debugger) how many such variables exist, how they appear in the input cell array, and how they are referenced by the script or function.
8. Try manually creating, transferring, and calculating with small-sized variables of the relevant type.
9. When manually performing the tasks, work with test data. Try troublesome syntax within the function or script file.
10. Save commands that successfully duplicate the troublesome tasks; try reusing these in the script or function file you are building.

Checking MATLAB Paths and References (Windows)

We can suggest the following for MATLAB installations on Microsoft Windows machines:

1. Check the matlab.ini file in your Windows directory (or equivalent) to ensure that it refers to the proper directories.
2. Check the Windows path parameter to ensure that the MATLAB root directory is listed (e.g., D:\ MATLAB r13).
3. We do not recommend running MATLAB or, for that matter, anything important on Windows operating systems that do not have the more stable kernel that exists in the versions designed for use in business environments.[418]

A Note on Operating Systems

A discussion of the merits and drawbacks of operating systems is outside the scope of this book. However, we use and recommend MATLAB partially because it is reliable, robust, and can be used on multiple platforms. We suggest that users consider these factors when selecting an operating system as well.

418 At the current time, the author has successfully, and fairly reliably, used MATLAB on Windows NT, XP Pro, and 2000 versions. We do not recommend use of the earlier (3.x, 95, 98, and Me) versions. Note that The MathWorks no longer supports installations on some Windows versions.

Troubleshooting Simulink Models

One of the advantages of Simulink is the graphical display of the underlying equations. This allows for a truly visual debugging of common problems stemming from the incorrect use of variables. However, Simulink has its quirks, and the following problem areas are discussed below:

1. Data problems, including missing data, wrong formats, incorrect sizes, and other problems.
2. Date problems, including problems with the periodicity of data or the parameters of the simulation.
3. Calculation problems such as intermediate variables having unexpected magnitudes.
4. Model structure problems, which often evidence themselves with final variables that are quite different from what they should be.

Data and Date Problems in Simulink Models

Symptoms

The following are symptoms of data problems:

1. Numbers that are astronomically higher than you suspect.
2. Data matrices that do not line up properly — in that the size of some matrices do not match those of other data structures used in the model.

Troubleshooting Steps

Try the following steps:

1. You may have a source that starts — unknown to you — growing in the year 0, and has experienced over 2000 years of growth by the time you expect to see the results. Check that the time vectors in your model, including those in initialization files and simulation parameters, match.
2. Check to make sure that all start dates are correct, including those in:
   - Model parameters
   - Ramps
   - Integrator blocks
   - Clocks

One way to ensure this is to use Tstart rather than a specific date on every block that includes time as a parameter. The use of the Tstart variable for the starting time period is discussed in Chapter 3, under "Workspace and Timespan" on page 50.

3. Beware of relying on Simulink's default setting of zero for parameters in many blocks. We suggest using either zero consistently (a difficult choice for economists but perhaps appropriate for engineers, scientists, and others who might consult this text) or Tstart.
4. For mismatches in size of the data vectors, carefully review the input data, then the parameters, and then the output data. Attempt to identify the portion of the model where the errors start occurring. Use test data, simulating just the portion of the model in which the problem appears to be arising.

Calculation Problems

One of the fundamental advantages of Simulink is the ability to visualize the model and see the data at various steps along the way. However, unless you carefully build into the model the correct data, and then check to see you are using the data properly, you will still make mistakes.

Symptoms

If the simulation model runs fine, but produces numbers that do not seem right, you may have calculation problems.

Troubleshooting Steps

Try the following steps:

1. Check all the source blocks to see if they reference the correct variables.
2. Check the values of the variables and their dimensions (scalar, vector, and matrix). You can do this in the MATLAB command window or with the data browser.
3. Check the time parameters and other simulation parameters.
4. Use scopes, plotting, and display blocks to trace the value of the signals along the entire path of the simulation.
5. Check to ensure that the inports and outports of each subsystem are correctly labeled and properly connected.

Model Structure Problems

Most models are created in an initial draft, then modified substantially along the way. Usually, at various points in the development of a model, the equations do not work as the creator intends. To troubleshoot such occurrences, we suggest the following:

1. Review your model structure from the top down. Of course, we assume you have already done this before consulting this text, but there is no substitute for looking at it again. We suggest you save your files (perhaps using a temporary name), shut down the program, take a short break, and then return. When you do, follow these steps:
    1. Check the input data.
    2. Check the simulation parameters.
    3. Check the overall flow of data through the model.
    4. Double-check that the information you *think* is coming from one subsystem is the same information the next subsystem receives.
    5. Review subsystem-by-subsystem all the intermediate calculations. Consult the subsection on "Calculation Problems" on page 453.
    6. If the preceding steps do not solve the problem, they should have at least isolated it to one or two areas.

In addition, you may wish to strictly apply style guidelines throughout the model. Using good style consistently will make your model easier to understand and may help you find problems. See Chapter 3, "MATLAB and Simulink Design Guidelines," on page 29.

Appendix B: Excel Link Debugging

Excel Link Debugging

For users comfortable with spreadsheets, Excel Link greatly increases the ability to import and report data. We suggest a strategy for importing and reporting data using Excel Link in Chapter 4, "Importing and Reporting Your Data," on page 63.

However, Excel Link is not as reliable as MATLAB. Furthermore, it has a number of quirks that often require debugging. We collect a number of troubleshooting suggestions into this appendix to assist you in using this tool.

Keep in mind that this is not a book on Excel, and there will be future versions of both Excel and MATLAB. Therefore, expect many of these suggestions to eventually become obsolete, and always check the help information for your specific version.

Common Excel Link Problems

Some of these problems are:

1. Excel indicates "Runtime error '429': Active X component can't create object."
2. Excel does not start MATLAB automatically.
3. Your commands show up in Excel with errors.
4. Excel Link shuts down entirely, taking Excel with it.

If you are having problems with Excel Link, try the following:

Check Excel Link Files and References

1. Check, and if necessary reinstall, the correct Excel Link files, using the Excel Tools | Add-Ins function. If you have switched MATLAB versions, try the steps listed under "Run the Program with A Newer Version of MATLAB" on page 456.
2. Check the exlink.ini file in your Windows directory (or equivalent) to ensure that it refers to the proper directories.

Check MATLAB *References (Windows)*

1. Most users of Excel Link will probably also use a Microsoft operating system, as Microsoft also sells Excel. Such users should review Appendix A, "Troubleshooting," on page 449.

Start the Program in Different Ways

1. MATLAB should start automatically with Excel if Excel Link is working properly. If it does not, a "brute force" method is to use the Excel command Tools | Macro, and then enter matlabinit.

Run the Program with a Newer Version of MATLAB

1. If you run both a current version and the R11 versions of MATLAB, note that the location of the excllink.xla file changes with R12. The R12, R13 and R14 installers put the Excel Link files within the \toolbox directory. Update any references such as those discussed below.
2. To ensure that starting Excel opens up the right version of Excel Link, point the Tools | Add-ins dialogue within Excel to the proper file.
3. Another approach is moving the old version of the excllink.xla to within the new MATLAB root directory and then pointing Excel to the relocated file.
4. A new version of the excllink.xla file may be available from The MathWorks; see their Web site.
5. Try starting the correct version of MATLAB, then Excel. This will sometimes force Excel to connect with the correct version. You may, however, end up with more than one instance of MATLAB running, and should check which one is connecting to Excel (by sending a matrix between the two) before closing one. This is a brute force method that should be used if more elegant, but time-consuming, methods have not yet succeeded.

Use Spreadsheets with Different MATLAB *Instances*

Some errors arise when a file is used by users from different machines, or with different versions of MATLAB. If this is the case, try these debugging tips when problems arise:

1. On Excel files that have been opened after switching from one version to another or from one machine to another, the simple

commands embedded in the cells such as MLEvalString(...) are sometimes replaced with path-specific strings such as D:\MATLABr11\exlink\mlevalstring.... Use the replace command to remove these additional path strings (and any quotes). Do this after correcting the add-in location using the Excel Tools | Add-Ins feature.

2. If the Excel worksheet shows something like:

   d:/MATLABr12/excllink/excllink.xla'!...

   then delete the file prefix entirely, so the command simply calls the Excel Link command without a file prefix.

3. This method is suggested by The MathWorks: use the Excel function Edit | Links and click on change source, and then point to the correct location for excllink.xla.

Errors in Formulas

If your Excel formula in a cell generates an error, try the following:

1. Make sure that the syntax begins with an equals sign and does not end in a semicolon.
2. Check that the entire MATLAB command is in quotes.
3. Compare the commands you are using with those in the sample Excel spreadsheet included in your MATLAB installation.

Excel Link Crashes

If Excel Link crashes after you attempt to exchange data between the two programs, it may be due to an attempt to transfer a data type that is prohibited. Check to see that you are attempting to send back and forth only data matrices and other data types that you can confirm are supported.[419]

Excel Link and Workspace Changes

If you upgrade to MATLAB R12 or later versions, you can use the workspace browser and other features. However, Excel Link may pull up the command window that was the main interface in older versions.[420] If you experience this problem and want MATLAB to start with its full functionality, you have

[419] In addition to matrices, we have successfully transferred strings before, although this sometimes causes crashes. We have not attempted other data types. Check the MATLAB Help and The MathWorks Web site for notes on this toolbox.

[420] Some users prefer using this interface, and The MathWorks, at least in the current version, allows those users to continue with the command window.

two options.

Option 1: Automatic through Excel — One suggestion from a MathWorks technical note is to include the following in a cell of an Excel worksheet:

```
=mlevalstring("desktop")
```

Option 2: Automatic through MATLAB — Add the command `desktop` to your `startup.m` file or type it manually.

Other Recommendations

We discuss in some detail different approaches to data in Chapter 4, "Importing and Reporting Your Data," on page 63. Excel Link has advantages for convenient collection and reporting of many data sources, but it is not an industrial-strength solution for all types of data. Therefore, consider acquiring data in other formats. For a summary of these, see Chapter 4, under "Appendix III: File Importing with MATLAB," on page 93.

Index

C

D

N

O

P

Q

R

S

T

U

V

W

X

Y

Z